Recent Progresses in Bioelectronics

生物电子学最新进展

顾忠泽　主编

科 学 出 版 社

北　京

内 容 简 介

美国国家标准与技术研究所(NIST)将生物电子学定义为生物学与电子学的融合领域。这主要包括两部分内容：一是生物系统中电子学现象，包括研究生物分子的电子学特性、信息的传输和储存；二是利用电子学的手段解决生物学问题，包括利用传感器获取和分析生物信息等。近几十年以来，生物电子学有了飞速的发展，其内涵和外延也不断变化。2015 年是东南大学生物电子学国家重点实验室成立 30 周年。本书汇集了该实验室成立以来的主要研究进展，由实验室的现任研究人员和曾经在实验室工作过的人员编写，其内容涵盖材料与界面、生物传感与芯片、生物信息学和生物医学影像等四个方面，主题包括生物分子相互作用、自组装、单细胞分析、生物芯片、生物信息学、神经接口、多模态成像技术等，这些也是生物电子学的重要内容。

本书适合生物医学工程、电子工程、分析化学、材料工程和生物工程，以及其他生物医学相关专业的本科生、研究生及研究人员阅读和参考。

图书在版编目（CIP）数据

生物电子学最新进展=Recent Progresses in Bioelectronics/顾忠泽主编.
—北京：科学出版社，2016

ISBN 978-7-03-047280-9

Ⅰ. 生…　Ⅱ. 顾…　Ⅲ. 生物电子学–英文　Ⅳ. Q-331

中国版本图书馆 CIP 数据核字（2016）第 025434 号

责任编辑：张淑晓　张翠霞 / 责任校对：贾伟娟
责任印制：肖　兴 / 封面设计：耕者工作室

科学出版社 出版
北京东黄城根北街 16 号
邮政编码：100717
http：//www.sciencep.com

中国科学院印刷厂 印刷
科学出版社发行　各地新华书店经销
*
2016 年 3 月第　一　版　　开本：720 × 1000　1/16
2016 年 3 月第一次印刷　　印张：21 1/2　插页：16
字数：410 000

定价：128.00 元

（如有印装质量问题，我社负责调换）

Editorial Board

编　委　会

主　编　顾忠泽

副主编　赵祥伟

编　委　顾　宁　肖守军　陈　战

刘凌云　孙　啸　艾竹茗

周晓林　吕晓迎　王志功

万遂人　徐　蓁

Contributors

Can Li
State Key Laboratory of Bioelectronics, Jiangsu Key Laboratory for Biomaterials and Devices, School of Biological Science and Medical Engineering, Southeast University, Nanjing 210096, China

Ning Gu
State Key Laboratory of Bioelectronics, Jiangsu Key Laboratory for Biomaterials and Devices, School of Biological Science and Medical Engineering, Southeast University, Nanjing 210096, China

Hongbo Liu
Nanjing National Laboratory of Microstructures, School of Chemistry and Chemical Engineering, Nanjing University, Nanjing 210093, China

Shoujun Xiao
Nanjing National Laboratory of Microstructures, School of Chemistry and Chemical Engineering, Nanjing University, Nanjing 210093, China

Zhan Chen
Department of Chemistry, 930 North University Avenue, University of Michigan, Ann Arbor, MI 48109, USA

Xiangwei Zhao
State Key Laboratory of Bioelectronics, School of Biological Science and Medical Engineering, Southeast University, Nanjing 210096, China

Junjie Yuan
State Key Laboratory of Bioelectronics, School of Biological Science and Medical Engineering, Southeast University, Nanjing 210096, China

Zhongze Gu
State Key Laboratory of Bioelectronics, School of Biological Science and Medical Engineering, Southeast University, Nanjing 210096, China

Wenchen Li
Department of Chemical and Biomolecular Engineering, the University of Akron, Akron, Ohio 44325, USA

Qingsheng Liu
Department of Chemical and Biomolecular Engineering, the University of Akron, Akron, Ohio 44325, USA

Lingyun Liu
Department of Chemical and Biomolecular Engineering, the University of Akron, Akron, Ohio 44325, USA

Xiao Sun
State Key Laboratory of Bioelectronics, School of Biological Science and Medical Engineering, Southeast University, Nanjing 210096, China

Yue Hou
State Key Laboratory of Bioelectronics, School of Biological Science and Medical Engineering, Southeast University, Nanjing 210096, China

Huan Huang
State Key Laboratory of Bioelectronics, School of Biological Science and Medical Engineering, Southeast University, Nanjing 210096, China

Yumin Nie
State Key Laboratory of Bioelectronics, School of Biological Science and Medical Engineering, Southeast University, Nanjing 210096, China

Yiru Zhang
State Key Laboratory of Bioelectronics, School of Biological Science and Medical Engineering, Southeast University, Nanjing 210096, China

Honde Liu
State Key Laboratory of Bioelectronics, School of Biological Science and Medical Engineering, Southeast University, Nanjing 210096, China

Zhuming Ai
Information Management and Decision Architectures, U.S. Naval Research Lab 4555 Overlook Avenue, SW Washington, D.C. 20375, USA

Li Zhang
Center for Brain and Cognitive Sciences and Department of Psychology, Peking University, Beijing 100871, China

Yuan Zhang
Center for Brain and Cognitive Sciences and Department of Psychology, Peking University, Beijing 100871, China

Philip R. Blue
Center for Brain and Cognitive Sciences and Department of Psychology, Peking University, Beijing 100871, China

Xiaolin Zhou
Center for Brain and Cognitive Sciences and Department of Psychology, Peking University, Beijing 100871, China
Key Laboratory of Machine Perception (Ministry of Education), Peking University, Beijing 100871, China

PKU-IDG/McGovern Institute for Brain Research, Peking University, Beijing 100871, China

Zhigong Wang
Institute of RF-& OE-ICs, Southeast University, Nanjing 210096, China

Xiaoying Lü
State Key Laboratory of Bioelectronics, School of Biological Science and Medical Engineering, Southeast University, Nanjing 210096, China

Boshuo Wang
Department of Biomedical Engineering, University of Southern California, Los Angeles, CA 90007, USA

Jisu Hu
The Laboratory for Medical Electronics, School of Biological Sciences and Medical Engineering, Southeast University, Nanjing 210096, China

Suiren Wan
The Laboratory for Medical Electronics, School of Biological Sciences and Medical Engineering, Southeast University, Nanjing 210096, China

Yu Sun
The Laboratory for Medical Electronics, School of Biological Sciences and Medical Engineering, Southeast University, Nanjing 210096, China

Bing Zhang
Department of Radiology, The Affiliated Drum Tower Hospital of Nanjing University Medical School, Nanjing 210096, China

Zhen Xu
Department of Biomedical Engineering and Department of Pediatrics, University of Michigan, Ann Arbor, MI 48105, USA

Preface

Bioelectronics was first published in 1912 and focused on measurement of electrical signals generated by the body, which is the basis of the electrocardiogram. In the 1960s two new trends appeared in bioelectronics. One was implantable electronic devices and systems, such as the pacemaker. The other was the study of electron transfer in electrochemical reactions. In 1970s-1980s, Erwin Neher and Bert Sakmann invented patch clamp and investigated the potential of cell membrane. In 1990s, bioelectronics went to single molecule level and molecular electronics appeared. In February 2009, the National Institute of Standards and Technology (NIST) of the U.S. Department of Commerce published *A Framework for Bioelectronics*: *Discovery and Innovation* and defined bioelectronics as "the discipline resulting from the convergence of biology and electronics". The convergence contains two aspects: one is to study the electronics phenomena in biological systems, including the electronic characters of biomolecules, the transmission and storage of information; the other is to solve biological problems with tools of electronics, such as acquisition and analysis of bioinformation, and the related biomedical detection techniques, sensors, instruments and therapeutic methods.

Today, the concept of bioelectronics is broaden and cross-disciplinary, especially when it is boosted by nanotechnology. In general, the content of bioelectronics covers acquisition system and detection system of bioinformation, modelling and simulation of biosystem, nanobiology, biochips and micro total analysis system (μTAS), and biomedical instrumentation. In the next five years, the rising of flexible bioelectronics and bioelectronics medicine as frontiers of bioelectronics will be anticipated. It allows wearable and implantable sensors, actuators and drug delivery systems, which will monitor the health of human, and intervene or treat diseases in real-time. Bioelectronics allows people to understand life more precisely and deeply, and engineering life more effectively and creatively. It has very close relationships with material science, electronics, nanotechnology, biology and medicine and thereby it is of great significance to the industry, medicine and security.

State Key Laboratory of Bioelectronics (SKLB), Southeast University of China, was founded in 1985 by Dr. Yu Wei, one of China's noted experts in electronics. The laboratory was conferred on the name of Chien-Shiung Wu Laboratory by the world-famous physicist Dr. Chien-Shiung Wu in 1992 and was approved as an open laboratory of molecular and biomolecular electronics by the Ministry of Education (MOE) of China in the same year. In 1997, the laboratory was renamed the Key Laboratory of Molecular and Biomolecular Electronics of the MOE. In 2005, the laboratory was promoted to the State Key Laboratory of Bioelectronics by the Ministry of Science and Technology of China, which represents the highest level of research in bioelectronics in China. The objective of the laboratory construction is to conduct frontier research in the light of the development of bioelectronics; to come up with innovative research methods and technologies by applying the rapid developments in information science with emphasis on the study of life process, related pathologensis and its medical application on the level of nanometers and molecules. Now, there are four main research directions—biomaterials and interfaces, biosensors and chips, genomics and bioinformatics, and biomedical imaging. The research activities in the laboratory focus on the fabrication of biomaterials and biodevices, acquisition and sensing of bioinformation, and application of bioinformation systems. The research fields cover the preparation of nano (molecular) materials, assembly and characterization of molecular ordered structures, application of bio/nano materials, nano (molecular) devices, implantable bioelectronic devices, detection of single molecule and single cell, biosensor, micro-array chip, microfluid biochip, bioinformatics, application of biomimic information processing system, modeling and using of brain information system, etc.

Over 30 years of fast development, lots of achievements are accomplished by SKLB. The faculties have published hundreds of articles in significant international academic journals like *Nature*, *Cell*, and *Nature Materials*, obtained dozens of provincial and ministerial awards, and held over 100 patents. The laboratory has organized many academic conferences and initiated fruitful cooperation with famous laboratories in United States, Japan, Germany, United Kingdom, France, Switzerland, etc. With all the endeavors of faculties, SKLB gains their leadership in gene sequencing, biochips and nano-medicine all over the country and pioneers in the frontier of bioelectronics in the world.

In order to fulfill the need of the national and local economic development, SKLB also has established five laboratories, i.e., Wuxi Biochip Key Laboratory, Suzhou Biomedical Materials and Technology Key Laboratory, Suzhou Environment and Biosafety Key Laboratory, and Nanjing Innovation Center of Strategic Biomedical Industry in collaboration with local governments, which contributed a lot to the innovation and development of cities in Jiangsu Province. In this year, SKLB also established Institute of Biomaterials and Device in alliance with six other province key laboratories in the field of biomedical engineering in Southeast University. This institute is affiliated with Jiangsu Industry Technology Research Institute, whose aim is to reform the construction of industry in Jiangsu Province and bridge the gap between fundamental research, application development and industrialization. By these affiliated laboratories and institutes, SKLB aims to transfer the original innovative research to the productivity, explores new models of innovation and addresses the innovation driven development.

2015 is the 30th anniversary of SKLB and the fast growing bioelectronics is also attracting more and more attention from people especially when "internet+health" is widely acknowledged and the fusion of "bio" and "electronics" is more and more deep. To celebrate this, this book was proposed and the recent progress of the aforementioned four directions was reviewed, which ranges from interfaces and materials to imaging and algorithms and covers all levels of bioelectronics. By this book, the readers could not only learn the research development of bioelectronics in China, but also the cutting edge technologies in the related fields of bioelectronics.

Zhongze Gu

December 30, 2015

Contents

Chapter 1 The Photoelectric Detection at the Single Cell Level

Can Li, Ning Gu

1.1 Introduction

Cells are the basic structural and functional units of living organisms. Highly efficient and sensitive detection of the components in cell will explain some important physiological processes such as metabolism, signal transduction and so on.

In the past years, increasing interest has evolved for single-cell analytical methods, which could give exciting new insights into genomics, proteomics, transcriptomics and systems biology. Therefore, single-cell analysis has become a "hot topic", various high sensitivity, high selectivity, high temporal resolution and high throughput techniques have been developed for single-cell analysis.

In the recent years, single-cell analysis has gone deep into subcellular (local of cytoplast, cellular membrane, vesicular) and single molecule level[1, 2], while the detection methods determine accuracy and degree of analysis. As it is known to all, optical and electrical methods are more direct and fundamental in single-cell detection. By optical or electrical detection, we can directly get the signal of single cell such as impedance, voltage, which may be more accurate and fast as there is no need for conversion. Here we review the single main optical and electrical methods for single-cell detection and analysis. There are many excellent research results. While it is not realistic to cover all of them, here we just list parts of them.

1.2 Dielectrophoresis

Positioning single cells is of utmost importance in areas of biomedical research as diverse as *in vitro* fertilization, cell-cell interaction, cell adhesion, embryology, microbiology, stem cell research, and single-cell transfection. Dielectrophoretic tweezers, a sharp glass tip with electrodes on either side, are capable of trapping single cells with electric fields. Mounted on a micromanipulator, dielectrophoresis tweezers can position a single cell in three dimensions, holding the cell against fluid flow of hundreds of microns per second with more than 10 pN of force[3]. Figure 1.1 (a) is an illustration of how dielectrophoretic tweezers work. Two electrodes a few microns apart provide a non-uniform electric field which polarizes a nearby cell. If the cell is more polarizable than the solution, it will be attracted to the field maximum at the tip of the electrodes. The movement of the cell due to the force on the induced dipole is called dielectrophoresic (DEP). In more detail, the DEP force on a spherical particle is given by

$$F_{\text{DEP}} = 2\pi\varepsilon_{\text{f}} r^3 Re[\text{CM}(\omega)]\nabla E_{\text{rms}}^2 \tag{1.1}$$

where r is the radius of the particle; E_{rms} is the root mean squared electric field; ε_{f} is the fluid permittivity; ω is the frequency of the electric field; and CM (ω) is the Clausius-Mossotti factor which depends on the permittivity and conductivity of the cell and the fluid. For a given cell, a positive CM (ω) can be achieved by adjusting the frequency of the electric field and the conductivity of the fluid, which will trap the cell at the electric field maximum at the tip of the DEP tweezers. The tweezers are also capable of being used when the CM factor is negative, in which case, cells will be pushed away from the tip. Using MHz AC voltages for DEP minimizes the induced potential across the trapped cell membrane and the electro-osmotic flow of the charged double layer[3].

T. P. Hunt and R. M. Westervelt[4] fabricated dielectrophoretic tweezers [Figure 1.1 (b)] by a standard method for fabricating micropipettes first, and then depositing electrodes. The sharpened glass rod was placed in a high-vacuum thermal evaporator. 7 nm Ti and 20 nm Au were evaporated on one side of the rod. Figure 1.1 (c) is an electron micrograph of the tweezer tip. With the dielectrophoretic tweezers, they show that cells are trapped without harm while they divide in the trap.

DEP with geometry of 2D microelectrode has limitations on the movement and placement of a particle at a desired position. The methods using 2D microelectrodes have shown results mainly on trapping of multiple particles. In general, the size of 2D microelectrode is larger than the objects such as a single bead or cell, which causes the difficulty in manipulation for a single particle. Kiha Lee et al.[5] proposed a localized and 3D movable electric field configuration, and analyzed for the functional requirements of dielectrophoretic tweezers. To achieve a steeply focused field, they developed an electrochemical machining method for a sharp probe electrode with the range of 200-300 nm in its radius of curvature (ROC), and used the developed dielectrophoretic tweezers to manipulate cells and beads.

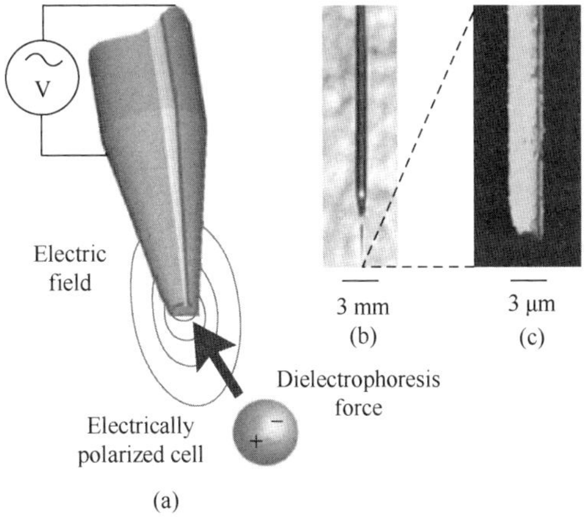

Figure 1.1 Dielectrophoretic tweezers for cell manipulation. (a) Schematic of dielectrophoretic tweezers in operation. A voltage across two electrodes on either side of a sharp glass tip creates an electric field which polarizes a cell and pulls the cell into the field maximum at the end of the tip; (b) Photograph of dielectrophoretic tweezers; (c) SEM image of the tweezer tip. Electrodes appear light while the insulating gap between electrodes is dark

Rupert S Thomas et al.[6] present a novel design of micron-sized particle trap that uses negative dielectrophoresis (nDEP) to trap cells in high conductivity physiological media. The design is scalable and suitable for trapping large numbers of single cells. Each trap has one electrical connection and the design can be extended to produce a large array. The trap consists of a metal ring electrode and a surrounding ground plane, which creates a closed electric field cage in the center. The operation of the device was demonstrated by trapping single latex spheres and HeLa cells against a moving fluid. The dielectrophoretic holding force was determined experimentally by measuring the displacement of a trapped particle in a moving fluid. Then they compared this with theory by numerically solving the electric field for the electrodes and calculating the trapping force, demonstrating good agreement. Analysis of the 80 mm diameter trap showed that a 15.6 mm diameter latex particle could be held with a force of 23 pN at an applied voltage of 5 V peak-peak.

Growth monitoring is the method of choice in many assays measuring the presence or properties of pathogens, e.g., in diagnostics and food quality. Established methods, relying on culturing large numbers of bacteria, are rather time-consuming, while in healthcare time often is crucial. Several new approaches have been published, mostly aiming at assaying growth or other properties of a small number of bacteria.

However, no method so far readily achieves single-cell resolution with a convenient and easy to handle set-up that offers the possibility for automation and high throughput. Ingmar Peitz and Rien van Leeuwen[7] demonstrated these benefits in their study by employing dielectrophoretic capturing of bacteria in microfluidic electrode structures, optical detection and automated bacteria identification and counting with image analysis algorithms. For a proof-of-principle experiment they chose an antibiotic susceptibility test with *E. coli* and polymyxin B. Growth monitoring is demonstrated on single cells and the impact of the antibiotic on the growth rate is shown. Their experiments demonstrated the feasibility of single-cell growth monitoring based on facilitated image analysis of immobilized but dividing bacteria in a time span of a few hours.

The ability to research individual cells has been seen as important in many kinds of biological studies. In Kung-Chieh Lan and Ling-Sheng Jang's[8] study, cell impedance analysis was integrated into a single-cell trapping structure. For the purpose of precise positioning, they developed a cell manipulation and measurement microchip, which uses an alternating current electrothermal (ACET) effect and an nDEP force to move a particle and cell on measurement electrodes. An ACET and an nDEP can be easily combined with subsequent analyses based on electric fields. The capture range of the microwell can be designed for cells of various sizes. In order to demonstrate the precision of the positioning, a particle is captured, measured, and released twice. The results show that the impedance error of the particle is about 3%. Finally, the developed structure is applied to trap and measure the impedance of a HeLa cell.

Different types of human cells repeatably undergo a first order transient motion response when subjected to a specific optically induced dielectrophoresis (ODEP) force field. Yuliang Zhao et al.[9] proposed an improved theoretical model for the motion of a cell under the influence of DEP force. Based on the trajectory tracking of various types of human cells, they measured cell displacement as a function of time and calculated the associated velocities and accelerations. Then, they analyzed the motion of the cells and distinguished between different types of cells based on their transient motion response under a specific DEP force field. The computer-vision based cell motion analysis described in this letter using an ODEP system could be used to track cell motion trajectory accurately, while enabling the simultaneous identification and sorting of different types of cells.

Mengxing Ouyang et al.[10] investigated the phenomenon of self-rotation observed in naturally and artificially pigmented cells under an applied linearly polarized alternating current (non-rotating) electrical field. The repeatable and controllable rotation speeds of the cells were quantified and their dependence on dielectrophoretic parameters such as frequency, voltage, and waveform was studied. Moreover, the rotation behavior of the pigmented cells with different melanin content was compared to quantify the correlation between self-rotation and the presence of melanin. Most importantly, macrophages, which did not originally rotate in the applied non-rotating electric field, began to exhibit self-rotation that was very similar to that of the pigmented cells, after ingesting foreign particles (e.g., synthetic melanin or latex beads). The discovery presented in this book will enable the development of a rapid, non-intrusive, and automated process to obtain the electrical conductivities and permittivities of cellular membrane and cytoplasm in the near future.

The density of a single cell is a fundamental property of cells. Cells in the same cycle phase have similar volume, but the differences in their mass and density could elucidate each cell's physiological state. Yuliang Zhao et al.[11] combined sedimentation theory, computer vision, and microparticle manipulation techniques in an optically induced electrokinetics (OEK) microfluidic platform, which enables the measurement of single-cell volume, density, and mass rapidly and accurately in a repeatable manner. They could rapidly measure the density and mass of a large quantity of cells rapidly using an OEK microfluidic platform which is not very easy before. The microparticles and cells could be lifted up by a digitally controlled DEP force in the medium of the OEK chip and then allowed to "free fall" to the OEK chip's surface (Figure 1.2, see the Color Inset, p.1). To be specific, a simple time-controlled projected light pattern is used to illuminate the selected area on the OEK microfluidic chip that contains cells to lift the cells to a particular height above the chip's surface. Then, the cells are allowed to "free fall" to the chip's surface, with competing buoyancy, gravitational and fluidic drag forces acting on the cells. By using a computer vision algorithm to accurately track the motion of the cells and then relate the cells' motion trajectory to sedimentation theory, the volume, mass, and density of each cell can be rapidly determined. Based on the sedimentation theory of microparticles in solution, they successfully implemented a sedimentation velocity detection scheme to measure the buoyant density and mass of microparticles and yeast cells. This new method could potentially be widely used for determining the density and mass of many types of cells rapidly.

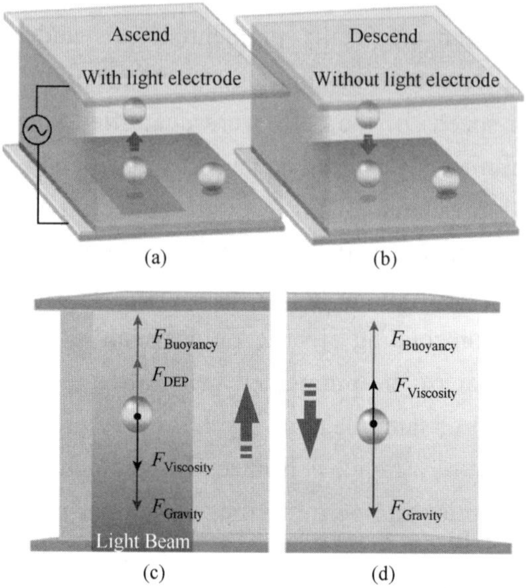

Figure 1.2　Illustration of the manipulation of microparticles. (a) When a light beam is projected onto the OEK chip, the particle is lifted up and (b) when there is no light projected, the particle starts to sediment. The forces act on the particle, when the beam is (c) rising and (d) falling

1.3　Optical Tweezers

Optical tweezers are tools that use optical pressure in trapping micro objects including living cells and microorganisms, and also in directionally rotating artificial microobjects fabricated by micromachining. Given their non-invasive nature, optical tweezers are useful particularly in biological processes. Nowadays, these optical tweezers are used to control and manipulate various types of micro/nanoobject in various research and industrial fields.

The field of optical tweezers has enjoyed a wide range of applications since its inception in the early 1970s. Optical tweezers (also known as optical traps) use a strongly focused beam of laser light to trap and manipulate particles/cells suspended in a medium very precisely within a three-dimensional space[12].

By using light to trap microscopic objects non-invasively, optical tweezers provide a flexible tool for ultrafine positioning, measurement, and control. In practice, forces up to 200 pN or thereabouts may be applied with sub-pN resolution on objects whose characteristic dimensions are similar to the wavelength of light. Particle positioning and detection capabilities are therefore on a spatial scale of micrometers down to angstroms. Optical tweezers used in combination with other methodologies, such as fluorescence spectroscopy, micropipettes, and optical microbeams, have all helped to make optical tweezers an extremely versatile tool[13].

The laser light in optical tweezers imparts two opposing forces on the trapped particle. A scattering force is caused by reflection of the laser beam, whereas a gradient force is caused by the refraction of the laser beam. As seen in Figure 1.3 (see the Color Inset, p.1), the scattering force moves the particle along the direction of beam propagation. Figure 1.4 (see the Color Inset, p.1) illustrates how the gradient force acts to restore the particle toward the focal point of the beam. For the particle to experience this restoring gradient, the refractive index of the particle has to exceed the refractive index of the surrounding medium, and a stable optical trap results when the gradient force is greater than the scattering force on the particle[14].

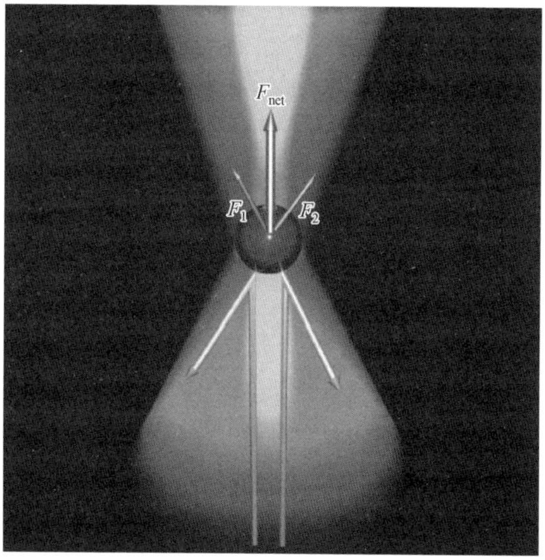

Figure 1.3 Scattering force. The reflection of rays produces momentum in the opposite direction, resulting in a net force along the direction of laser propagation

Trapped particles still undergo Brownian motion, and the trap stiffness, which is directly proportional to the laser power and the volume of the trapped particle, keeps the particles from escaping. A high trap stiffness is desirable to keep the particle as still as possible when making precise measurements. To achieve the stiffness sufficient for trapping single biomolecules, a detrimentally high laser power would be required to compensate for the small volumes of the individual biomolecules. Therefore, molecules of interest are most often linked to micrometer-sized beads, which can be stably trapped by the piconewton forces relevant for biomolecular measurements.

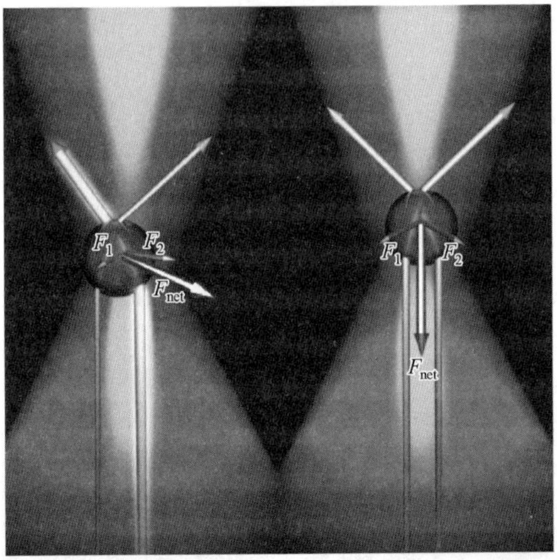

Figure 1.4 Gradient force. (left) When the bead is not in the beam's center, the larger momentum change of the more intense rays causes a net force that pulls the bead back toward the center of the trap; (right) when the bead is laterally centered in the beam, the net force points toward the focal point of the beam

Although the first optical trap was built with two counter propagating lasers, currently, the most common configuration has a single trapping laser. The key parts of an optical trap are the trapping laser and the high-numerical-aperture microscope objective that focuses the laser to a tiny spot. Trapping lasers in the near-IR band of the spectrum (~800-1100 nm) are frequently used to minimize damage to biological samples. Many experiments are conducted in an aqueous medium to more closely resemble the native environment of biological systems. Most optical traps incorporate oil-immersion objectives to achieve the maximum numerical aperture, but water-immersion objectives, which minimize spherical aberration and can be used farther from the surface, are increasing in popularity.

Optical tweezers have emerged as an essential tool for manipulating single biological cells and performing sophisticated biophysical/biomechanical characterizations. Distinct advantages of using tweezers for these characterizations include non-contact force for cell manipulation, force resolution as accurate as 100 aN and amiability to liquid medium environments[15].

Optical tweezers has been widely employed as a tool for actively manipulating and positioning biological objects at the nano/microscale. One of the most popular applications is to apply tweezers to confine or constrain single cells, as well as to organize, assemble, locate, sort and modify them.

The development of a large variety of imaging techniques allowed detailed information to be gained from investigations of single cells to be possible. The use of multiple optical traps has high potential within single-cell analysis since parallel measurements provide good statistics. Multifunctional optical tweezers are, for instance, used to study cell heterogeneity in an ensemble, and force measurements are used to investigate the mechanical properties of individual cells.

Optical manipulation has prospects within the field of cell signaling and tissue engineering. When combined with microfluidic systems the chemical environment of cells can be precisely controlled. Hence the influence of pH, salt concentration, drugs and temperature can be investigated in real time. Fast advancing technical developments of automated and user-friendly optical manipulation tools and cross-disciplinary collaboration will contribute to the routinely use of optical manipulation techniques within the life science[16].

Yuichi Wakamoto et al.[17] described an improved on-chip microcultivation system using an on-chip microculture system and optical tweezers for continuous obser vation of isolated single cells, which enables genetically identical cells to be compared.

To be specific, they use photolithography to construct microchambers with 5-μm-high walls made of thick photoresist (SU-8) on the surface of a glass slide. These microchambers are connected by a channel through which cells are transported, by means of optical tweezers, from a cultivation microchamber to an analysis microchamber, or from the analysis microchamber to a waste microchamber. The microchambers are covered with a semi-permeable membrane to separate them from nutrient medium circulating through a "cover chamber" above. Differential analysis of isolated direct descendants of single cells showes that this system could be used to compare genetically identical cells under contamination-free conditions. It should thus help in the clarification of heterogeneous phenomena, for example unequal cell division and cell differentiation.

Raman spectroscopy is a vibration spectroscopic technique that has been widely used to probe biochemical changes of biological sample such as tumor tissue, blood cells, bacteria and yeast. Haiyang Tang et al.[18] applied near-infrared Raman spectroscopy to analyze the chemical composition changes of intact or swollen mitochondria induced by calcium ions. Using a confocal Laser Tweezers Raman Spectroscopy (LTRS) system that combined optical trapping and near-infrared Raman spectroscopy, they confined a single mitochondrion and consequently measured its Raman spectra following the addition of calcium ion solution. Then Raman spectra of mitochondria isolated from rat liver, heart muscle and kidney was analyzed respectively. The major Raman peaks at 1654 cm^{-1}, 1602 cm^{-1}, 1446 cm^{-1}, 1301 cm^{-1} and 1226 cm^{-1} were observed from individual intact mitochondria. The differences in near-infrared spectra between intact and Ca^{2+} damaged mitochondria were examined. It was found that after the exposure of the intact mitochondria to the 100 μM Ca^{2+} solution the band of 1602 cm^{-1} decreased very rapidly in the first period and then disappeared after 30 minutes, while the intensities of the phospholipids and protein bands changed slowly in the first period and then suddenly disappeared, corresponding to the Ca^{2+} induced swelling process. These results demonstrate the potential of LTRS technique as a valuable tool for the study of bioactivity and molecular composition of mitochondria.

Optical tweezers have undergone a resurgence of interest because computer addressable spatial light modulators (SLMs) make it possible to create multiple traps that can be steered individually in recent years[19-22]. The ability to introduce focal power with the SLM also means that the traps can be manipulated axially, allowing controlled 3D positioning within the sample volume, typically over a range of several 10's microns[23].

Pamela Jordan et al.[24] used holographic optical tweezers to manipulate individual *E. coli*, creating permanent 3D configurations of cells at predefined positions within a gelatin matrix. Within a solution comprising monomer precursors (Figure 1.5), the matrix was then set, and after the laser beam was removed the structures remained intact for many days. In the presence of appropriate nutrients, the *E. coli* survived within the gelatin matrix for several days. The technique could have a number of potential future applications, including the arrangement of a variety of different cell types in complex architectures, as motifs for promoting tissue differentiation and growth within the field of cell engineering.

It is possible to collect the immobilized target cell for analysis or culture by switching off the micro heater and releasing the cell from the entrapment. Fumihito Arai et al.[25] proposed a novel approach appropriate for rapid separation and immobilization of a single cell by concomitantly utilizing laser manipulation and locally thermosensitive hydrogelation. A single laser beam is employed as optical tweezers for separating a target cell and locating it adjacent to a fabricated, transparent micro heater. Simultaneously, the target cell is immobilized or partially entrapped by heating the thermosensitive hydrogel with the micro heater. The state of the thermosensitive hydrogel can be switched from sol to gel and gel to sol by controlling the temperature through heating and cooling by the micro heater. After other unwanted cells are removed by the high-speed cleaning flow in the microchannel, the entrapped cell is successfully isolated. The proposed approach is feasible for rapid manipulation, immobilization, cleaning, isolation and extraction of a single cell.

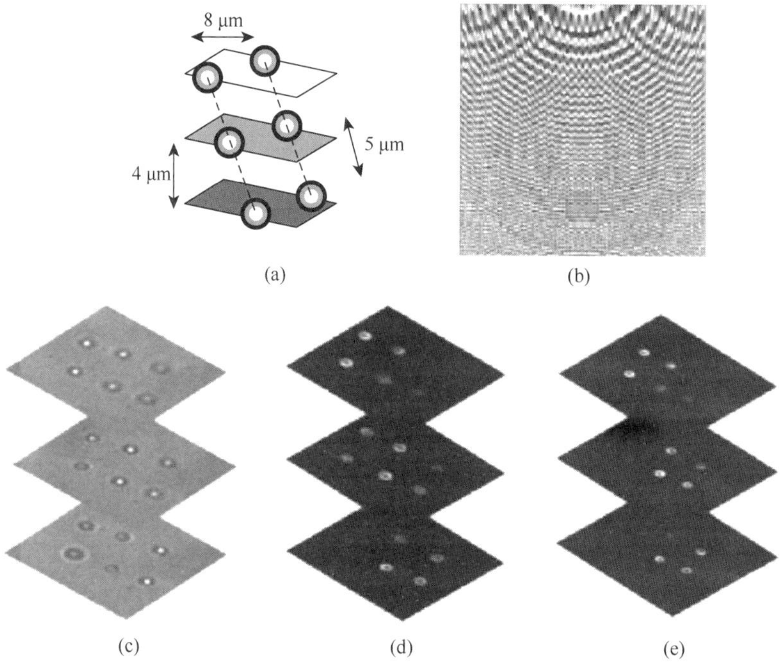

Figure 1.5 Permanently set 3D structure of fluorescent spheres. A schematic of the structure (a), the hologram used to create the traps (b), and images taken in optical tweezers (c), and then removed from the tweezers and imaged within a confocal (d) and multi-photon (e) microscope

Cells naturally exist in a dynamic chemical environment, and therefore it is necessary to study cell behaviour under dynamic stimulation conditions in order to understand the signaling transduction pathways regulating the cellular response. However, most experiments looking at the cellular response to chemical stimuli have mainly been performed by adding a stress substance to a population of cells and thus only varying the magnitude of the stress. Emma Eriksson et al.[26] demonstrate an experimental method for time-resolved single-cell analysis using a combination of microfluidics, fluorescence microscopy and optical tweezers, enabled acquisition of data on the behaviour of single cells upon reversible environmental perturbations; the cells are individually selected and positioned in the measurement region on the bottom surface of the microfluidic device using optical tweezers. The optical tweezers were introduced to actively select and position cells within the measurement region. Consequently, the number of cells in each experiment can be optimized while clusters of cells, that render subsequent image analysis more difficult, can be avoided. The active selection made it possible to favor a specific category of cells from the population as well as avoiding clusters of cells and cells with a deviating morphology, and cellular regulation in *S. cerevisiae* was quantitatively studied by monitoring the real-time changes in the spatio-temporal distribution of GFP-tagged proteins in response to reversibly changing glucose levels.

Sorting (or isolation) and manipulation of rare cells with high recovery rate and purity are of critical importance to a wide range of physiological applications. Xiaolin Wang et al.[27] reported a generic single-cell manipulation tool that integrates optical tweezers and microfluidic chip technologies for handling small cell population sorting with high accuracy. The design is based on dynamic fluid and dynamic light pattern, in which single as well as multiple laser traps are employed for cell transportation, and a recognition capability of multiple cell features. The laminar flow nature of microfluidics enables the targeted cells to be focused on a desired area for cell isolation. To recognize the target cells, an image processing methodology with a recognition capability of multiple features, e.g., cell size and fluorescence label was developed. The target cells can be moved precisely by optical tweezers to the desired destination in a non-invasive manner. The unique advantages of this sorter are its high recovery rate and purity in small cell population sorting. Yeast cells and Hesc respectively with the established cell sorter prototype were isolated by the sorter.

Deformation imposed by optical tweezers provides a useful means for the study of single-cell mechanics under a variety of well-controlled stress-states[28]. Measuring the induced deformation allows researchers to investigate cell viscoelastic properties. Optical tweezers are used to study the elasticity of red cell membranes[29] and apply calibrated forces to human erythrocytes[30-32]. Simultaneous measurements of forces and deformations exerted on a red blood cell enable the inferration of the elastic shear modulus of the membrane. The cell is seized and deformed by means of two small silica beads bound to the membrane. The beads are trapped in split optical tweezers and used like handles to pull on the membrane with two opposite forces. The mechanical responses of the cell during loading and upon release of the optical force are analyzed to extract the elastic properties of the cell membrane by recourse to several different constitutive formulations of the elastic and viscoelastic behavior within the framework of a fully three-dimensional finite element analysis.

Cell fusion, a process by which two or multiple cells merge to form a single entity, plays an important role in numerous biological events and applications such as tissue regeneration. The combination of an optical trap and a pulsed UV laser was allowed for very selective and precise fusions of two preselected cells in suspension[33]. Recent stem cell research has discovered that stem cell could reprogram somatic cell through cell fusion. Shuxun Chen et al.[34] reported they use optical tweezers and pulsed UV laser to accomplish efficient fusion of human embryonic stem cells (hESCs) and also hESC with somatic cell. Results indicated that two or multiple hESCs could be fused by pulsed UV laser irradiation with a high efficiency ($>50\%$). This targeting laser-induced cell fusion could be an efficient tool for studying stem cell differentiation, maturation, and reprogramming.

1.4 Microfluidics

Methods for single-cell analysis are critical to revealing cell-to-cell variability in biological systems, especially in cases where relevant minority cell populations can be obscured by population-averaged measurements. For single-cell studies, there is a considerable need for a platform that allows integrated multiparameter measurement and manipulation of cells. However, to date single-cell studies have been limited by the cost and throughput required to examine large numbers of cells and the difficulties associated with analyzing small amounts of starting material. Microfluidic technologies are increasingly bringing new and improved protocols to single-cell analysis strategies. Microfluidic approaches are well suited to resolving these issues by providing increased sensitivity, economy of scale, and automation. After many years of development microfluidic systems are now finding traction in a variety of single-cell analytics including gene expression measurements, protein analysis, signaling response, and growth dynamics. With newly developed tools now being applied in fields ranging from human haplotyping and drug discovery to stem cell and cancer research, the long-heralded promise of microfluidic single-cell analysis is now finally being realized[35].

A single cell weighs a few ng, has a volume of ~1 pL, a size of ~10 μm and consists mainly of water. Inorganic ions and small organic molecules (sugars, vitamins and fatty acids) make up most of the cellular content, and less than 25% (by weight) are made up by proteins, DNA, and RNA. The molar concentration of a gene is only in the order of 10-12 M while the total protein content is as high as 109 molecules per cell (hundreds of pg). It has been estimated that a cell contains more than 100000 different proteins, ranging from <100 copies of many receptors, 1000-10000 copies of signaling enzymes, to 108 copies of some structural proteins, according to the book *The Cell*[36]. With those basic facts in mind, the suitability of Lab-on-chip (LOC) devices for single-cell analyses can be envisioned. Common guidelines for analytical single-cell techniques are the aim of time-efficient, systematically arranged and accurate analysis (often referred to as high-throughput), instrument compatibility, transparency (imaging/optical properties), cell accessibility (manipulation and handling), stability and robustness (cell tracking), sensitivity, mimicking of *in vivo* conditions, user friendliness, and gentle cell handling[37].

Lab-on-chip technologies are being developed for multiplexed single-cell assays. Impedance offers a simple non-invasive method for counting, identifying and monitoring cellular function. A number of different microfluidic devices for single-cell impedance have been developed. Individual cells can be identified on the basis of differences in size and dielectric properties using electrical techniques which are non-invasive and label-free. Characterization of the dielectric properties of biological cells is generally performed in two ways using AC electrokinetics or impedance spectroscopy. AC electrokinetic techniques are used to study the behaviour of particles (movement and/or rotation) and fluids subjected to an AC electric field.

The electrical forces act on both the particles and the suspending fluid and have their origin in the charge and electric field distribution in the system. Electrical impedance spectroscopy measures the AC electrical properties of particles (in suspension) from which the dielectric parameters of the particles can be obtained.

Impedance micro-cytometry is an evolution of the micro-Coulter counter[38-40], but instead of fabricating an aperture with sensing electrodes either side, microelectrodes are integrated into the walls of the microchannel[40].

David Holmes et al.[41] have used a microfluidic impedance cytometer to accurately and precisely measure the dielectric properties of single cells in minimally prepared whole blood. As is shown in Figure 1.6, the microfluidic chip is fabricated from glass and contains a microchannel through which cells flow. Micro electrodes are fabricated in the channel, and these are connected to a small signal AC voltage. The ratio of this voltage to the measured current is used to measure the impedance of a single cell. Simultaneously, the cell is illuminated by laser light that excites fluorescence from any fluorescent antibodies. The emitted light is collected through an objective lens, spatially and spectrally filtered and measured using photomultipliers. The simultaneous collection of both the impedance and fluorescence signals allows independent identification and verification of cell phenotype or physiological state using fluorescent probes (anti-CD antibodies).

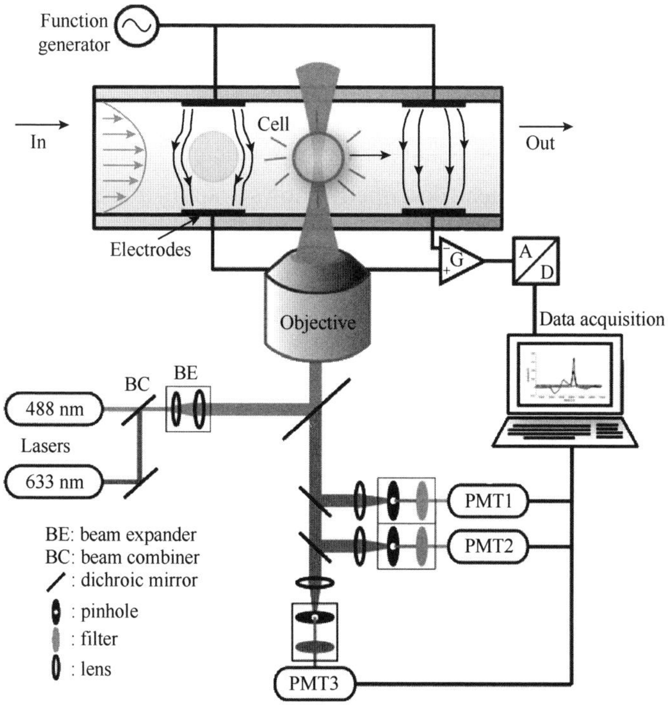

Figure 1.6 Schematic diagram of the micro impedance cytometer system, including the confocal-optical detection set-up. Dual-laser excitation and three-colour detection were implemented along with dual frequency impedance measurement. The cell flows through the microchannel and passes between two pairs of electrodes and the optical detection region. The fluorescence properties of the cell are measured simultaneously with the impedance allowing comparison of the electrical and optical properties on a cell-by-cell basis

Miniature high speed label-free cell analysis systems have the potential to deliver fast, inexpensive and simple full blood cell analysis systems that could be used routinely in clinical practice. The microfluidic single-cell impedance cytometer was used to perform a white blood cell differential count and measure the impedance of single cells at two frequencies. Human blood, treated with saponin/formic acid to lyse erythrocytes, flows through the device and a complete blood count is performed in a few minutes. Verification of cell dielectric parameters was performed by simultaneously measuring fluorescence from CD antibody-conjugated cells. This enabled direct correlation of impedance signals from individual cells with phenotype. Tests with patient samples showed 95% correlation against commercial (optical/Coulter) blood analysis equipment, demonstrating the potential clinical utility of the impedance microcytometer for a point-of-care blood analysis system.

Mengsu Yang have developed a microfluidic device that integrates cell docking and concentration gradient functions. The ATP-dependent calcium uptake reaction of HL-60 cells was used as a model for on-chip measurement of the threshold ATP concentration that induces significant intracellular calcium signal.

The microfluidic device described allows rapid transportation and gentle immobilization of particles, particularly mammalian cells, in a controllable manner. The integration of the cell manipulation and solution manipulation on a microchip provides an efficient and stable platform for the measurement of concentration-dependent biological responses within a confined microscale feature with improved reproducibility and reliability[42].

The microfluidic chip with electrochemical detection (ED) is a versatile technique for the analysis of biochemical electroactive substances within single cells. The selectivity of the analytes of interest detected for the different electrode material makes the electropherograms simple. The microfluidic chip-ED can integrate the whole process in single-cell analysis including single-cell loading, lysis, separation and detection of intracellular species.

Fangquan Xia developed a method for single-cell analysis by using electrochemical detection (ED) combining a microfluidic chip with a double-T injector. When the chip with a double-T injector is used, the single-cell loading is controlled easily by using electric field. The DC electric field of 220 V/cm can lyse single protoplasts docked in the double-T injector. They have used the system with an Au/Hg electrode to determine glutathione in single human hepatocarcinoma cells. After taking into consideration leakage effects during injection of standard solutions, the external standardization can be used to quantify the analytes of interest in individual cells. The system and the approach can be also applicable to the other chemical species[43].

A simple and effective means for the directed capture and analysis of single cells is important for cell detection and analysis. A novel combination of cell surface modification and electric field-directed adhesion is developed for the rapid capture and chemical activation of living single cells[44]. The device uses direct labeling of the cells, instead of the more common approach of tailoring the substrate for cell adhesion. The combination of the driving electric field and cell thiolation provides adhesion sufficient to withstand subsequent flow used for rinsing or reagent introduction. Additionally, the ability to individually direct the capture of single cells from multiple populations on neighboring electrodes in an integrated microfluidic chip presents the advantage of multicell type patterning. This device also provides a platform for single-cell genetic studies as well as the development of a bioelectronic interface for fundamental studies of cell activity.

Biological analyses traditionally probe cell ensembles in the range of 103-106 cells, thereby completely averaging over relevant individual cell responses, such as differences in cell proliferation, responses to external stimuli or disease onset. In past years, this fact has been realized and increasing interest has evolved for single-cell analytical methods, which could give exciting new insights into genomics, proteomics, transcriptomics and systems biology.

Microfluidic or lab-on-a-chip devices are the methods of choice for single-cell analytical tools as they allow the integration of a variety of necessary process steps involved in single-cell analysis, such as selection, navigation, positioning or lysis of single cells as well as separation and detection of cellular analytes. Along with this advantageous integration, microfluidic devices confine single cells in compartments near their intrinsic volume, thus minimizing dilution effects and increasing detection sensitivity[45].

Although other choices for cell-docking, such as trapping in channel recesses, by dielectrophoretic approaches, or employing adherents on channel walls, might be used, the experiences of Jian Gao et al.[46] showed that "switching the HV on and off" is a most effective, robust and convenient approach for cell docking. They have demonstrated the feasibility of integrating the whole process for single-cell analysis on a microfluidic chip including single-cell injection, lysis, separation and detection of cellular constituents by capillary electrophoretic (CE), employing the docked-cell-electrolysing approach. As demonstrated by the determination of Glutathione (GSH) in human erythrocytes, the proposed microfluidic system provides an effective and efficient platform for qualitative analysis and quantitative analysis of intracellular constituents in a single cell, which might be of significance in biological and medical research. This work also demonstrates that well-controlled docking and lysing of single cells before CE separation is a vital factor for ensuring reproducibility and separation efficiency. The high separation efficiency achieved with single cells in this work shows favorable potential for the analysis of multi-intracellular components in the single cell. The challenge lies in the development of suitable fluorimetric labeling reagents, which could penetrate the cell membrane under physiological conditions and form derivatives with multi-components of biological interest. This work also showed that good performance in single-cell analysis need not always be related to complicated microstructures, since the performance achieved in this work was achieved using a microfluidic chip of the simplest design.

The real-time detection of *E. coli* through latex immunoagglutination using a microfluidic device with proximity optical fibers in phosphate buffered saline (PBS) was demonstrated in a Y-channel polydimethylsiloxane (PDMS) microfluidic device[47]. This method is essentially one-step and requires no sample pre-treatment or cell culturing. The detection limit was as low as 40 cfu/mL or 4 cfu per device (viable cells only), or <10 cfu/mL or <1 cfu per device (including dead cells and free antigens), which are superior to the other results of the *E. coli* detection performed in a microfluidic device.

The idea of using discrete droplets to maintain a high concentration of sub-cellular contents for digital detection at single molecule or cell level is highly promising as it can greatly enhance the detection sensitivity and quantification accuracy. However, realization of this idea has been hampered by the need for complicated target amplification and off-chip sample processing. Tushar D. Rane et al.[48] have extended the repertoire of digital microfluidic applications to include a versatile "sample-to-answer" platform for pathogen detection by introducing a hybridization based assay within droplets and demonstrating its functioning using *E. coli* as a model organism. The proposed platform can perform integrated sample preparation and amplification-free pathogen detection on a microfluidic chip. The integration enables digital detection in a continuous flow format and thereby high-throughput measurement of single cells.

Resistive pulse detection is based on the transient resistance modulation when a particle/cell passes through a small sensing aperture and displaces a volume of electrolyte equivalent to its own volume. Characterization of individual cells by electrical impedance spectroscopy has many advantages over the suspension technique with cell populations. The single-cell approach does not require time-consuming consideration for a uniform population of cells in buffer solution, which is required for the control of the suspension technique. Moreover, the theoretical analysis becomes relatively simple and easy because it is not necessary to take into account electrical interactions between cells.

Younghak H. Cho et al.[49] developed a novel device for the characterization of a single red blood cell which has the microchannels for filtering cells and the twin microcantilever array for measuring electrical impedance of single cell. The device was designed to cause red blood cells to be under condition similar to actual capillary circulations. The electrical properties of a suspended single cell will be explored by the analysis of the frequency-dependent response to electric fields.

The electrical impedance of normal and abnormal red blood cell were measured over the frequency range from 1 Hz to 10 MHz from the electrical impedance experiment of normal and abnormal red blood cells, it was examined that the electrical impedance between normal and abnormal red blood cells was significantly different in magnitude and phase shift.

Thus, the normal cell can be taken apart from the abnormal cell by electrical impedance measurement. The applicability of this technology can be used in cellular studies such as cell sorting, counting or membrane biophysical characterization.

Tien Anh Nguyen et al.[50] present a new cell-based impedance sensor chip with integrated microfluidics for cancer-cell detection with single-cell resolution. By integrating the passive pumping method with the proposed sensor chip, single cells can be efficiently trapped on the microelectrode array for sequential 2D cell culture and impedance measurement without the requirement of physical connections to off-chip syringe pumps or off-chip pneumatics. The impedance spectrums along with the equivalent circuit model indicated that the cellular activities such as cell adhesion and cell spreading at single-cell level can be identified by the electrical parameters in the circuit. The proposed single-cell based sensor chip has the potential for further detection and quantification of cancer cells from various tumor stages and different cell lines.

To simultaneously count, size and measure the density of a yeast or *E. coli* population, Jiashu Sun et al.[51] have presented a microfluidic resistive pulse-based cell sensor. A unique feature of the design is to combine the pressure driven flow and the electric field across the small sensing aperture of the microfluidic device. The small size of the sensing aperture enabled the highly sensitive detection of yeast and bacteria, making it suitable for clinical use and on-site disease diagnostics. The precise control of the flow rate through the syringe pump also enables the measurement of cell density in a short time period. This miniaturized microfluidic cell sensor can be easily integrated with cell culture systems and realize on-line cell analysis, and also has the potential to be an alternative to conventional Coulter counters for on-site sample measurements.

Cell mass depends on the synthesis of proteins, DNA replication, cell wall stiffness, cell cytoplasm density, cell growth, ribosome and other analogous of organisms[52]. Chronic diseases like cancer and tumour affect intracellular physiological properties of cells[53], subsequently cell mass and density will be changed as well[54, 55].

Therefore, it is strongly believed that studying single-cell mass and its measurement techniques will enhance our understanding of physiological properties of cell and perhaps it may provide new tools for disease diagnosis through the variation of single-cell mass property of identical cells at different health conditions[56].

The suspended microchannel resonator (SMR), which consists of an embedded microfluidic channel inside a cantilever that resonates in an on-chip vacuum, enables single-cell measurements of buoyant mass with femtogram-level resolution. Cells or particles with a different density than the surrounding fluid cause a small change in the cantilever's resonant frequency as they flow through the cantilever, and their buoyant mass can be determined from the magnitude of the frequency change. The buoyant mass, or mass of a particle in fluid, is defined as

$$m_{\text{buoyant}} = V(\rho_{\text{p}} - \rho_{\text{f}}) \tag{1.2}$$

where V is the particle volume, and ρ_{p} and ρ_{f} are the particle and fluid densities, respectively.

Mechanical trapping structures integrated with the SMR can effectively load and unload a single cell while its buoyant mass and the density of the surrounding fluid are continuously monitored. Weng et al.[57] augment current SMR capabilities with single-cell manipulation techniques based on mechanical trapping structures. Two types of mechanical trap were evaluated: the first, three-channel SMRs capture single cells and rapidly exchange the surrounding buffer for a new fluid [Figure 1.7 (a)(see the Color Inset, p.2) and Figure 1.7 (b)(see the Color Inset, p.2)] proved to be most suitable for single-cell density measurements. Consecutive buoyant mass measurements in different fluids are rapidly acquired without the need for microfluidic mixing. This approach is not optimal for monitoring growth behavior of the cells prior to and after drug delivery as the presence of a third channel introduces fluidic pressure variations that prevent stable dynamic flow trapping. To address this limitation, the second type, columned SMRs [Figure 1.7 (c)(see the Color Inset, p.2)] that enable a complete fluid exchange throughout the system by temporarily capturing a cell was developed. Dynamic flow trapping can be resumed without hindrance following fluid exchange, thereby allowing for effective growth monitoring before and after drug delivery.

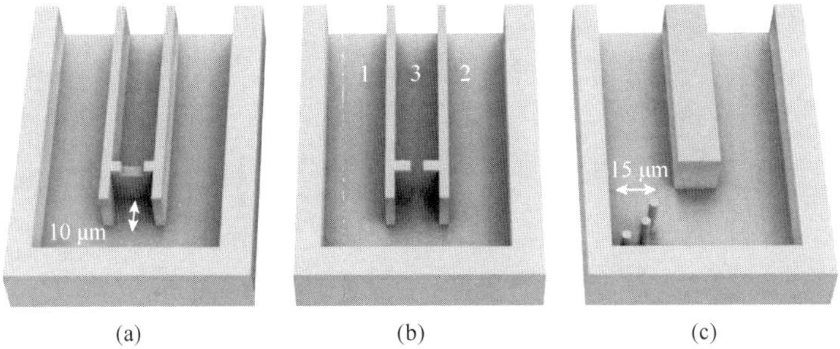

Figure 1.7 Top perspective of SMRs with mechanical traps. Three-channel SMRs with different third-channel dock geometries: (a) 3 mm×8 mm device with a 200 nm horizontal slit and (b) 8 mm×8 mm device with a vertical 2 mm wide opening (c) columned SMR

The ability to measure the buoyant mass of an individual cell in two fluids allows its density as well as its response to a drug to be measured. They measured the density, mass, and volume of individual yeast cells in their culture medium and in PBS: Percoll solution, as well as the dynamics of buoyant mass accumulation and loss in mouse lymphoblast cells before and after complete buffer replacement with and without the presence of a drug or stimulus.

Even though suspended micro channel resonator has a great contribution to the advancements of single-cell mass measurement techniques, this method is limited to non-adherent cell only[58] and cell stiffness data remained elusive.

To measure adherent cell mass, Park et al.[52] have demonstrated a new platform for the mass measurement and optical observation of single cells in their physiological conditions using a cantilever mass sensor (Figure 1.8, see the Color Inset, p.2). The "living cantilever arrays" cantilevers were submerged into the L-15 growth medium and cells were cultured. Hence, cells remain alive and adherent cell mass was measured.

Pedestal measurement sensor has an excellent geometrical shape that enables cell to be trapped within the mass sensing region.

Park et al.[58] have developed an array of micro-electro-mechanical systems (MEMS) resonant mass sensors that can be used to directly measure the biophysical properties, mass, and growth rate of single adherent cells. Unlike conventional cantilever mass sensors, these sensors retain a uniform mass sensitivity over the cell attachment surface.

By measuring the frequency shift of the mass sensors with growing (soft) cells and fixed (stiff) cells, and through analytical modeling, the Young's modulus of the unfixed cell and the dependence of the cell mass measurement on cell stiffness can be derived.

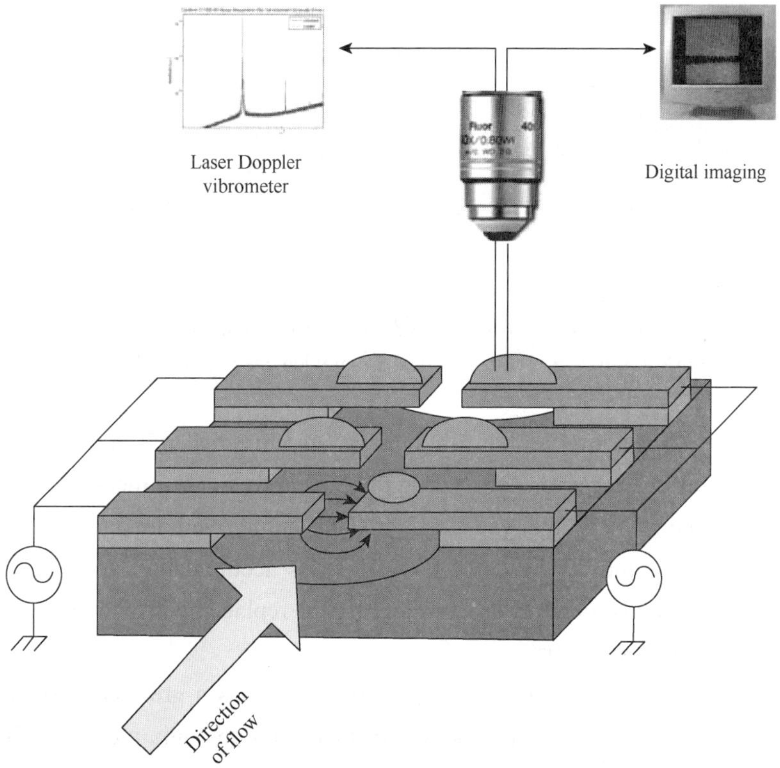

Figure 1.8　Schematic diagram of living cantilever array. Target cells in suspension are captured and immobilized on the cantilever. Then the cells are cultured and the mass of a cell on a cantilever is measured via the resonance frequency shift of a cantilever

The results demonstrate that of individual cells grew on the mass sensors, adherent human colon epithelial cells have increased growth rates with a larger cell mass, and the average growth rate increases linearly with the cell mass, at 3.25% per hour. The sensitive mass sensors with a position-independent mass sensitivity can be coupled with microscopy for simultaneous monitoring of cell growth and status, and provide an ideal method to study cell growth, cell cycle progression, differentiation, and apoptosis.

1.5 Microelectrodes

Intracellular thermal signal is closely related with disease. Living cells can change their intramembranous temperature during cell activities such as division, gene expression, enzyme reaction, and metabolism[59, 60]. Temperature change inside cells is usually at a small scale and is of transient nature due to the thermo-influence by the extracellular environment, rendering it rather difficult to measure using the conventional temperature detection methods.

Thermocouple (TC) is widely used in settings that require detection of temperature changes. The TC-based detection method has a number of advantages, including the capacity for achieving high precision and rapid response[61].

Changling Wang et al.[62] designed a novel TC device for detecting intracellular temperature [Figure 1.9 (a) and Figure 1.9 (b)]. Briefly, our TC probe is made of a sandwich structure consisting of the tungsten (W) substrate, an insulating layer made of polyurethane (PU; except at the tip), and a platinum (Pt) film.

They calculated the temperature resolution of the TC based on the derived thermo-electric power temperature equation and then calculated the Seebeck coefficient. The temperature resolution of their measurement was below 0.1℃ incorporating the resolution limit of the digital multimeter (Agilent 34410A, 0.1 μV).

Then the designed TC was used to detect the temperature of U251 and the temperature change of the cell after two drugs' (camptothecin, CPT; doxorubicin) stimulation. They found that CPT treatment led to an acute increase in intracellular temperature [Figure 1.9 (c) and Figure 1.9 (d)], while doxorubicin did not cause any obvious intracellular temperature change [Figure 1.9 (c) and Figure 1.9 (d)].

Substantial change in impedance for an electrode with a size analogous to the cell leads to possibilities for high accuracy, label free and automated counting of cancer cells, and for improved cancer prognosis, diagnosis and therapy monitoring.

Microelectrodes (Figure 1.10, see the Color Inset, p.3) are desirable for applications such as in rare cell counting, wherein higher resolution is required. Sunil K. Arya et al.[62] have used cyclic voltammetry (CV) and electrochemical impedance spectroscopic (EIS) techniques to analyze gold microelectrodes with different diameter size (25 mm, 50 mm, 75 mm, 100 mm and 250 mm) for breast cancer MCF-7 cell capture response. Cells were successfully and specifically captured using an anti-EpCAM modified surface and the optimum electrode size (25 mm diameter) was established for the precise detection of single MCF-7 cells with an average diameter of (18±2) mm. The 25 mm electrode in the EIS investigation exhibited an impedance change of 2.2×10^7 V at 20 Hz in response to a single MCF-7 cell captured on its surface, whereas other electrodes (250 mm, 100 mm, 75 mm and 50 mm) showed responses between 3.6×10^5 V and 1.1×10^5 V for a single MCF-7 cell (Figure 1.11, see

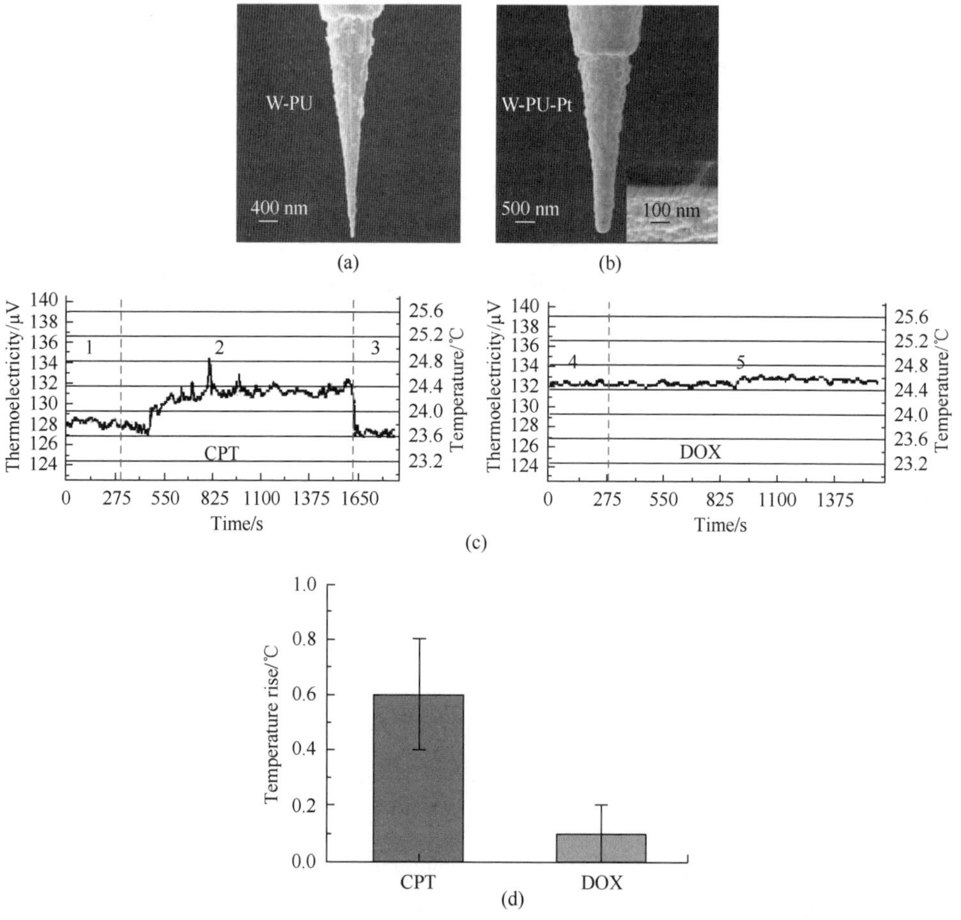

Figure 1.9 (a) SEM image of tungsten probe coated by polyurethane (PU; except at the tip which is uncoated). (b) SEM image of the thin platinum film as an outermost layer. (c) Upper panel, a typical curve showing the changes of the intracellular temperature of a single U251 cell after CPT treatment. 1-TC inserted into the cell; 2-addition of CPT; 3-TC withdrawn from the cell—lower panel, a typical curve showing no temperature changes after addition of DOX; 4-TC inserted into the cell; 5-addition of DOX. (d) A column figure indicating the average temperature changes for total cells after two drug treatment

the Color Inset, p.3). Based on observed binary "on/off" type responses upon the presence and absence of a single MCF-7 tumor cell on 25 mm surfaces, efforts are in progress to fabricate and use a microelectrode array of 25 mm electrodes to sense and count MCF-7 cells over a wide concentration range with single-cell precision. The results of present investigations have shown the promise for fabrication of a lab-on-a-chip device for simultaneous counting and testing of CTCs with single-cell precision. Specific electrodes can be designed for specific cell types having electrode size analogous to cell size to result in optimum response for their testing at a multicell level in microarrays.

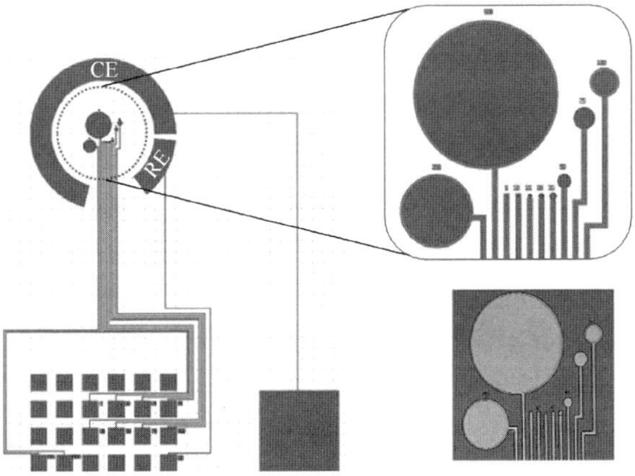

Figure 1.10 Design of microelectrodes of various sizes and their arrangement. Inset on the lower right is the optical image of the actual fabricated electrode

Needle shaped probes in submicron size have been developed by R. J. Fasching et al.[63] for electrochemical analyses of living single cells. The sharpness of the probe tips and the mechanical robustness of the needle structures allow for a controlled penetration of cell membranes and electrochemical probing of the cytosolic cell environment. Measurements of cell membrane potentials and cell membrane impedances demonstrated the probes' analytical capability in biological environments.

The birth of nanotechnology has provided numerous interesting ways which may overcome the limitation of the current conventional methods of any possible field. Single-cell analyses which before seemed difficult to be realized are now showing rapid growth in research endeavors. The ability to perform characterization task at single-cell level gives much better understanding of the biological cells which was previously realized on the data from the population averaging-based approach.

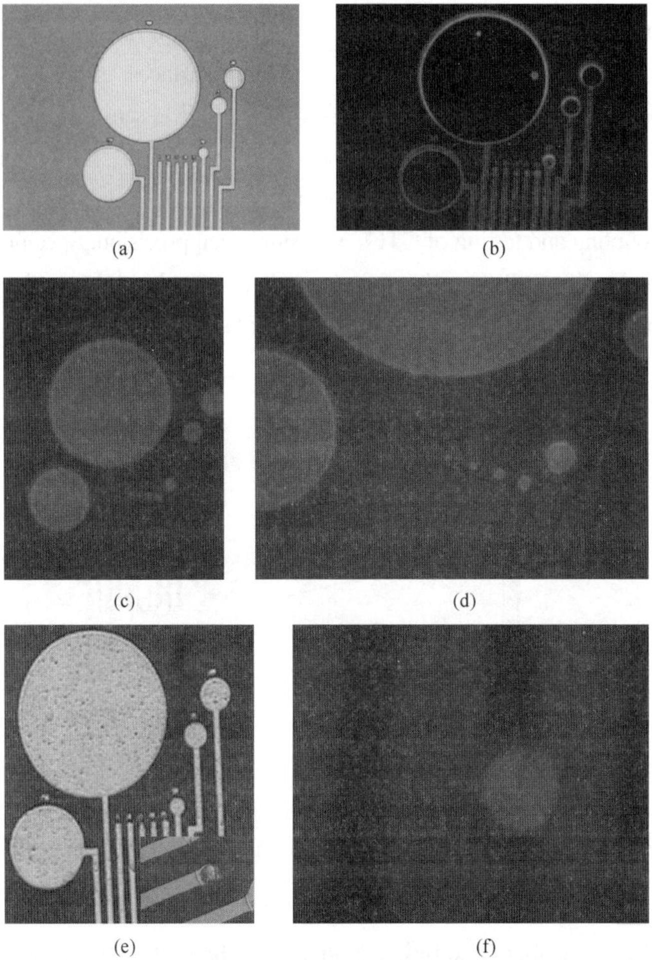

Figure 1.11　(a) Optical image of blank microelectrodes; (b) fluorescent image of blank microelectrodes; (c) and (d) fluorescence image of microelectrodes modified with fluorescent tagged antibody; (e) optical image of a cell bound on microelectrodes, inset (e) optical image of a cell bound on 25 mm diameter electrode; and (f) fluorescent image of cells bound on 25 mm diameter electrode

Ahmad et al.[64] have introduced a new method that employs a direct electrical measurement to instantly and quantitatively determine the viability of single cells. The ability to characterize the electrical property of single cells can be used as a novel method for cell viability detection in quantitative and instantaneous manners. Single pulses current measurement on single cells was performed using dual nanoprobe through environmental scanning electron microscope nanomanipulator system (Figure 1.12). The nanoprobe was successfully fabricated using focused ion beam tungsten deposition and etching processes. The characteristics of the nanoprobe were examined from the energy dispersion spectrometry and noise analyses. The electrical property of single cells under their native condition was presented. In order to apply this method for cell viability detection, two types of cells were used, i.e., dead cells and live cells. The results showed that there is a significant difference on the electrical measurement data between dead and live cells.

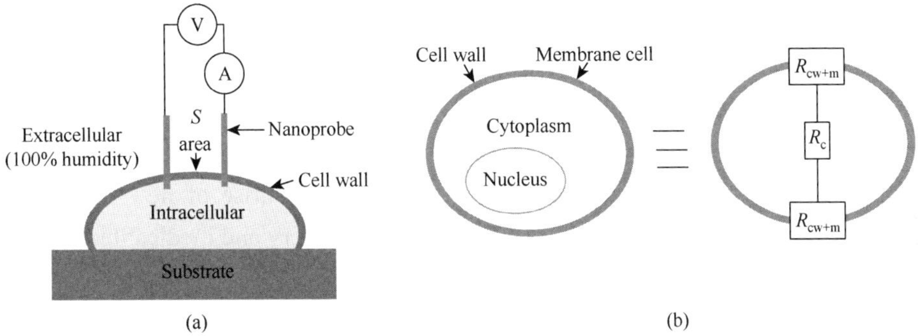

Figure 1.12 (a) Schematic of the single-cell electrical measurement using dual nanoprobe; (b) Schematic of the electrical model of a single biological cell

1.6 Conclusions

Detection and analysis of single cells have important influence upon morphology, function, metabolism, proliferation, differentiation and survival of single cells. Individual cells have discrete molecular, metabolic, and proteomic identities[65].

In traditional approaches, cellular parameters are represented by populations. They are averages and cannot exactly represent individual cells. Additionally, heterogeneity cannot be evaluated in bulk cultures. Multiple parameters must be measured in real time in single living cells to correlate cellular events with genomic information and thus understand complex cellular processes[66].

This book reviews several optical and electrical methods for single-cell detection and analysis.

Dielectrophoresis is a promising tool for rapid detection of single cells' impedance, mass, density, etc. The method is non-invasive, does not require the use of antibodies or other labelling (although dielectric labelling to increase specificity can be an option) and can be employed at either the single-cell or multicell level. While optical tweezers have emerged distinct advantages including non-contact force for cell manipulation, force resolution as accurate as 100 aN and amiability to liquid medium environments, their wide range of applications, such as transporting foreign materials into single cells, delivering cells to specific locations and sorting cells in microfluidic systems, are reviewed in this article. The implementation of microfluidic technologies in single-cell analysis is one of the most promising approaches that not only offers information rich, high throughput screening but also enables the creation of innovative conditions that are impractical or impossible by conventional means. The possibilities for distinguishing the difference between individual cells and the benefits from miniaturisation (e.g.,confinement) have led to many discoveries.

The arise of different micro-probes designed leads to new methods for single-cell detection, which may be more convenient, low cost and principle simple as described before. In the future, combination of several methods or arise of new method will inevitably bring new insight into single-cell level's discovery and medical technology's progress.

References

[1] Lu X, Huang W H, Wang Z L, et al. Recent developments in single-cell analysis. *Analytica Chimica Acta*, 2004, 510 (2): 127-138.

[2] Andersson H, van den Berg A. Microtechnologies and nanotechnologies for single-cell analysis. *Current Opinion in Biotechnology*, 2004, 15 (1): 44-49.

[3] Jones T B. *Electromechanics of Particles*. New York: Cambridge University Press, 1995.

[4] Hunt T P, Westervelt R M. Dielectrophoresis tweezers for single cell manipulation. *Biomedical Microdevices*, 2006, 8 (3): 227-230.

[5] Lee K, Kwon S G, Kim S H, et al. Dielectrophoretic tweezers using sharp probe electrode. *Sensors and Actuators A: Physical*, 2007, 136 (1): 154-160.

[6] Thomas R S, Morgan H, Green N G. Negative DEP traps for single cell immobilisation. *Lab on a Chip*, 2009, 9 (11): 1534-1540.

[7] Peitz I, van Leeuwen R. Single-cell bacteria growth monitoring by automated DEP-facilitated image analysis. *Lab on a chip*, 2010, 10 (21): 2944-2951.

[8] Lan K C, Jang L S. Integration of single-cell trapping and impedance measurement utilizing microwell electrodes. *Biosensors and Bioelectronics*, 2011, 26 (5): 2025-2031.

[9] Zhao Y L, Liang W F, Zhang G L, et al. Distinguishing cells by their first-order transient motion response under an optically induced dielectrophoretic force field. *Applied Physics Letters*, 2013, 103 (18): 1-4.

[10] Ouyang M X, Cheung W K, Liang W F, et al. Inducing self-rotation of cells with natural and artificial melanin in a linearly polarized alternating current electric field. *Biomicrofluidics*, 2013, 7 (5): 054112.

[11] Zhao Y L, Lai H S S, Zhang G, et al. Rapid determination of cell mass and density-using digitally-controlled

electric field in a microfluidic chip. *Lab on a Chip*, 2014, 14 (22): 4426.

[12] Thomas C R, Stenson J D, Zhang Z B. Measuring the mechanical properties of single microbial cells//Müller S, Thomas B. High Resolution Microbial Single Cell Analytics. Berlin, Heidelberg: Springer, 2011: 83-98.

[13] Lang M J, Block S M. Resource letter: LBOT-1: Laser-based optical tweezers. *American Journal of Physics*, 2003, 71 (3): 201-215.

[14] Piggee C. Optical tweezers: not just for physicists anymore. *Analytical Chemistry*, 2008, 81 (1): 16-19.

[15] Zhang H, Liu K K. Optical tweezers for single cells. *Journal of the Royal Society Interface*, 2008, 5 (24): 671-690.

[16] Ramser K, Hanstorp D. Optical manipulation for single-cell studies. *Journal of Biophotonics*, 2010, 3 (4): 187-206.

[17] Wakamoto Y, Inoue I, Moriguchi H, et al. Analysis of single-cell differences by use of an on-chip microculture system and optical trapping. *Fresenius Journal of Analytical Chemistry*, 2001, 371 (2): 276-281.

[18] Tang H Y, Yao H L, Wang G W, et al. NIR Raman spectroscopic investigation of single mitochondria trapped by optical tweezers. *Optics Express*, 2007, 15 (20): 12708-12716.

[19] Grier D G. A revolution in optical manipulation. *Nature*, 2003, 424 (6950): 810-816.

[20] Neuman K C, Block S M. Optical trapping. *Review of Scientific Instruments*, 2004, 75 (9): 2787.

[21] Reicherter M, Haist T, Wagemann E U, et al. Optical particle trapping with computer-generated holograms written on a liquid-crystal display. *Optics Letters*, 1999, 24 (9): 608-610.

[22] Curtis J E, Koss B A, Grier D G. Dynamic holographic optical tweezers. *Optics Communications*, 2002, 207 (1): 169-175.

[23] Sinclair G, Jordan P, Leach J, et al. Defining the trapping limits of holographical optical tweezers. *Journal of Modern Optics*, 2004, 51 (3):409-414.

[24] Jordan P, Leach J, Padgett M, et al. Creating permanent 3D arrangements of isolated cells using holographic optical tweezers. *Lab on a Chip*, 2005, 5: 1224-1228.

[25] Arai F, Ng C, Maruyama H, et al. On chip single-cell separation and immobilization using optical tweezers and thermosensitive hydrogel. *Lab on a Chip*, 2005, 5 (12): 1399-1403.

[26] Eriksson E, Sott K, Lundqvist F, et al. A microfluidic device for reversible environmental changes around single cells using optical tweezers for cell selection and positioning. *Lab on a Chip*, 2010, 10 (5): 617-625.

[27] Wang X L, Chen S X, Kong M, et al. Enhanced cell sorting and manipulation with combined optical tweezer and microfluidic chip technologies. *Lab on a Chip*, 2011, 11 (21): 3656-3662.

[28] Mills J P, Qie L, Dao M, et al. Nonlinear elastic and viscoelastic deformation of the human red blood cell with optical tweezers. *Mechanics and Chemistry of Biosystems*, 2004, 1 (3): 169-180.

[29] Sleep J, Wilson D, Simmons R, et al. Elasticity of the red cell membrane and its relation to hemolytic disorders: An optical tweezers study. *Biophysical Journal*, 1999, 77 (6): 3085-3095.

[30] Henon S, Lenormand G, Richert A, et al. A new determination of the shear modulus of the human erythrocyte membrane using optical tweezers. *Biophysical Journal*, 1999, 76 (2): 1145-1151.

[31] Dao M, Lim C T, Suresh S. Mechanics of the human red blood cell deformed by optical tweezers. *Journal of the Mechanics and Physics of Solids*, 2003, 51 (11): 2259-2280.

[32] Mejean C O, Schaefer A W, Millman E A, et al. Multiplexed force measurements on live cells with holographic optical tweezers. *Optics Express*, 2009, 17 (8): 6209-6217.

[33] Steubing R W, Cheng S, Wright W H, et al. Laser induced cell fusion in combination with optical tweezers: The laser cell fusion trap. *Cytometry*, 1991, 12 (6): 505-510.

[34] Chen S X, Cheng J, Kong C W, et al. Laser-induced fusion of human embryonic stem cells with optical tweezers. *Applied Physics Letters*, 2013, 103: 033701.

[35] Lecault V, White A K, Singhal A, et al. Microfluidic single cell analysis: from promise to practice. *Current Opinion in Chemical Biology*, 2012, 16 (3): 381-390.

[36] Cooper G M, Hausman R E. *The Cell: A Molecular Approach*. 3rd ed. Washington D.C.: ASM Press, 2007.

[37] Lindström S, Andersson-Svahn H. Overview of single-cell analyses: Microdevices and applications. *Lab on a Chip*, 2010, 10 (24): 3363-3372.

[38] Larsen U D, Blankenstein G, Braneb J. Microchip Coulter particle counter//Solid State Sensors and Actuators, TRANSDUCERS'97 Chicago. 1997 International Conference on IEEE, 1997, 2: 1319-1322.

[39] Koch M, Evans A G R, Brunnschweiler A. Design and fabrication of a micromachined Coulter counter. *Journal of Micromechanics and Microengineering*, 1999, 9 (2): 159.

[40] Saleh O A, Sohn L L. Quantitative sensing of nanoscale colloids using a microchip Coulter counter. *Review of Scientific Instruments*, 2001, 72 (12): 4449-4451.

[41] Holmes D, Pettigrew D, Reccius C H, et al. Leukocyte analysis and differentiation using high speed microfluidic Single cell impedance cytometry. *Lab on a Chip*, 2009, 9 (20): 2881-2889.

[42] Yang M S, Li C W, Yang J. Cell docking and on-chip monitoring of cellular reactions with a controlled concentration gradient on a microfluidic device. *Analytical Chemistry*, 2002, 74 (16): 3991-4001.

[43] Xia F Q, Jin W R, Yin X F, et al. Single-cell analysis by electrochemical detection with a microfluidic device. *Journal of Chromatography A*, 2005, 1063 (1): 227-233.

[44] Toriello N M, Douglas E S, Mathies R A. Microfluidic device for electric field-driven single-cell capture and activation. *Analytical Chemistry*, 2005, 77(21): 6935-6941.

[45] Chao T C, Ros A. Microfluidic single-cell analysis of intracellular compounds. *Journal of the Royal Society Interface*, 2008, 5 (Suppl 2): S139-S150.

[46] Gao J, Yin X F, Fang Z L. Integration of single cell injection, cell lysis, separation and detection of intracellular constituents on a microfluidic chip. *Lab on a Chip*, 2004, 4 (1): 47-52.

[47] Han J H, Heinze B C, Yoon J Y. Single cell level detection of Escherichia coli in microfluidic device. *Biosensors and Bioelectronics*, 2008, 23 (8): 1303-1306.

[48] Rane T D, Zec H, Puleo C, et al. High-throughput single-cell pathogen detection on a droplet microfluidic platform//Micro Electro Mechanical Systems (MEMS), 2011 IEEE 24th International Conference on IEEE, 2011: 881-884.

[49] Cho Y H, Yamamoto T, Sakai Y, et al. Development of microfluidic device for electrical/physical characterization of single cell. *Journal of Microelectromechanical Systems*, 2006, 15 (2): 287-295.

[50] Nguyen T A, Yin T I, Urban G. A cell impedance sensor chip for cancer cells detection with single cell resolution//SENSORS. 2013 IEEE. IEEE, 2013: 1-4.

[51] Sun J S, Kang Y J, Boczko E M, et al. A microfluidic cell size/density sensor by resistive pulse detection. *Electroanalysis*, 2013, 25 (4):1023-1028.

[52] Park K, Jang J, Irimia D, et al. "Living cantilever arrays" for characterization of mass of single live cells in fluids. *Lab on a Chip*, 2008, 8 (7): 1034-1041.

[53] de Flora S, Izzotti A, Randerath K, et al. DNA adducts and chronic degenerative diseases. Pathogenetic relevance and implications in preventive medicine. *Mutation Research/Reviews in Genetic Toxicology*, 1996, 366 (3): 197-238.

[54] Cooper S. Distinguishing between linear and exponential cell growth during the division cycle: Single-cell studies, cell-culture studies, and the object of cell-cycle research. *Theoretical Biology and Medical Modelling*. 2006, 3: 10-25.

[55] Mitchison J M. Single cell studies of the cell cycle and some models. *Theoretical Biology and Medical Modelling*, 2005, 2 (1): 4.

[56] Rahman M H, Ahmad M R. Lab-on-chip microfluidics system for single cell mass measurement: A comprehensive review. *Jurnal Teknologi*, 2014, 69 (8):85-93.

[57] Weng Y, Delgado F F, Son S, et al. Mass sensors with mechanical traps for weighing single cells in different fluids. *Lab on a Chip*, 2011, 11 (24): 4174-4180.

[58] Park K, Millet L J, Kim N, et al. Measurement of adherent cell mass and growth. *Proceedings of the National Academy of Sciences of the United States of America*, 2010, 107 (48): 20691-20696.

[59] Lowell B B, Spiegelman B M. Towards a molecular understanding of adaptive thermogenesis. *Nature*, 2000, 404 (6778): 652-660.

[60] Tanaka E, Yamamura M, Yamakawa A, et al. Microcalorimetric measurements of heat production in isolated rat brown adipocytes. *Biochemistry International*, 1992, 26 (5): 873-877.

[61] Wang C L, Xu R Z, Tian W J, et al. Determining intracellular temperature at single-cell level by a novel thermocouple method. *Cell Research*, 2011, 21 (10): 1517-1519.

[62] Arya S K, Lee K C, Dah'alan D B, et al. Breast tumor cell detection at single cell resolution using an electrochemical impedance technique. *Lab on a Chip*, 2012, 12 (13): 2362-2368.

[63] Fasching R J, Bai S J, Fabian T, et al. Nanoscale electrochemical probes for single cell analysis. *Microelectronic Engineering*, 2006, 83 (4): 1638-1641.

[64] Ahmad M R, Nakajima M, Kojima M, et al. Instantaneous and quantitative single cells viability determination using dual nanoprobe inside ESEM. *IEEE Transactions on Nanotechnology*, 2012, 11 (2): 298-306.

[65] Sweedler J V, Arriaga E A. Single cell analysis. *Analytical and Bioanalytical Chemistry*, 2007, 387 (1): 1-2.

[66] Jang L S, Huang P H, Lan K C. Single-cell trapping utilizing negative dielectrophoretic quadrupole and microwell electrodes. *Biosensors and Bioelectronics*, 2009, 24 (12): 3637-3644.

Chapter 2 Development of Multiple Transmission-Reflection Infrared Spectroscopy

Hongbo Liu, Shoujun Xiao

2.1 Introduction of Infrared Absorption Methodologies for Thin Films

Infrared spectroscopy is one of the most important analytical techniques available to today's scientists. Using various sampling accessories, IR spectrometers can accept a wide range of sample types such as gas, liquid, powder, and thin film on a supporting substrate. IR is powerful in surface analyses, for its capability to determine the molecular structure on a surface, compared to other surface measurements such as auger electron spectroscopy (AES), ultraviolet or X-ray photoelectron spectroscopies (UPS or XPS), and grazing-incidence X-ray diffraction (GIXD)[1]. According to the optical properties of the substrate, the IR measurement configurations can be classified as reflection on a metallic surface, transmission on a semiconductor surface, multiple internal reflections on a high refractive index prism, etc. External reflection at the grazing angle incidence or so-called infrared reflection absorption spectroscopy (IR-RAS) is well-known for detection of molecules on a metallic surface. The molecular layer on an absorbing substrate can only be measured by reflection. While for a transparent substrate, more IR methods are available: external reflection[2, 3], transmission[4-7], attenuated total reflection (ATR)[8-17], and our recently developed grazing angle mirror-backed reflection (GMBR)[18] and multiple transmission-reflection infrared spectroscopy (MTR-IR)[19, 20].

Silicon is a group IV semiconductor, transparent in the mid-IR region of 4000-1500 cm^{-1} (2.5-6.7 μm), and is widely used in electronic industry. Fabrication of organic monolayers on silicon attracts much attention in molecular, nano-, and micro-electronics[21-23]. With the development of monolayers based on Si—C and Si—O—C bonds, more potential applications are emerging, such as biochips, (bio-) chemical sensors, molecular recognition, solar cells, chemical and electrical surface passivation, and control of photopatterning[24-27].

Because of the potential applications of organic-modified silicon surfaces, many researchers are endeavoring to develop new IR methods for analyses of molecules on the silicon surface[12, 13, 15, 17, 28, 29]. Apart from the chemical structural identification of adsorbed species, the determination of molecular orientation is another advantage of IR analyses.

IR-RAS is well-known on a metal surface for its powerful and non-destructive illumination of structure, quality, and orientation of molecular monolayers. However, its application on infrared-transparent substrates such as silicon is limited because of the angular dependence of reflectance, which differs substantially from metals. For the p-polarized radiation, the existence of the Brewster angle φ_B, where the reflectance is zero, makes it difficult to get good spectra at the grazing angle of 80°, which is close to φ_B of 73.6° for the air/silicon interface. The angular dependence of reflectance on the air/silicon interface, both for p- and s-polarized radiation, is shown in Figure 2.1. The reflectance of p-polarization is zero at the Brewster angle 73.6°, while for s-polarization the reflectance increases with the angle of incidence.

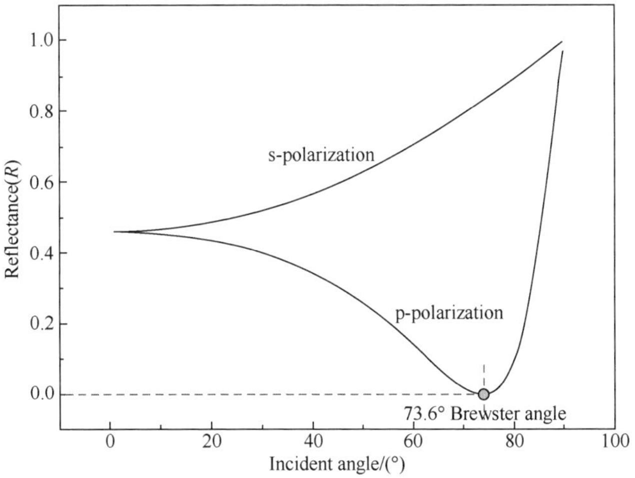

Figure 2.1 Dependence of reflectance (R) on incident angle of the air/silicon interface with p- and s-polarization: n_{Si}=3.42, n_{air}=1.0, v=2918 cm^{-1}. n_{Si} is the refractive index of silicon, n_{air} is the refractive index of air, v represents the wavenumber

Since the external reflection at the grazing angle of incidence is not applicable for IR measurements on the silicon substrate, how about the transmission mode? Actually, transmission is the simplest IR measurement method for molecules chemisorbed on an infrared-transparent surface, because no additional accessory is needed. However, the normal incident transmission cannot get enough sensitivity for molecular monolayers on the silicon surface. As shown in Figure 2.2, the absorbance goes up with the increment of incident angle for both p- and s-polarizations. The maximum absorption depth of p-polarization gets the highest value near the Brewster angle because 100% light transmits through the silicon layer and reaches the detector, while for s-polarization, it decreases smoothly with increasing the angle of incidence. The contrary increment direction of absorbance and absorption depth for s-polarization makes it difficult to get good spectra,

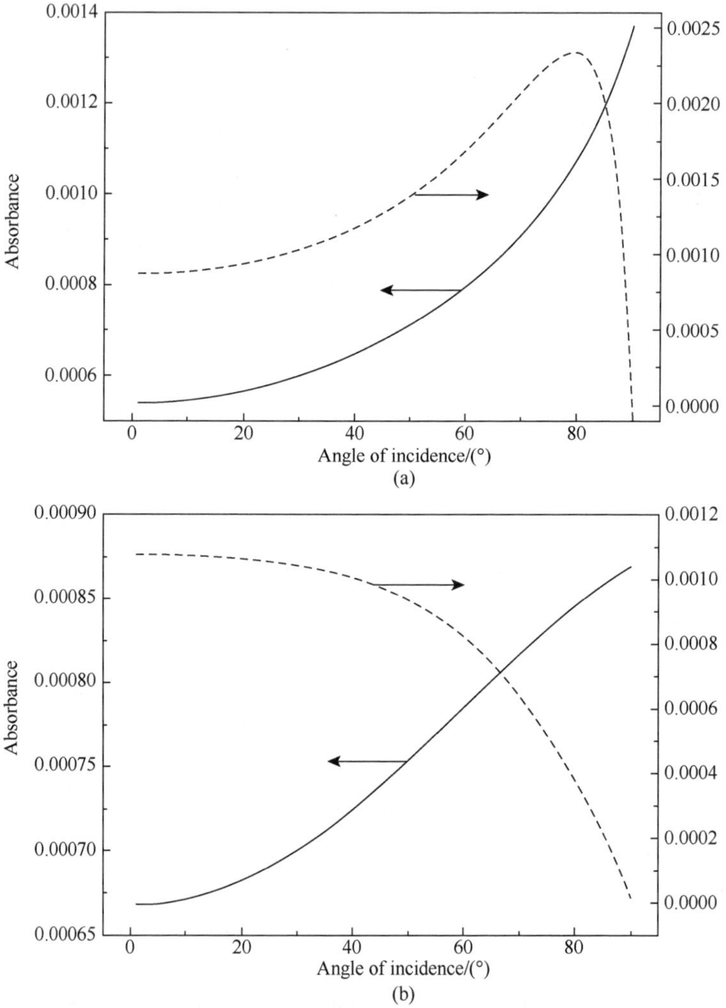

Figure 2.2　Calculated absorbance (A, definition see Equation (2.33), solid line) and absorption depth (ΔR, definition see Equation (2.34), dashed line) versus angle of incidence for (a) p-polarization and (b) s-polarization at the transmission mode with a 1 nm layer (a complex refractive index of $\hat{n}=1.5+0.5i$) at 2918 cm^{-1} on Si (n_{Si}=3.42)

and hence p-polarized transmission at the Brewster angle is usually employed to measure silicon-based molecular monolayers. Si is excellent for mid-IR measurements because it is transparent in the range of 4000-1500 cm^{-1}, but its absorption is strong below 1500 cm^{-1} because of the multi-phonon absorption (silicon lattice) and impurities inside the silicon wafer, such as Si—O, Si—C, doped phosphor or boron.

Among all the IR measurements for monolayers on the infrared-transparent surface, p-polarized transmission at the Brewster angle is the simplest configuration. To get good optical transmission and to avoid scattering, the double-side-polished silicon wafer is necessary for measurements. For the single Brewster angle transmission, although the strong absorption peaks such as CH_2 stretchings from long alkyl chains can be detected with great care, its sensitivity for weak absorptions is not high enough to meet the demands of academia and industry. The weak peaks are easily to be covered by the interference fringes coming from multiple reflections inside a thin Si substrate. For all optical configurations, the absorption from Si can be removed by subtraction of a reference background from the sample. Although practically the elimination of the absorption at low wavenumbers cannot be perfect, the weak peaks at around 600 cm^{-1}, for example the bending mode of Si—H at 627 cm^{-1} and the stretching and bending modes of Si—Cl at 583 and 528 cm^{-1} respectively[4, 5, 7], can be seen with great care.

As we discussed above, an external reflection with p-polarized incident light cannot obtain enough reflection signals because the Brewster angle of air/silicon interface (73.6°) is close to the grazing angle and most of the light passes through the silicon wafer and escapes away from the detector. Transmission with p-polarized light at the Brewster angle is by far the simplest configuration for detection of monolayers on silicon. However, the sensitivity of the single transmission, especially for monolayer samples, is still not high enough to allow quantitative analyses. The most popular configuration for detection of silicon-based monolayers is multiple internal reflection (MIR)[8], in which the monolayer is fabricated on an attenuated total reflection (ATR) silicon crystal with 45° bevel cuts at both ends and the sensitivity is increased by multiple reflections (Figure 2.3). This technique can provide excellent spectra for molecular monolayers for both p- and s-polarizations, but the costly ATR crystal restricts its use as a routine laboratory method. Technically, it has to be recycled and the monolayer/silicon interface cannot be used for further device-fabrication. The Ge-ATR measurement[13, 14] was developed by directing the IR beam through a higher refractive index ATR crystal, Ge, to a sample monolayer supported by another higher refractive index Si material (briefly as Ge/monolayer/Si) at an incident angle larger than the critical angle (58° for the Ge/Si interface) to ensure a total reflection (Figure 2.4). The signal enhancement up to two orders of magnitude was reported to be possible for perfect optical contact between the incidence phase and the sample. Since the spectra of sandwiched sample layers strictly follow the surface selection rule, where only the perpendicular vibration modes are enhanced, the molecular orientation can be deduced from the relative intensities of different orientational dipole moments. Practically a big challenge exists for the Ge-ATR measurement, i.e., the peak intensity depends on individual operations because the balance among the loaded external pressure, the damage of the sample, and the conformal contact between the sample and the Ge prism surface cannot be easily found out.

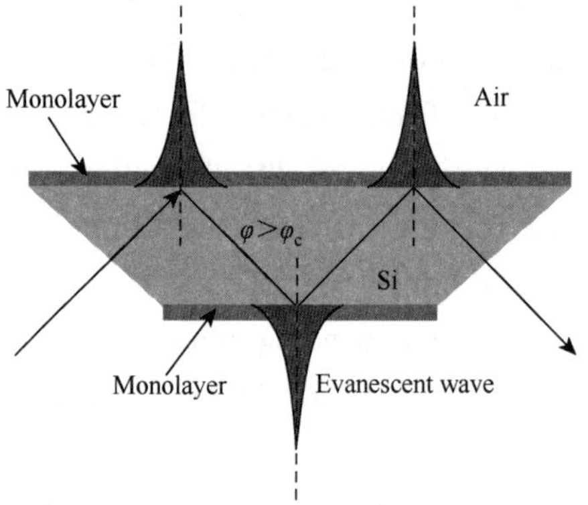

Figure 2.3　Schematic representation of total internal reflection for detection of a monolayer on the Si—ATR crystal. IR radiation was directed inside the crystal from the side face with a 45° bevel cut. Monolayers were fabricated on both sides of the Si—ATR crystal and can be detected by the evanescent wave

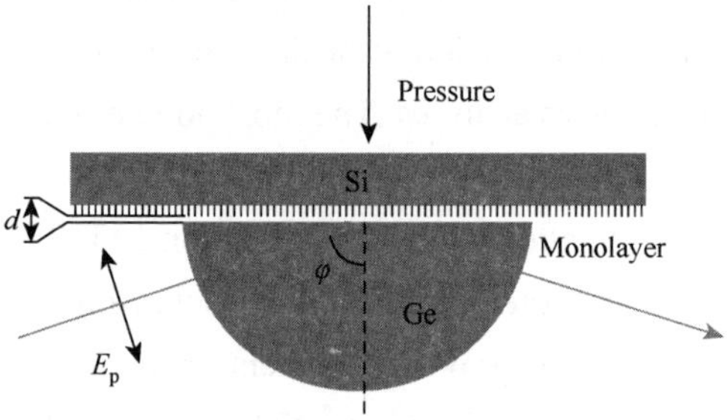

Figure 2.4　Schematic of the p-polarized Otto's ATR configuration at an incidence angle φ for a three-phase system (Ge/monolayer/Si). The sample was pressed on the Ge surface. The thickness of the air gap between Ge and monolayer was d. Above the critical angle of approximately 58°, total internal reflection occurs at the Ge/monolayer/Si interface while, below this angle, regular reflection and transmission occur

The aim of our work is to establish an IR method for detection of the monolayer on the semiconductor surface with a high sensitivity like the MIR measurement, but with easy handling and at much lower cost. The obtained spectra should be reproducible and can be used for quantitative analyses. In the following sections, we will discuss our inventions in detail: optical theory, grazing angle mirror-backed reflection (GMBR), especially its upgraded technique—multiple transmission-reflection infrared spectroscopy (MTR-IR), and finally orientation analysis. We shall demonstrate these powerful technologies with examples of covalently bound molecular monolayers on silicon. Not limited only to the molecular monolayers, MTR-IR can be applied in many other scientific fields: impurities of oxygen and carbon measured at room temperature, shallow impurities of boron, tin, phosphorus, aluminum, etc. at cryo-temperature, charge transfer for opto-electronic devices, etc., you name it.

2.2 Optical Theory

Optical simulation is an important and useful tool to interpret IR spectra and choose optimal conditions for IR measurements. IR spectra of a surface are influenced by the measurement conditions and optical properties of the film, the substrate, and the surroundings. For example, the sign of peaks (positive + or negative −) in IR-RAS spectra of a monolayer on silicon depends on the angle of incidence and the orientation of transitional dipole moment (TDM). Furthermore, orientation analysis of molecules on a surface can only be obtained from the combination of simulation analyses and experimental spectra. Here the optical simulation method will be established to calculate the optimal measurement conditions for thin films on infrared-transparent substrates.

The optical thin-film theory is essentially based on the Maxwell theory, which summarizes all the empirical knowledge on electromagnetic phenomena. But here we will only introduce the equations used in IR analyses, without any introduction of the original optical theory and deduction of equations, which can be found easily in text books for film optics[30].

2.2.1 Law of reflection

The basic optical phenomena are the reflection and refraction of light when light passes through the boundary between two different dielectric media. The incident ray, the reflected ray, and the normal to the interface at the point of incidence all lie in *xz* plane (the plane of incidence). The angle of incidence φ_1 is equal to the angle of reflection φ_1' (see also Figure 2.5).

$$\varphi_1 = \varphi_1' \tag{2.1}$$

2.2.2 Snell's law of refraction

The angle of incidence φ_1 is related to the angle of refraction φ_2 by

$$\frac{\sin\varphi_1}{\sin\varphi_2} = \frac{n_2}{n_1} \tag{2.2}$$

where n_1 and n_2 are the refractive indices of the first medium, from which the ray passes, and the second medium, in which the ray refracts respectively. The refracted ray lies in the plane of incidence. If the light passes through several refracting media with parallel boundaries, the angle of refraction in each medium is simultaneously the angle of incidence of the following. Snell's law of refraction can then be written in the following form

$$n_1 \sin\varphi_1 = n_2 \sin\varphi_2 = n_3 \sin\varphi_3 = \cdots \tag{2.3}$$

where n_1, n_2, $n_3\cdots$ are the refractive indices of the different media, φ_1, φ_2, $\varphi_3\cdots$ are the angles of incidence and angles of refraction in their corresponding media. The result, that the product of the given refractive index and the sine of the angle of incidence (refraction) is constant for all refracting media, is very important.

The reflection from the boundary is called the external or specular reflection when the light is incident from an optically rare medium to a denser medium ($n_1 < n_2$). On the other hand, if the light passes from a denser to a rarer medium ($n_1 > n_2$), an internal reflection takes place. In the later case refraction of light can take place parallel to the boundary (angle of refraction φ_2=90°), and the angle of incidence is called critical angle φ_c.

$$\sin\varphi_c = \frac{n_2}{n_1} \tag{2.4}$$

If the angle of incidence is greater than this critical angle, there is no refraction of the light and total reflection occurs.

2.2.3 Linearly polarized light

If the light is constrained to vibrate in only one plane, the light is so-called plane polarized or linearly polarized. In IR measurement, linear polarizer is usually used to get pure p- and s-polarizations. For isotropic bounding media, the incident IR ray to a surface can be resolved into two orthogonal and, therefore, independent components (Figure 2.5): the s-component (s-polarization) and the p-component (p-polarization). In Figure 2.5, these components are schematically denoted by the arrows lying in the plane of incidence (*xz* plane, p) and by circles pointing out of the plane of the drawing (*y* direction, s). The s-polarized component has the electric field vector $\boldsymbol{E}_{s,}$ oriented perpendicular to the plane of incidence (*xz* in Figure 2.5). The p-polarized component has the electric field vector $\boldsymbol{E}_p$ parallel to the *xz*-plane.

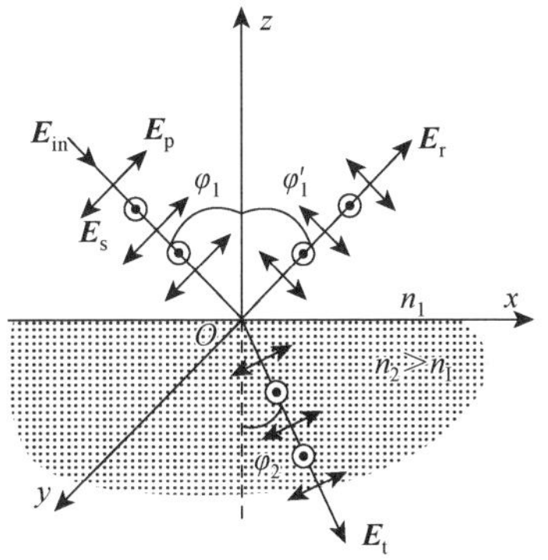

Figure 2.5 Reflection and refraction of light at a plane boundary between two media. The incident electric field vectors in two media were shown with arrows

2.2.4 Fresnel's formulae

Fresnel's formulae describe the reflection and transmission of electro magnetic waves at an interface, i.e.,the reflection and transmission coefficients of both p- and s-components. For the boundary of two isotropic media, these relations can be written as the ratio coefficients:

$$r_{ij}^{s} = \frac{\xi_i - \xi_j}{\xi_i + \xi_j}; \quad r_{ij}^{p} = \frac{\hat{n}_j^2 \xi_i - \hat{n}_i^2 \xi_j}{\hat{n}_j^2 \xi_i + \hat{n}_i^2 \xi_j} \tag{2.5}$$

$$t_{ij}^{s} = \frac{2\xi_i}{\xi_i + \xi_j}; \quad t_{ij}^{p} = \frac{2\hat{n}_i \hat{n}_j \xi_i}{\hat{n}_j^2 \xi_i + \hat{n}_i^2 \xi_j} \tag{2.6}$$

$$\xi_i = \hat{n}_i \cos\varphi_i = \sqrt{\hat{n}_i^2 - \hat{n}_0^2 \sin^2\varphi_0} \tag{2.7}$$

where $\hat{n} = n + ik$ is the complex refractive index. The subscript 0 denotes incident medium, and superscripts p and s for p- and s-polarized radiations respectively. If an isotropic layer with finite thickness is located at the planar interface of two semi-infinite media, the Fresnel coefficients can be rewritten in the form as

$$r_{123}^{\mathrm{p,s}} = \frac{r_{12}^{\mathrm{p,s}} + r_{23}^{\mathrm{p,s}}\mathrm{e}^{2i\beta}}{1 + r_{12}^{\mathrm{p,s}} r_{23}^{\mathrm{p,s}}\mathrm{e}^{2i\beta}} \qquad t_{123}^{\mathrm{p,s}} = \frac{t_{12}^{\mathrm{p,s}} t_{23}^{\mathrm{p,s}}\mathrm{e}^{i\beta}}{1 + r_{12}^{\mathrm{p,s}} r_{23}^{\mathrm{p,s}}\mathrm{e}^{2i\beta}} \tag{2.8}$$

$$\beta = 2\pi k d \xi_2 \tag{2.9}$$

where $r_{ij}^{\mathrm{p,s}}$ and $t_{ij}^{\mathrm{p,s}}$ are the two-phase Fresnel coefficients calculated by Equation (2.5) and Equation (2.6) respectively; β is the phase shift of the electromagnetic wave after one pass through the film; k is the wavenumber of incident light; d is the thickness of film. The transmittance and reflectance related to Fresnel coefficients are described as below:

$$R_{1(2)3}^{\mathrm{p,s}} = \left|r_{1(2)3}^{\mathrm{p,s}}\right|^2; \quad T_{1(2)3}^{\mathrm{s}} = \frac{\mathrm{Re}\,\xi_3}{\xi_1}\left|t_{1(2)3}^{\mathrm{s}}\right|^2; \quad T_{1(2)3}^{\mathrm{p}} = \frac{\mathrm{Re}(\xi_3/\hat{n}_3^2)}{\xi_1}\left|\hat{n}_3 t_{1(2)3}^{\mathrm{p}}\right|^2 \tag{2.10}$$

To get the anisotropic reflectance and transmittance, the refractive angle inside the monolayer (φ_2) is determined with $\hat{n}_{2z}$ ($n_1 \sin\varphi_1 = \hat{n}_{2z} \sin\hat{\varphi}_2$), especially for p-polarization. In other equations, $\hat{n}_2$ was replaced by $\hat{n}_{2x}(\hat{n}_{2y})$ [28, 31].

All these equations completely express the relationship between the reflecting and transmitting properties of an interface and the refractive indices of the substances on both sides of that interface. With these equations, the reflectance and transmittance of each boundary can be calculated, and the absorption of a layer can be shown with the difference of reflectance and transmittance with and without the layer between two isotropic media.

2.2.5 Approximate calculation of reflectance and transmittance for a transparent substrate

Since the thickness of the monolayer (~1 nm) is much smaller than the IR wavelength (~10000 nm), the approximate equations of reflectance and transmittance can be used in the spectral simulations[31]:

$$R^{\mathrm{s}} \approx \left|\frac{n_1\cos\varphi_1 - n_3\cos\varphi_3}{n_1\cos\varphi_1 + n_3\cos\varphi_3}\right|^2 \left(1 - \frac{8\pi k d_2 n_1 \cos\varphi_1}{(n_1 - n_3)(n_1 + n_3)}\mathrm{Im}(\hat{\varepsilon}_{2y})\right) \tag{2.11}$$

$$R^{\mathrm{p}} \approx \left|\frac{n_3\cos\varphi_1 - n_1\cos\varphi_3}{n_3\cos\varphi_1 + n_1\cos\varphi_3}\right|^2 \times \left\{ \begin{aligned} & 1 - \frac{8\pi k d_2 n_1 \cos\varphi_1}{(n_3\cos\varphi_1 + n_1\sin\varphi_1)(n_3\cos\varphi_1 - n_1\sin\varphi_1)(n_1^2 - n_3^2)} \\ & \times \left[(n_3^2 - n_1^2\sin^2\varphi_1)\operatorname{Im}(\hat{\varepsilon}_{2x}) - n_1^2 n_3^4 \sin^2\varphi_1 \operatorname{Im}\left(-\frac{1}{\hat{\varepsilon}_{2z}}\right)\right] \end{aligned} \right\} \tag{2.12}$$

$$T^{\mathrm{s}} \approx \frac{4n_1 n_3\cos\varphi_1\cos\varphi_3}{(n_1\cos\varphi_1 + n_3\cos\varphi_3)^2}\left(1 - \frac{4\pi k d_2}{n_1\cos\varphi_1 + n_3\cos\varphi_3}\operatorname{Im}(\hat{\varepsilon}_{2y})\right) \tag{2.13}$$

$$T^{\mathrm{p}} \approx \frac{4n_1 n_3\cos\varphi_1\cos\varphi_3}{(n_3\cos\varphi_1 + n_1\cos\varphi_3)^2} \times \left\{1 - \frac{4\pi k d_2 n_1 n_3}{n_3\cos\varphi_1 + n_1\cos\varphi_3} \times \left[\frac{\cos\varphi_1\cos\varphi_3}{n_1 n_3}\operatorname{Im}(\hat{\varepsilon}_{2x}) + n_1^{\,2}\sin^2\varphi_1 \operatorname{Im}\left(-\frac{1}{\hat{\varepsilon}_{2z}}\right)\right]\right\} \tag{2.14}$$

where $\hat{\varepsilon} = \hat{n}^2$ is the complex dielectric constant, and d_2 is the thickness of the monolayer.

2.2.6 The reflection and refraction of light from multiple layers

Equation (2.8) gives the Fresnel's formulae for only one thin layer located between two isotropic media. If there are two or more layers, Fresnel's formulae or matrix will be more complicated. Therefore the accumulative equations for multiple layers deduced from the Fresnel's formulae[30] are rarely used nowadays. Here we only introduce the matrix method[32].

The multilayer structure shown in Figure 2.6 is composed of (N–1) layers with complex refractive indexes. The subscripts of the field amplitudes in Figure 2.6 indicate the layer and surrounding medium; the + and – superscripts signify right- and left-going waves, respectively. The field amplitudes in each layer are related by a product of 2×2 matrices in sequence.

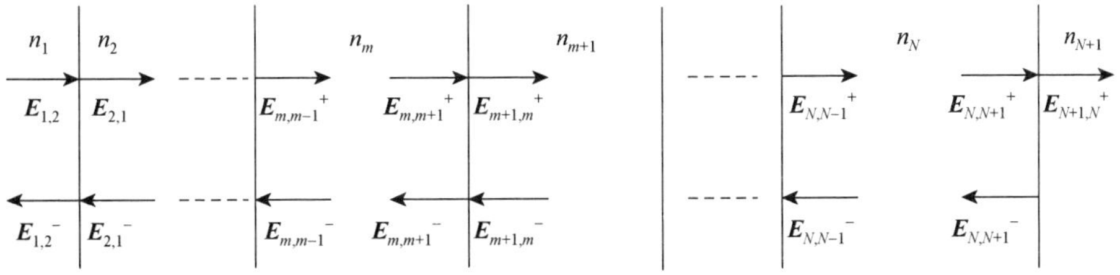

Figure 2.6 Notation of electric field amplitudes within an arbitrary multilayer. n represents the refractive index. The subscripts 2 to N indicate optical layers, while 1 and N+1 indicate semi-infinite media. The + and – signs distinguish between left- and right-going waves

The electric field amplitudes on each side of an interface can be related by the transmission matrix $\boldsymbol{W}$:

$$\begin{pmatrix} \boldsymbol{E}_{m-1,m}^{\ +} \\ \boldsymbol{E}_{m-1,m}^{\ -} \end{pmatrix} = W_{m-1,m} \begin{pmatrix} \boldsymbol{E}_{m,m-1}^{\ +} \\ \boldsymbol{E}_{m,m-1}^{\ -} \end{pmatrix} = \frac{1}{t_{m-1,m}} \begin{bmatrix} 1 & r_{m-1,m} \\ r_{m-1,m} & 1 \end{bmatrix} \begin{pmatrix} \boldsymbol{E}_{m,m-1}^{\ +} \\ \boldsymbol{E}_{m,m-1}^{\ -} \end{pmatrix} \tag{2.15}$$

where $t_{m-1,\,m}$ and $r_{m-1,\,m}$ are the complex Fresnel transmission and reflection coefficients of the $(m-1)/m$ interface which can be obtained from Equation (2.5) to Equation (2.7).

The field amplitudes on the left- and right-hand sides of mth layer are related by the propagation matrix $\boldsymbol{U}_m$:

$$\begin{pmatrix} \boldsymbol{E}_{m,m-1}^{\ +} \\ \boldsymbol{E}_{m,m-1}^{\ -} \end{pmatrix} = U_m \begin{pmatrix} \boldsymbol{E}_{m,m+1}^{\ +} \\ \boldsymbol{E}_{m,m+1}^{\ -} \end{pmatrix} = \begin{bmatrix} \mathrm{e}^{-j\beta_m} & 0 \\ 0 & \mathrm{e}^{j\beta_m} \end{bmatrix} \begin{pmatrix} \boldsymbol{E}_{m,m+1}^{\ +} \\ \boldsymbol{E}_{m,m+1}^{\ -} \end{pmatrix} \tag{2.16}$$

where $\beta = 2\pi k d \xi_m$ is the phase shift of the electromagnetic wave after one pass through the film; k is the wavenumber of incident light; and d is the thickness of the film. So the incident wave in the first layer ($\boldsymbol{E}_{1,\,2}^{\ +}$) can be related to reflected wave ($\boldsymbol{E}_{1,\,2}^{\ -}$) and to transmitted wave ($\boldsymbol{E}_{N+1,\,N}^{\ +}$) by the transfer matrix $\boldsymbol{Q}_{1,\,N+1}$:

$$\begin{pmatrix} \boldsymbol{E}_{1,2}^{\ +} \\ \boldsymbol{E}_{1,2}^{\ -} \end{pmatrix} = \boldsymbol{Q}_{1,N+1} \begin{pmatrix} \boldsymbol{E}_{N,N+1}^{\ +} \\ \boldsymbol{E}_{N,N+1}^{\ -} \end{pmatrix} = \boldsymbol{W}_{1,2}\boldsymbol{U}_2\boldsymbol{W}_{2,3}\boldsymbol{U}_3 \cdots \boldsymbol{W}_{N-1,N}\boldsymbol{U}_N\boldsymbol{W}_{N,N+1} \begin{pmatrix} \boldsymbol{E}_{n+1,n}^{\ +} \\ 0 \end{pmatrix} \tag{2.17}$$

The reflection and transmission coefficients can be obtained from Q:

$$r_{1,N+1} = \frac{\boldsymbol{Q}_{1,N+1}(2,1)}{\boldsymbol{Q}_{1,N+1}(1,1)} \tag{2.18}$$

$$t_{1,N+1} = \frac{1}{\boldsymbol{W}_{1,N+1}(1,1)} \tag{2.19}$$

$$r_{N+1,0} = -\frac{\boldsymbol{Q}_{N+1,0}(1,2)}{\boldsymbol{Q}_{N+1,0}(1,1)} \tag{2.20}$$

$$t_{N+1,0} = \frac{\boldsymbol{W}_{N+1,0}(1,1)\boldsymbol{W}_{N+1,0}(2,2) - \boldsymbol{W}_{N+1,0}(1,2)\boldsymbol{W}_{N+1,0}(2,1)}{\boldsymbol{W}_{N+1,0}(1,1)} \tag{2.21}$$

2.2.7 The reflection and refraction of light from a thick film

When silicon is double-side-polished, the optical reflection and refraction are limited by two parallel boundaries, it can be considered as thick film without interference to calculate the reflectance and transmittance.

The essential calculation difference between thin and thick films is that in a thick film only the intensity or the square of the amplitude A^2 proportional to the intensity I is taken into consideration, while in a thin film not only the amplitude but also the corresponding path or phase difference of the light rays in the film are considered.

As shown in Figure 2.7, multiple reflections occur inside the thick film, and the reflectance ($R_{a,b}{}^{+,-}$) and transmittance ($T_{a,b}{}^{+,-}$) on each boundary (a and b) are used to calculate the total reflectance (R) and transmittance (T) of the thick film:

$$R = R_a^+ + T_a^+ R_b^+ T_a^- + T_a^+ R_b^+ R_a^- R_b^+ T_a^- + \cdots = R_a^+ + \frac{T_a^+ R_b^+ T_a^-}{1 - R_a^- R_b^+} \tag{2.22}$$

$$T = T_a^+ T_b^+ + T_a^+ R_b^+ R_a^- T_b^+ + T_a^+ (R_b^+ R_a^-)^2 T_b^+ + \cdots = \frac{T_a^+ T_b^+}{1 - R_a^- R_b^+} \tag{2.23}$$

where the subscripts a and b denote the two boundaries of the thick film. The superscripts of + and – indicate the direction of incident light. The reflectance and transmittance of each boundary can be obtained from Equation (2.5) to Equation (2.10).

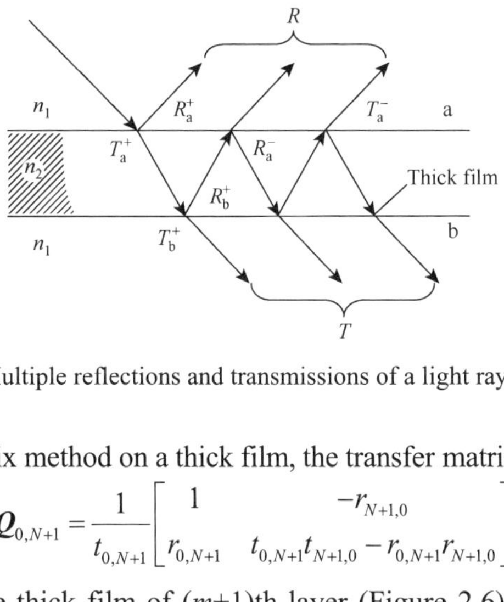

Figure 2.7 Multiple reflections and transmissions of a light ray in a thick film

To apply the matrix method on a thick film, the transfer matrix can be generalized as

$$\boldsymbol{Q}_{0,N+1} = \frac{1}{t_{0,N+1}} \begin{bmatrix} 1 & -r_{N+1,0} \\ r_{0,N+1} & t_{0,N+1} t_{N+1,0} - r_{0,N+1} r_{N+1,0} \end{bmatrix} \tag{2.24}$$

In the case of the thick film of (m+1)th layer (Figure 2.6), the coefficients were replaced with the absolute square amplitudes. The transfer matrices from 1st layer to (m+1)th and from (m+1)th to (N+1)th can be changed as

$$\boldsymbol{Q}_{1,m+1}{}^{\text{incoh}} = \frac{1}{t_{1,m+1}} \begin{bmatrix} 1 & -|r_{m+1,1}|^2 \\ |r_{1,m+1}|^2 & |t_{1,m+1} t_{m+1,1}|^2 - |r_{1,m+1} r_{m+1,1}|^2 \end{bmatrix} \tag{2.25}$$

$$\boldsymbol{Q}_{m+1,N+1}{}^{\text{incoh}} = \frac{1}{t_{m+1,N+1}} \begin{bmatrix} 1 & -|r_{N+1,m+1}|^2 \\ |r_{m+1,N+1}|^2 & |t_{m+1,N+1} t_{N+1,m+1}|^2 - |r_{m+1,N+1} r_{N+1,m+1}|^2 \end{bmatrix} \tag{2.26}$$

Then the transfer matrix of all the layers can be obtained by multiplying the squared propagation matrix, $\boldsymbol{U}_{3,m+1}^{\text{incoh}}$:

$$\boldsymbol{U}_{m+1}^{\text{incoh}} = \begin{bmatrix} \left|e^{-j\beta_{m+1}}\right|^2 & 0 \\ 0 & \left|e^{j\beta_{m+1}}\right|^2 \end{bmatrix} \tag{2.27}$$

$$\boldsymbol{Q}_{1,N+1}^{\text{incoh}} = \boldsymbol{Q}_{1,m+1}^{\text{incoh}} \boldsymbol{U}_{m+1}^{\text{incoh}} \boldsymbol{Q}_{m+1,N+1}^{\text{incoh}} \tag{2.28}$$

The reflectance and transmittance were described as

$$\boldsymbol{R} = \frac{\boldsymbol{Q}_{1,N+1}^{\text{incoh}}(2,1)}{\boldsymbol{Q}_{1,N+1}^{\text{incoh}}(1,1)} \tag{2.29}$$

$$\boldsymbol{T} = \frac{1}{\boldsymbol{Q}_{1,N+1}^{\text{incoh}}(1,1)} \tag{2.30}$$

2.3 Grazing Angle Mirror-Backed Reflection

As we discussed, external reflection with p-polarized incident light cannot give enough sensitivity for detection of the silicon-based monolayers because the Brewster angle of the air/silicon interface (73.6°) is close to the grazing angle and most of the light passes through the silicon wafer and escapes away from the detector. Compensation of the energy loss can be performed by reflecting back the transmitted infrared rays with a metal mirror. An example was given for analyses of an oxide layer in the metal-oxide-semiconductor (MOS) structure[33]. However, metal deposition is destructive to samples[34], which cannot be reused for other purposes. A simple alternate approach is to place a mirror behind the double-side-polished silicon wafer in the grazing angle external reflection configuration. Energy compensation simultaneously increases the band intensity and the signal-to-noise ratio (SNR) for both p- and s-polarization spectra.

Figure 2.8 shows the layout of the GMBR measurement with a grazing angle (80°) incidence. The incident infrared light is split into two parts as reflection and transmission when it reaches the silicon surface. Subsequently, the transmission light passing through silicon is reflected back by the gold mirror, and then traverses silicon again. Following the increment of the air spacer between gold mirror and silicon to 1 mm, the distance between the first reflection spot (point *A*) and the double-transmission spot (point *B*) on the guiding mirror is about 23 mm, and the span width (*L*) of the IR beam to enter the detector is about 2 mm. Since the guiding mirrors in the grazing angle accessory are 70 mm long, most reflection and double-transmission light can reach the detector. As a result, a relatively higher sensitivity can be obtained.

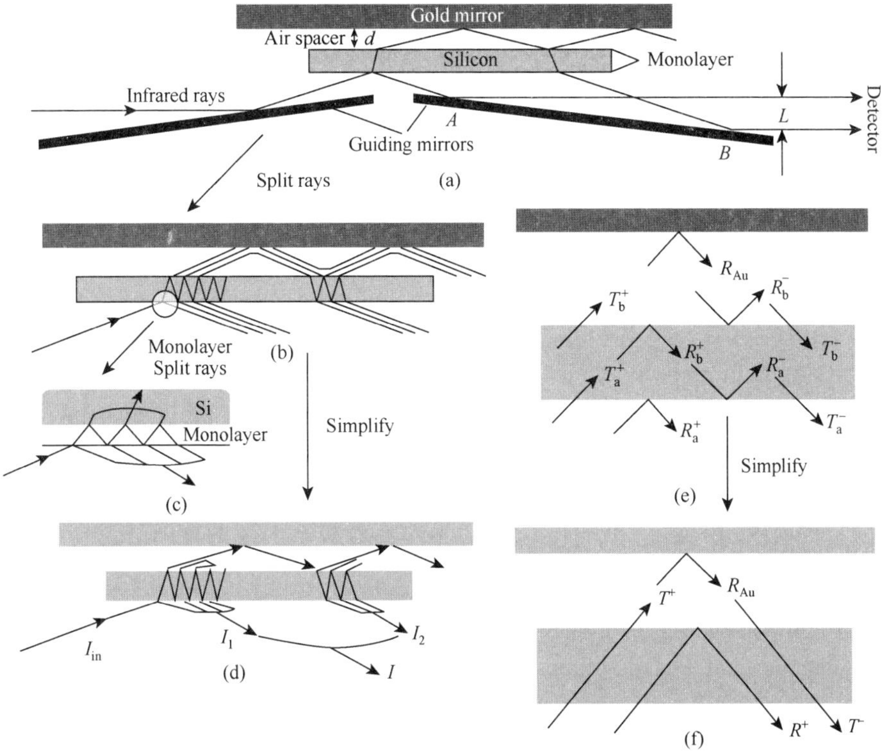

Figure 2.8 (a) Experimental set-up for GMBR. (b) Multiple-reflections and multiple-transmissions taken place both in the silicon wafer and a monolayer. (c) Amplification of multiple-reflections and multiple-transmissions in a monolayer. (d) Simplified three-media model for calculation with air/Si (monolayer)/air, I_{in} and I denote the incident and emergent light intensity, I_1 and I_2 are the reflection and double-transmission light respectively. (e) Symbols of reflectance (R) and transmittance (T) at the interface for lower (subscript a) and upper (subscript b) surface of the silicon wafer, the superscript + indicates the light traveling direction from the air/Si_a interface to the Si_b/air interface, and – for the opposite direction (from air/Si_b to Si_a/air interface). Si_a and Si_b denote the lower and upper silicon surface respectively. R_{Au} denotes the reflectance of gold mirror. (f) Sum reflectance (R^+) and transmittance (T^+, T^-) for the entire silicon wafer

Numerical simulation was carried out to analyze the sensitivity of GMBR measurements, and the simulated optimum parameters were examined. As shown in Figure 2.8 (b) and Figure 2.8 (c), all multiple-reflections and multiple-transmissions in both monolayer and silicon are taken into account. Calculation was simplified with a three-media model air/Si/air in Figure 2.8 (d), where in this model the monolayer-modified silicon [monolayer/Si in Figure 2.8 (c)] is integrated as a single interface. The reflectance [$R^{+(-)}{}_{a(b)}$] and transmittance [$T^{+(-)}{}_{a(b)}$] at the interface of air/Si and air/monolayer/Si shown in Figure 2.8 (e) can be obtained from Equation (2.5) to Equation (2.10). The sum reflectance (R^+) and transmittance (T^+, T^-) of the entire silicon wafer with and without the monolayer [Figure 2.8 (f)] can be calculated with $R^{+(-)}_{a(b)}$ and $T^{+(-)}_{a\ (b)}$ by Equation (2.22) and Equation (2.23). The total output light intensity can be written as

$$I = RI_{in} = I_1 + I_2 = R^+ I_{in} + (T^+ R_{Au} T^-) I_{in} \tag{2.31}$$

$$R = R^+ + T^+ R_{Au} T^- \tag{2.32}$$

where I_{in}, I_1 and I_2 are the light intensity of the incidence, reflection and double-transmission light respectively [Figure 2.8 (d)]. R is the reflectance of the output light, which equals the ratio of emergent light intensity to incident light intensity. R_{Au} is the reflectance of gold mirror. The absorbance (A) and absorption depth (ΔR) are defined through:

$$A = -\lg(I/I_0) = -\lg(R/R_0) \tag{2.33}$$

$$\Delta R = R_0 - R \tag{2.34}$$

where I and R with and without subscript 0 correspond to the output light intensity and the reflectance calculated without and with the monolayer respectively.

Stepwise modifications of the silicon hydride surface (1 in Figure 2.9) in Figure 2.9 were applied as an example to demonstrate the feasibility of GMBR. An undecylenic acid (UA) monolayer (2 in Figure 2.9) was prepared under microwave irradiation[35, 36]. The succinimidyl ester (NHS ester) (3 in Figure 2.9) was prepared by DCC/NHS (dicyclohexylcarbodiimide/N-hydroxysuccinimide) treatment at room temperature.

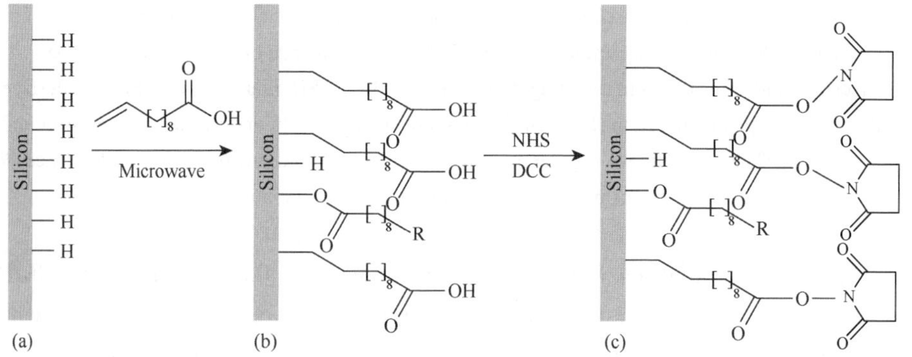

Figure 2.9　Schematic illustration of surface modification

(a) represents the freshly prepared hydrogen-terminated Si (111). (b) represents UA monolayer but with possible molecular structures: dominant terminal carboxylic acid and side product of Si-ester. R represents terminal vinyl group or bridged bulk—Si—C bond. (c) represents the NHS ester monolayer

For numerical analyses, the following parameters were selected as typical: n_1=1.0

(air), n_3=3.42 (silicon substrate), n_{Au}=3.2+20i (gold mirror), n_∞=1.46[29]. Since our goal is to characterize the molecular monolayer on silicon using the GMBR method, we choose the UA monolayer as an example for numerical analyses. The thickness of the UA monolayer is d=0.7 nm[28].

The complex refractive index of the monolayer can be determined from experimental spectra using the iterative fitting method of Buffeteau[28]. Briefly, the optical constants in the monolayer plane (n_x=n_y and k_x=k_y for uniaxial film) can be determined from the s-polarized spectra and the constants out of the monolayer plane (n_x and k_x) come from the p-polarized spectra. As shown in Figure 2.10, using the approximate equations [Equation (2.11)-Equation (2.14)] for reflectance and transmittance instead of the exact Fresnel equations [Equation (2.5)-Equation (2.10)], the approximate imaginary parts of the dielectric functions ($\hat{\varepsilon}_{x,y,z}$) were deduced from the experimental spectra at p- or s-polarization, and the real parts were calculated by Kramers-Kronig transformation[37]. The approximate real and imaginary parts of the dielectric functions were treated as initial parameters for simulating the polarized spectra with the exact Fresnel formulae. Comparing the calculated and experimental spectra, the imaginary parts of each wavenumber was perturbed until they were close enough to each other (spectra fitting was performed using the trust-region method in Matlab). Since the real part was not perturbed, this procedure was repeated until there was no obvious change between the real and imaginary parts from two adjacent iterative calculations. Because both parallel (x direction) and normal (z direction) components were included in the approximate reflectance equations for p-polarization, firstly, the approximate imaginary part of dielectric functions regarding the parallel component $[\mathrm{Im}(\hat{\varepsilon}_y) = \mathrm{Im}(\hat{\varepsilon}_x)]$ should be deduced from s-polarization, and then the approximate $\mathrm{Im}(\hat{\varepsilon}_z)$ can be determined. From $\hat{\varepsilon}_x$ and $\hat{\varepsilon}_z$, the anisotropic refractive index can be calculated with the relation of $\hat{\varepsilon} = \hat{n}^2$.

The UA monolayer is composed of 10 methylenes and 1 carboxyl group, thus we select the overwhelming methylene species as the monolayer for numerical analyses. The normalized experimental p- and s-polarized spectra of the ν_{as} (CH_2) mode (2927 cm^{-1}) were used for iterative simulation. The monolayer was treated as a uniaxially oriented layer and their optical properties can be described by in-plane ($\hat{n}_{2x} = \hat{n}_{2y}$) and out-of-plane ($\hat{n}_{2z}$) complex refractive indices. From Buffeteau et al.'s[28] iterative simulation at a fixed 80° of incidence, the complex refractive indices were determined as follows: $\hat{n}_{2x} = \hat{n}_{2y} = 1.52 + 0.39i$ and $\hat{n}_{2z} = 1.44 + 0.23i$.

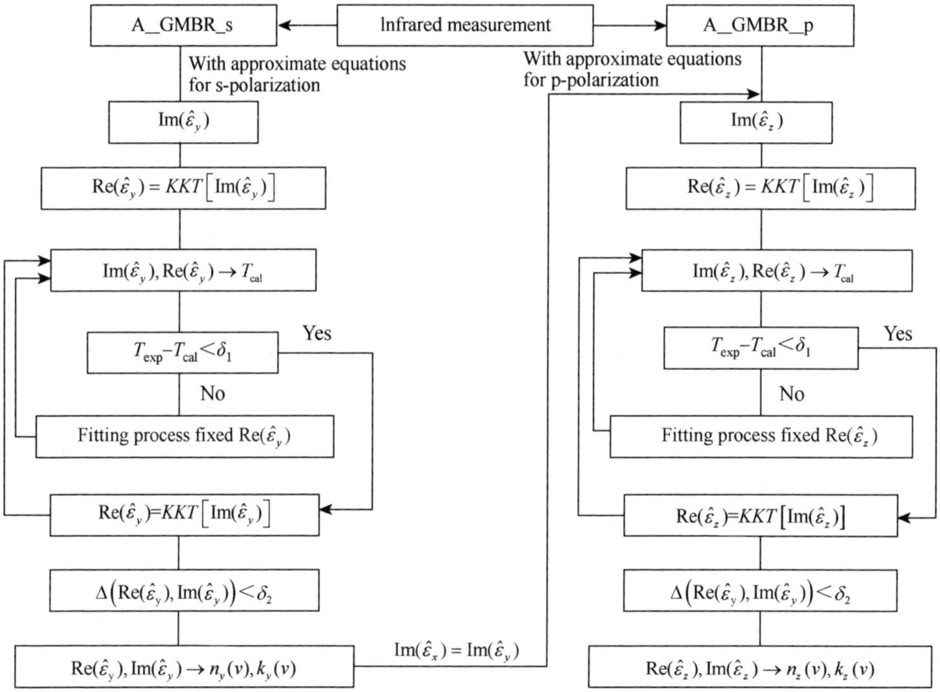

Figure 2.10 Flowchart of the computation method for determination of anisotropic constants

The result of numerical analyses for GMBR is shown in Figure 2.11, where the absorbance and absorption depth of the monolayer are drawn against the incident angle. Since the absorbance determines the band intensity and SNR is proportional to the absorption depth in a detector-noise-limited spectrometer[29], the simulated spectra allow for the distinguishment of the optimum conditions for measurements. The angular dependent absorbance for p-polarized radiation in Figure 2.11 (a) exhibits the highest absorbance at 78.1°, close to the maximum absorption depth, 75.4°. The coincidence of maximum spectral contrast and SNR facilitates the choice of the optimum angle of incidence. For s-polarization, both absorbance and absorption depth decrease smoothly with increasing the angle of incidence and the band intensity is far less than p-polarization. Given the calculation result, the optimum analyses can be achieved in the mirror-backed system by using p-polarized radiation at grazing angles of incidence. This coincidence occurred due to the energy conversation by placing a mirror behind the silicon wafer.

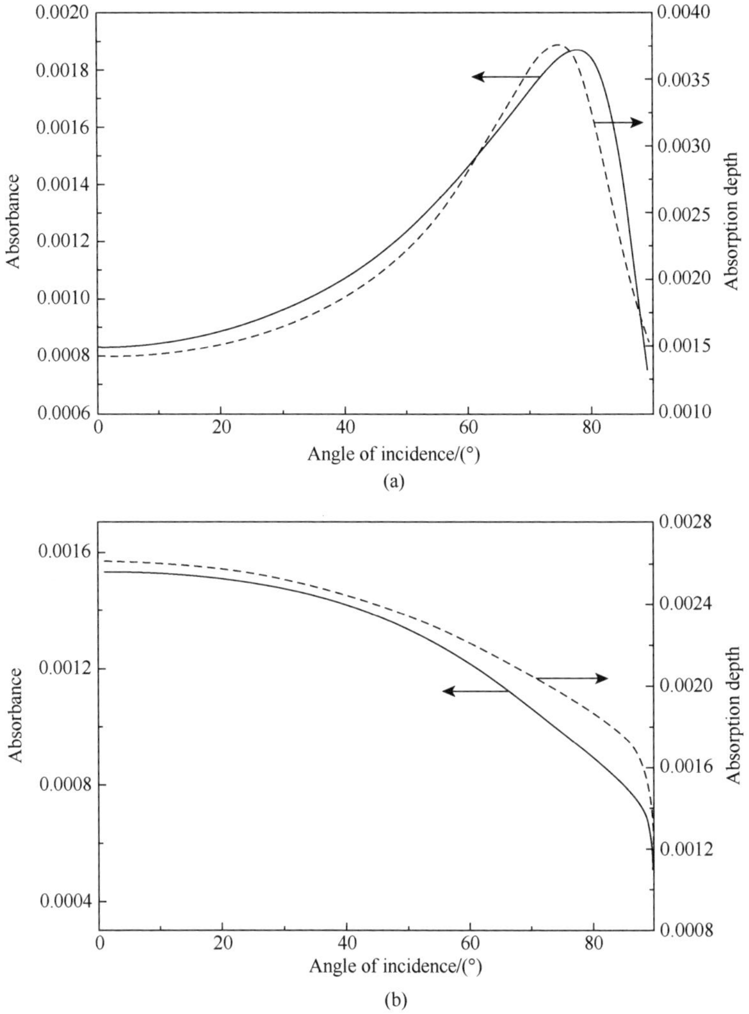

Figure 2.11 Calculated absorbance (A, solid line) and absorption depth (ΔR, dashed line) of thin film for (a) p-polarized and (b) s-polarized radiation with GMBR measurement for the UA monolayer at ν_{as} (CH_2) =2927 cm^{-1}

Experimental spectra of GMBR and Brewster angle transmission are given in Figures 2.12 (a) and 2.12 (b) for comparison. Three typical structures of Si—H [curves A and B in Figure 2.12 (a)], UA [curves C and D in Figure 2.12 (b)] and NHS ester [curves A and B in Figure 2.12 (b)] monolayer species on silicon are exemplified. The hydrogen terminated silicon surface is the basis for fabrication of Si—C bonded monolayers and its strong IR stretching mode (~2100 cm^{-1}) provides a suitable marker for testing the sensitivity of an IR measurement. The carboxylic acid terminated

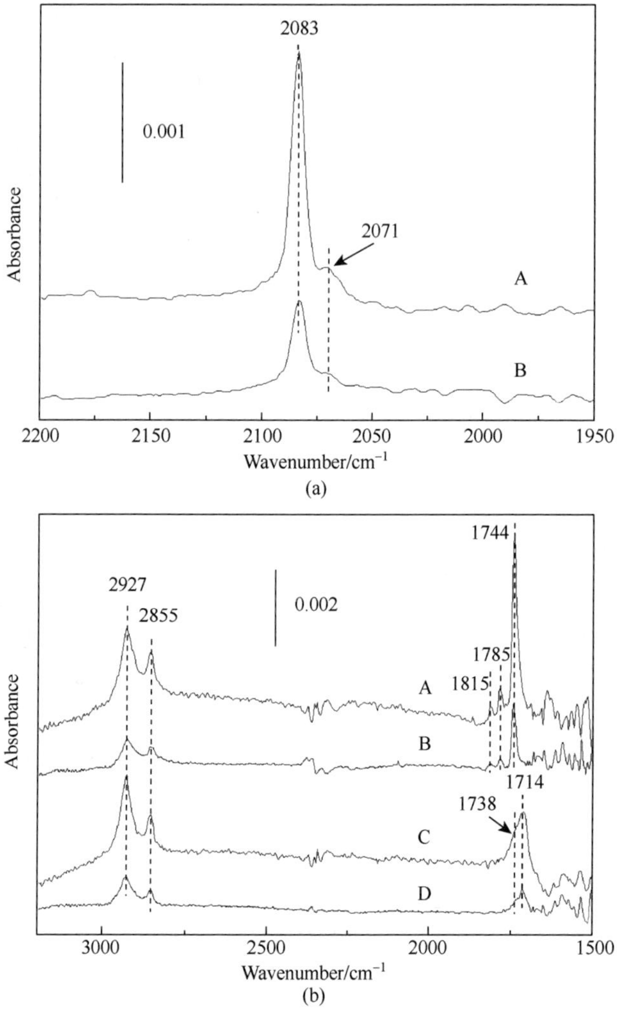

Figure 2.12 Experimental p-polarized spectra: (a) monohydrogen-terminated Si (111) surface by (A) GMBR and (B) Brewster angle transmission at 74°; (b) surfaces modified with (A, B) NHS ester and (C, D) UA monolayer by (A, C) GMBR and (B, D) Brewster angle transmission. In the measurement, the air gap [(d) in Figure 2.8] between silicon and mirror was 1.0 mm. The regions of 2400-2280 cm^{-1} and 1500-1800 cm^{-1} are affected by CO_2 and H_2O respectively. Spectra are shown after subtraction of H_2O and CO_2

monolayer (UA), and its subsequently derivatized NHS ester monolayer, frequently used for immobilization of biomolecules and fabrication of biochips[26, 35], are chosen as examples for testing organic monolayers with carbonyl stretchings at ~1700 cm^{-1}. The vibration bands in this region can exclusively distinguish the ambiguity of the C—H stretchings at 2800-2980 cm^{-1}, where the contaminated hydrocarbons also contribute to the absorption. The NHS ester species possesses a specifically coordinated triplex peaks,

1815 cm^{-1}, 1785 cm^{-1}, and 1744 cm^{-1}[26, 35]. It is also an excellent marker for testing the sensitivity of IR measurements. The Si—H band[38] intensity in Figure 2.12 (a) at 2083 cm^{-1} by GMBR [curve A in Figure 2.12 (a)] is about 2 times higher than that by the Brewster angle transmission [curve B in Figure 2.12 (a)], due to the contribution of both reflection and double-transmissions to the absorbance. It is consistent with the result of numerical analyses.

To get a good spectrum, the distance of the air spacer is critical because a smaller spacer will generate interference fringes, but a larger one will lose the double-transmission light and thus lose the intensity. In the GMBR measurement, if the mirror is placed directly on the silicon wafer, the spectra will be distorted. This problem can be solved by increasing the air spacer distance. When the spacer is increased to 0.5 mm, good spectra can be obtained, and spectra can be even better with 1 mm spacer. If an MCT detector is used, the interference fringes become stronger, while the sensitivity is higher. Although they can be attenuated to a large extent when the spacer is increased to 1.5 mm, the band intensity of alkyl peaks in the region of 2800-2980 cm^{-1} decreases due to partly the double-transmission light not reaching the detector.

2.4 Multiple Transmission-Reflection Infrared Spectroscopy

GMBR can enhance the measurement sensitivity 2-3 times higher than the p-polarized single transmission at the Brewster angle. But the sensitivity is still not high enough for molecules with weak absorption and subsequently affects the accuracy of quantitative analyses. We modified GMBR further to multiple transmission-reflection infrared spectroscopy (MTR-IR), for characterization of silicon-based monolayers, by placing the sample between two parallel gold mirrors. When light reaches the silicon surface, it will be split into transmitted and reflected parts leading to multiple transmissions and reflections between two mirrors (Figure 2.13). The absorbance of monolayers on a silicon surface will, therefore, be enhanced much higher than by GMBR. A pioneer work with this configuration was used by V. P. Tolstoy et al. for detection of the ultra-thin SiO_2 layer on silicon with p-polarized light at the Brewster incident angle about 20 years ago[29, 39]. Unfortunately, further development of this configuration for conformational characterization of organo- and bio-molecular monolayers are not yet investigated, especially for its wide application on silicon-based monolayers which are attracting much interest in molecular and nano-electronics and in bio-interfaces. Here, we show that this geometry is powerful for detection of monolayers on a silicon surface. Theoretical calculation demonstrates that the sensitivity of MTR is comparable to MIR. The most important advantage of MTR measurement, worth of special attention, is the replacement of the expensive ATR crystal with the ordinary double-side-polished silicon wafer. With optimized parameters, both p- and s-polarized spectra can be firmly obtained, from which more molecular structure information such as surface packing and orientation can be revealed.

2.4.1 Set-up

A schematic of the MTR set-up and the optical geometry is shown in Figure 2.13 (see the Color Inset, p.4).

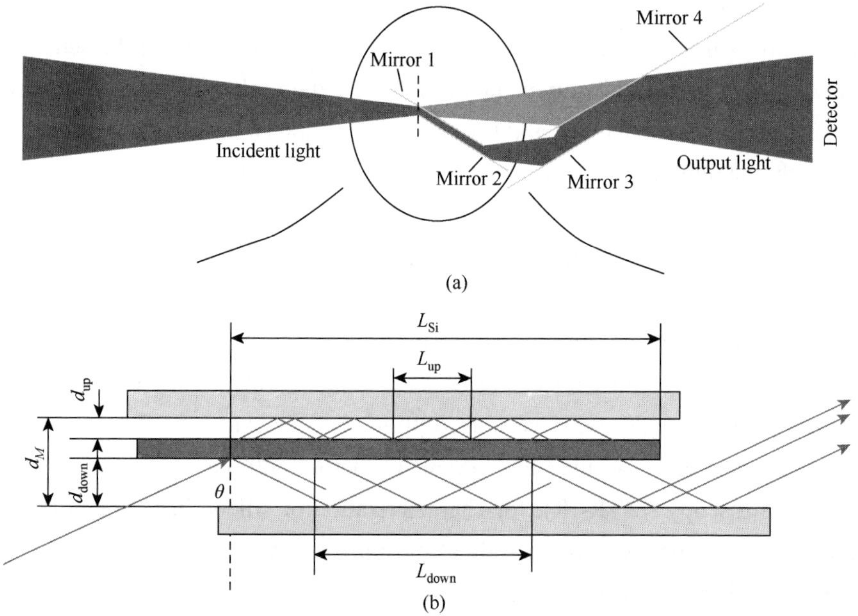

Figure 2.13 (a) Scheme of the MTR set-up. The incident light from the interferometer, after multiple transmissions and reflections on the silicon wafer placed between gold mirrors 1 and 2, is guided to the detector by mirrors 3 and 4. Red part denotes that the infrared light goes through the accessory, and the pink one is the original light path of instrument. (b) Detailed view of the sampling mirrors and the parameters used in the calculation. The effective length of silicon wafer (L_{Si}) is the distance from the incident spot to the other end of the wafer. L_{up} and L_{down} are defined as a periodic horizontal length from a spot on the upper face of silicon that a single ray of light shines on to the next spot that it comes back and hits on, in the upper and in the lower media respectively. d_M is the mirror distance, and d_{up} and d_{down} are the thicknesses of air gaps respectively

The set-up is composed of four mirrors, among which two are sampling mirrors (mirrors 1 and 2), and the other two are guiding mirrors (mirrors 3 and 4) to redirect the beam into the detector. The sample, silicon wafer, is inserted between the two sampling mirrors, with one end of the silicon wafer protruding out at about 5 mm of the incident spot, in order to make sure that the first reflection is on the silicon surface. The distance between the silicon wafer and two sampling mirrors can be adjusted within two millimeters. The two guiding mirrors can be moved back and forth to get the maximal amplitude. The incident angle was controlled by two step-motors with a minimal angle of 0.225°.

2.4.2 Numerical analyses of absorbance

The MTR configuration can be simplified as five layers for simulation calculation in Figure 2.14: two air gaps (1 and 5), two monolayers (2 and 4), and silicon (3). The reflectance (R_{Si}) and transmittance (T_{Si}) are obtained each time when the infrared beam interacts with the sample. Considering the silicon wafer as an incoherent thick film between two-infinite air media, according to Equation (2.5) to Equation (2.10), Equation (2.22), and Equation (2.23), the three layers, (monolayer)/Si/(monolayer), are treated as one medium interface during further calculations to get the final intensity of the output light.

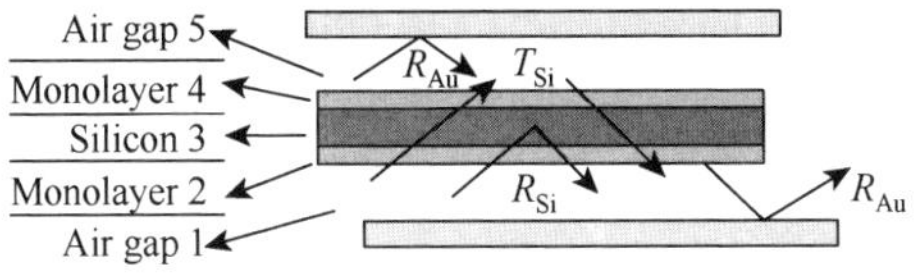

Figure 2.14 Simplified five-layer media in the MTR configuration for simulation calculations. T_{Si} and R_{Si} are the transmittance and reflectance of (monolayer)/Si/(monolayer) respectively. R_{Au} is the reflectance of gold surface

The step-by-step calculations are illustrated in Figure 2.15 to obtain the accumulative total intensities of the output light ($I_{0,\,out}$ and I_{out}). When light reaches the sample interface, (monolayer)/Si/(monolayer), it is split into reflection and transmission components, and both come back to this interface again after reflection on gold mirrors. This process defines one periodic horizontal length passed by the light in the upper media (L_{up}) and in the lower media (L_{down}) in Figure 2.13. The split light intensity is multiplied with T_{Si}, R_{Si}, or R_{Au} (reflectance on gold surface) respectively each time when the light meets these interfaces. Simultaneously, after each step, the horizontal length of light passed is added with L_{down} or L_{up} respectively (Figure 2.15). The relay calculation will be stopped if L is larger than the effective length of silicon wafer (L_{Si}) or the intensity is less than $I_0 \times 10^{-4}$. The sum of all terminal intensities of each branch in Figure 2.15 is the accumulative total intensity of the output light. The absorbance A and the absorption depth ΔR are obtained using the simple relations:

$$A = -\lg\left(\frac{I_{out}}{I_{0,out}}\right) \tag{2.35}$$

$$\Delta R = \frac{I_{0,out}}{I_{in}} - \frac{I_{out}}{I_{in}} \tag{2.36}$$

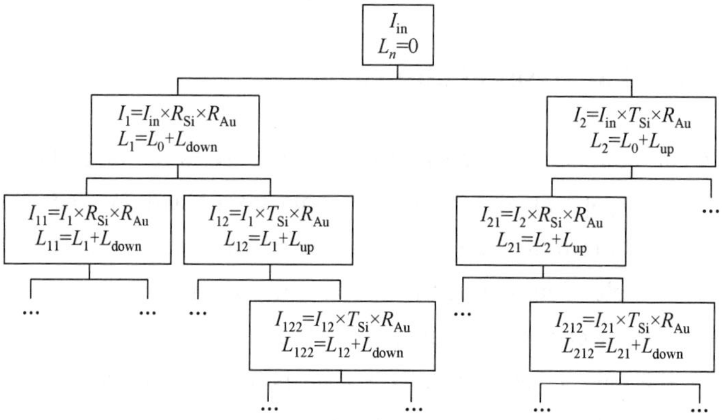

Figure 2.15　Flow chart for the calculation of accumulative total intensities of output light in the MTR configuration. I_{in} is the intensity of incident light. R_{Si} and T_{Si} are the reflectance and transmittance of (monolayer)/Si/(monolayer) respectively. The intensity in each branch is obtained by multiplying with R_{Si} and T_{Si} for reflection and transmission at the (monolayer)/Si/(monolayer) interface respectively and with R_{Au} for reflection on the gold mirror. The subscriptions indicate accumulative reflections and transmissions at the (monolayer)/Si/(monolayer) interface, where 1 and 2 represent one reflection and one transmission respectively

2.4.3　Optimal measurement conditions

The MTR-IR spectra are affected by several parameters, the most important amongst them being the incident angle (θ), the distances between two mirrors (d_M) and between silicon and upper and down mirrors (d_{up} and d_{down}) respectively, and the effective length of silicon wafer (L_{Si}). Since the number of reflections and transmissions through the silicon wafer changes when these parameters are perturbed, the absorbance and the signal to noise ratio (SNR) will be influenced. The band intensity is determined by the absorbance, and SNR is proportional to the absorption depth in a noise-limited spectrometer[29]. Simulation analysis will provide optimum parameters for high quality spectra. We define the number of simplified transmission times (N) as the passing times of the entrance beam through silicon until it leaves for the guiding mirror, assuming no reflection on the silicon surface. However, during the simulation process, we count all kinds of transmissions and reflections on silicon, and reflections on gold surfaces, as shown in Figure 2.15. Therefore the number of simplified transmission times (N) is just a parameter for simplicity and clarity in calculation.

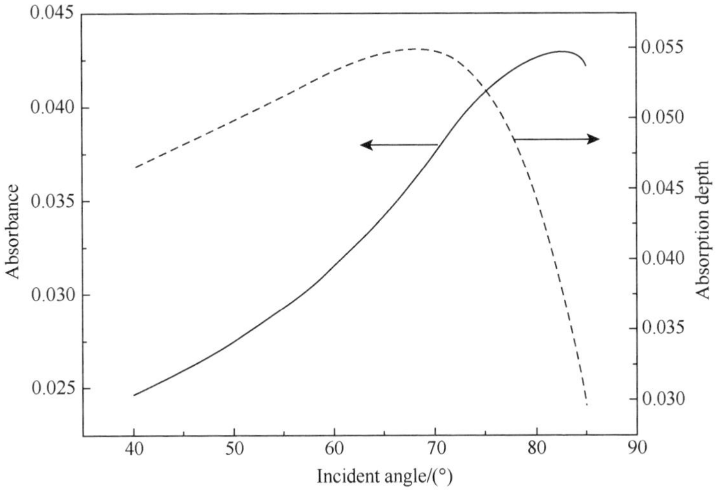

(a)

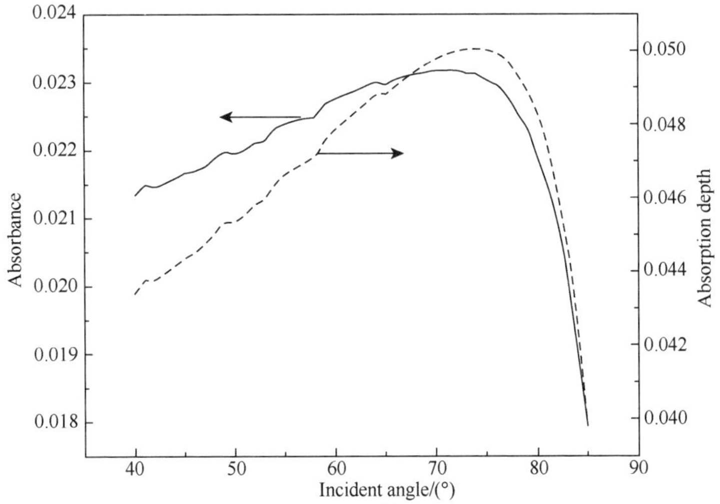

(b)

Figure 2.16 Calculated absorbance (A, solid line) and absorption depth (ΔR, dashed line) versus incident angle for a 2.6 nm isotropic hydrocarbon adlayer (n=1.5, k=0.5) by (a) p-polarization, and (b) s-polarization at 2918 cm^{-1}. MTR parameters for calculation are N=8, d_{up}=0.05 mm and L_{Si}=25 mm

First we consider the effect of the incident angle on the sensitivity of measurements. The goal is to figure out the optimal incident angle for obtaining both high absorbance and SNR at the same time. Figure 2.16 shows the curves of absorbance and absorption depth drawn against incident angle for p- and s-polarizations, where N is fixed to 8. The calculations have been done for an isotropic hydrocarbon layer at 2918 cm^{-1} with typical optical constants of n=1.5 and k=0.5. The angular dependence of absorbance for p-polarization exhibits the maximum around 80°, but the absorption depth goes down steeply after 74°. For s-polarization, both absorbance and absorption depth show maximum of 70°-74°. This similarity of maximum absorbance and absorption depth of both p- and s-polarizations facilitates the choice of the optimal angle of incidence to be 70°-74° in our measurements.

We now fix the angle of incidence at 74° to study the effect of the number of simplified transmission times (N) on the sensitivity. The distance between the two sampling mirrors and the size of the silicon sample determine the number of transmission-reflection times and consequently the measurement sensitivity. The values of N used in the calculation have been limited to a narrow range, 1 to 14, considering the unreasonably larger silicon sample size and the unrealizable small mirror distance that would be required to achieve larger values of N in practice. The absorbance and the absorption depth calculated for different values of N are plotted in Figure 2.17.

The absorbance of p-polarization goes up proportionally to N. This can be expected since no loss in intensity occurs for p-polarization at an incident angle close to the Brewster angle of air/silicon (73.6°). Therefore it is equivalent to measuring a stack of N identical samples in parallel. While this is true for the absorbance, the calculated absorption depth increases steeply only up to a value of N smaller than 6, and then begins to saturate at higher numbers of N [Figure 2.17 (a)]. The limitation of the absorption depth of p-polarization is due to the decrease of output intensity with increasing N, similar to what has been shown for multiple reflections between two metallic surfaces[41]. Because we consider Si to be transparent in all calculations, the only energy loss comes from the reflection on gold mirrors. The complex refractive index of gold is 2.018+21.087i at 2918 cm^{-1}, and the corresponding reflectance is 0.94 and 0.99 for p- and s-polarizations respectively[42]. For s-polarization, where the light has a high reflectance on gold, both absorbance and absorption depth increase with increasing N [Figure 2.17 (b)]. But an exception for N=2 exists. As shown in Figure 2.17 (b), the absorbance at N=2 is lower than that of the single transmission, which means that the MTR measurement for s-polarization, unlike the p-polarization, is a combination of transmission and reflection components, due to its high reflectance on air/silicon at the Brewster incident angle (R_{Si} is around 0.83[43]).

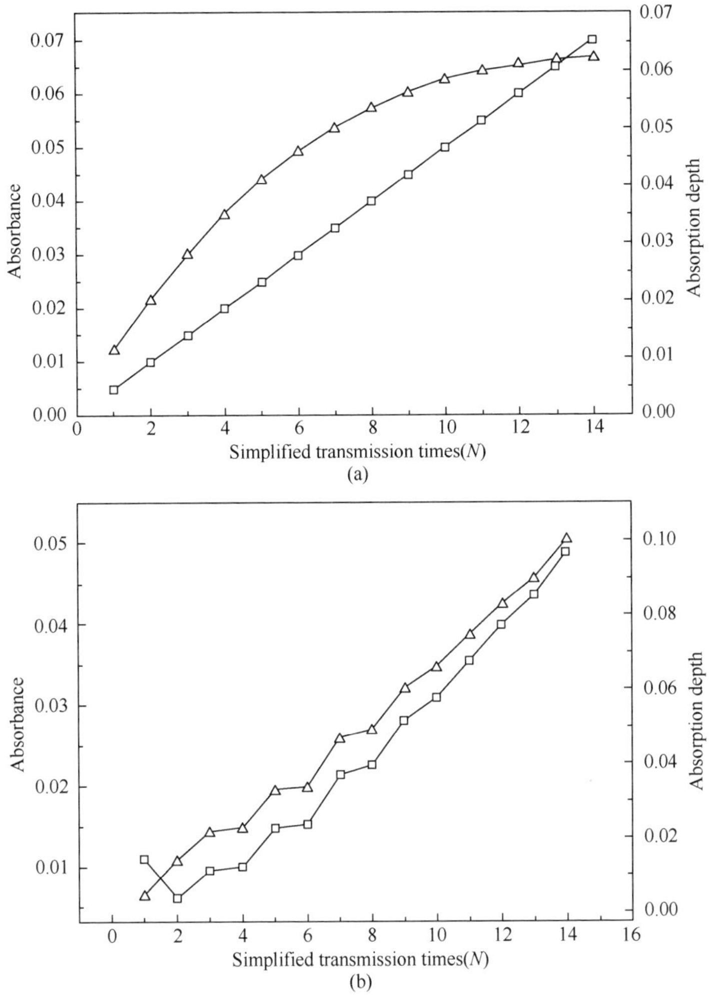

Figure 2.17 Calculated absorbance (A, –□–) and absorption depth (ΔR, –Δ–) versus simplified transmission times for a 2.6 nm layer (n=1.5, k=0.5) at 2918 cm^{-1} by (a) p-polarization, and (b) s-polarization at an incident angle of 74°. The mirror distance d_M between two gold mirrors is changed with different transmission times, while d_{up} is fixed with 0.3 mm

It is evident from the data presented in Figure 2.16 and Figure 2.17 that the MTR measurement is a unique combination of transmission and reflection. The above theoretical consideration, taken together with the practical limitation for the silicon wafer size and the precise control of mirror distances, has led to a set of optimal conditions for spectra acquisition: N=6, d_M=2 mm, d_{up}=0.3 mm, L_{Si}=25 mm, and an incident angle of 74°. All the experimental spectra discussed below are recorded with these optimized parameters. The high quality infrared spectra of adsorbed monolayers by MTR are demonstrated on the double-side-polished silicon substrate derivatized with carbonized (Si—C) and silanized (Si—O—C) organic monolayers.

Spectra measured with p-polarization on H-terminated silicon and its derivatized UA and NHS ester monolayers are shown in Figure 2.18. The peak at 2083 cm^{-1} is typical for monohydrogen-terminated Si (111) surface. Upon the reaction with the end double bond (C=C) of UA, this peak disappears completely, while a new set of bands appear at 2926 and 2854 cm^{-1}, for ν_{as} (CH_2) and ν_s (CH_2) respectively, and at 1714 cm^{-1}, for the C=O stretching band of carboxylic acid group. After further functionalization with DCC/NHS, the specifically coordinated three peaks (1815 cm^{-1}, 1785 cm^{-1}, and 1744 cm^{-1}), assigned to NHS ester, are clearly shown. For clarity, we also included the GMBR spectrum of H-terminated silicon in Figure 2.18 (bottom trace). Obviously, a substantial improvement in the spectral SNR is observed for the same monolayer on silicon. The higher sensitivity of the MTR methodology to study the surface species on a silicon surface compared to GMBR is clearly demonstrated. However, the real advantage of MTR over GMBR and Ge/monolayer/Si lies in its ability to measure the spectra of adsorbed layers with high sensitivity in both s- and p-polarizations, which can be used to infer information concerning the orientation of the adsorbed species. To demonstrate the feasibility of MTR for the revelation of the molecular orientation on Si from p- and s-polarized spectra, we used the well-known

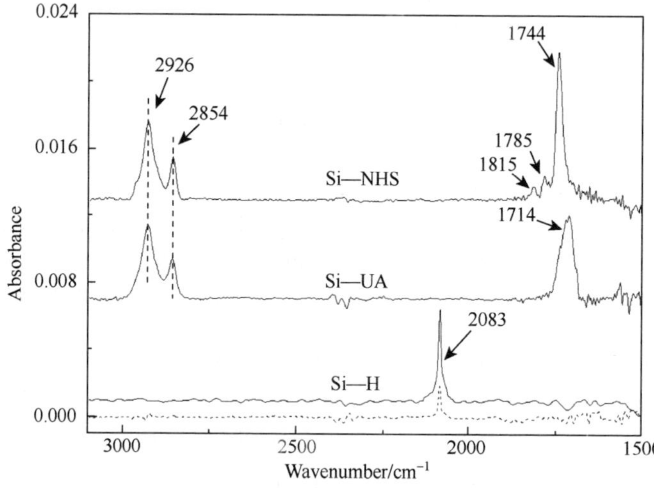

Figure 2.18 Experimental p-polarized spectra of Si—H, undecylenic acid (UA), and *N*-hydroxysuccinimidyl (NHS) ester monolayers measured with the MTR methodology. The region of 2400-2280 cm^{-1} is the absorption disturbance of CO_2 and of 1500-1800 and 3500-4000 cm^{-1} is due to H_2O. Spectra are shown after baseline correction and subtraction of H_2O and CO_2

self-assembled monolayers of OTS and decyl trichlorosilane (DTS) on the commercial silicon wafer surface with an ultrathin SiO_2 passivation layer. Here, we only show the ability of MTR to illustrate the surface molecular packing and orientation change by p- and s-polarized spectra. The quantitative orientation analysis will be discussed in details in Section 2.5.

Spectra of OTS and DTS with p- and s-polarizations in the C—H stretching region are given in Figure 2.19. The position of methylene vibration is sensitive to the molecular order. From OTS to DTS, the peaks of ν_{as} (CH_2) and ν_s (CH_2) shift from 2918 cm^{-1} and 2850 cm^{-1} to 2923 cm^{-1} and 2853 cm^{-1} respectively, but no shift of the wavenumber of any stretching mode of CH_3 is observed, which indicates a disordered gauche structure in the DTS monolayer. On the silicon surface, both normal (z) and parallel (x) components of transition dipole moment (TDM) contribute to the absorption for p-polarization, while only the parallel (y) component for s-polarization. This is evident from the spectra wherein both in-plane (2966 cm^{-1}) and out-of-plane (2958 cm^{-1}) modes of ν_{as} (CH_3) can be seen in p-polarized spectra, while only the out-of-plane mode is clearly visible in s-polarization. Assuming an extended all-trans conformation of the alkyl chains, the TDMs of the methylene stretching modes are perpendicular to the alkyl chain. The stronger intensity of methylene in the OTS s-polarized spectrum indicates that its alkyl chain is closer to the normal of the surface than that of DTS.

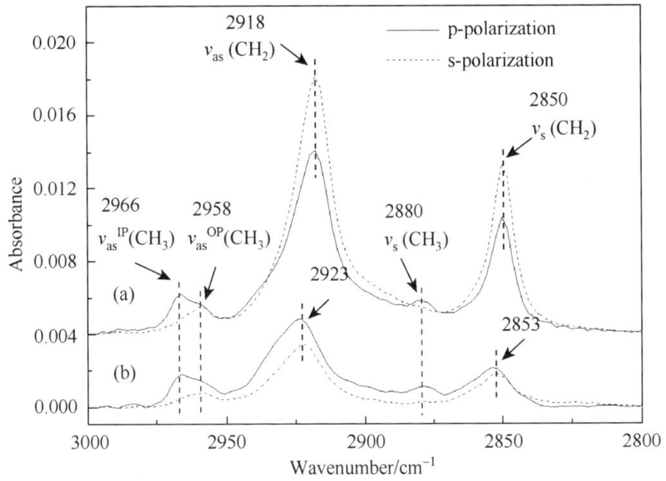

Figure 2.19 Experimental spectra of (a) OTS and (b) DTS monolayers with MTR for p- (solid line) and s-polarization (dashed line)

2.4.4 Comparison with other measurements

Compared with the single transmission (the case of N=1 in Figure 2.17), the absorbance of p-polarization of MTR is plotted against transmission times linearly at Brewster angle for the multiple transmissions, while the absorbance of s-polarization at N=2 to 4 is lower than that of N=1. The absorption depths of MTR for both p- and s-polarizations are much higher than the single transmission (Figure 2.17). The GMBR measurement is a special example of MTR with N=2. In Figure 2.18, we compared the experimental spectra of Si—H between MTR and GMBR. The absorbance peak of Si—H from MTR is about 3 times higher than that from GMBR. In Section 2.3, we have shown that its absorbance measured from GMBR is 2-3 times higher than that from the p-poalrized transmission at Brewster angle. These results are coincident with the calculated ones.

Hitherto, MIR is thought as the only measurement that can get good spectra with both p- and s-polarizations for detection of silicon-based monolayers. Theoretically, the absorbance of the MTR measurement can be compared as high as MIR. Calculated comparison results of MTR with MIR in Figure 2.20 (a) show the absorbance of p-polarization from MTR at 74° with N=10 is equivalent to that from MIR at 45° (the normally used incident angle) with reflection times of 16, while the s-polarized absorbance of MTR with N=10 is close to MIR with reflection times of 11. As we discussed above, the absorption depth of MTR for p-polarization is saturated at higher transmission times, while that for s-polarization goes up linearly with the increment of transmission times. Obviously the absorption depth of MTR is lower than MIR at higher transmission (reflection) times [Figure 2.20 (b)].

The high reflectance of s-polarization on silicon makes it possible that multiple reflections with s-polarized radiation between a gold mirror and a silicon surface could get considerable absorbance and SNR simultaneously. So it's necessary to compare it with the MTR measurement. As shown in Figure 2.21, the s-polarized absorbance of multiple reflections between a metal mirror and a silicon is linear against reflection times but the absorption depth decreases steeply when reflection times are more than 5. In the calcualtion, the corresponding simplified transmission times N are given by fixing L_{Si} and d_{up}, but changing d_{down} to get different reflection times in air gap 1. If we compare the simulated absorbance of s-polarization between one gold surface and one silicon surface (Figure 2.21) with the calculated MTR absorbance of s-polarization in Figure 2.17 (b), the absorbance could be comparable at high reflection times, but its absorption depth is much lower than MTR's.

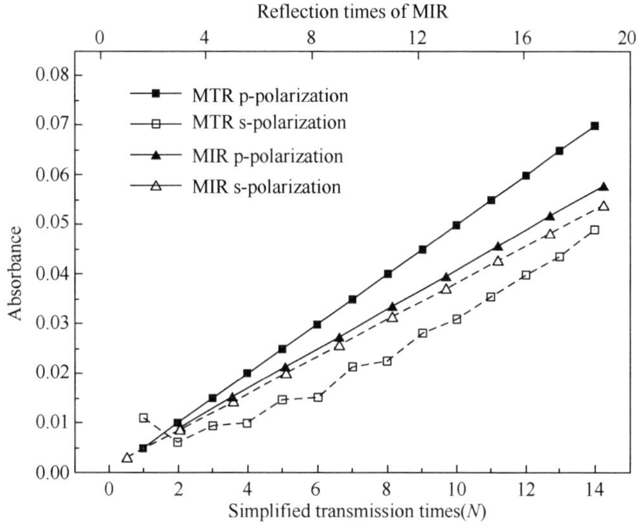

(a)

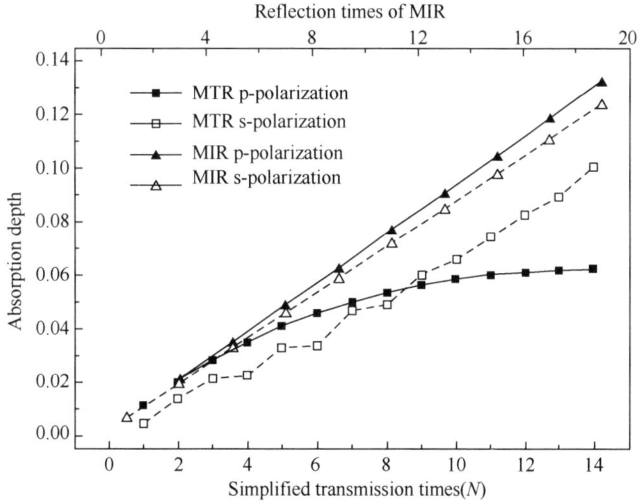

(b)

Figure 2.20 Comparison of calculated (a) absorbance and (b) absorption depth of MTR (incident angle is 73.6°) and MIR (incident angle is 45°) measurements with 2.6 nm layer (n=1.5, k=0.5) at 2918 cm^{-1}. MTR parameters for calculation are d_{up}=0.3 mm, and L_{Si}=25 mm

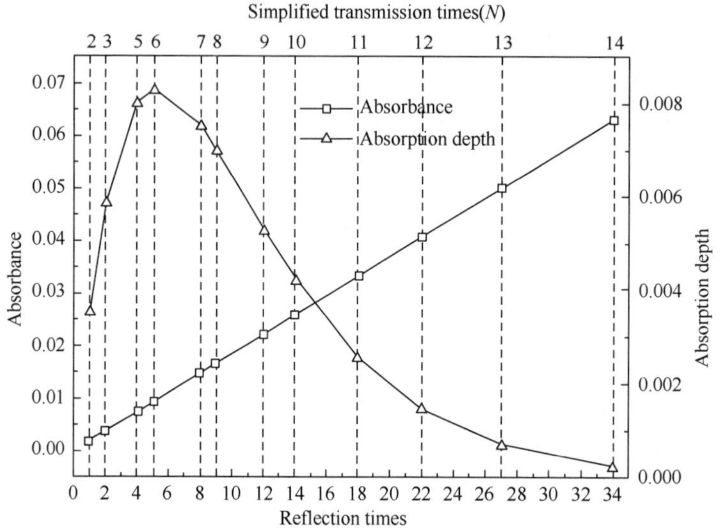

Figure 2.21　Calculated absorbance (A, –□–) and absorption depth (ΔR, –Δ–) for multiple reflections between a silicon surface and a gold mirror versus reflection times for a 2.6 nm layer (n=1.5, k=0.5) at 2918 cm^{-1} by s-polarization at an incident angle of 74°. The corresponding simplified transmission times (N) were obtained by changing d_{down} with fixed L_{Si}=25 mm and d_{up}=0.3 mm. The multiple reflections include only the reflections in air gap 1 between Si and gold mirror surfaces

2.4.5 Conclusions

We have investigated in detail the multiple transmission-reflection methodology to obtain high quality infrared spectra of molecular monolayers chemisorbed on silicon. The method is based on allowing for multiple transmissions and reflections to occur on a silicon surface by placing a double-side polished silicon sample between two parallel gold mirrors within a distance of 2 mm. Optimal conditions for spectral acquisition such as the number of transmission-reflections, incident angle, and mirror distance are chosen from theoretical calculations. Our experiments confirm that the MTR method can reach nearly the same high sensitivity and high quality spectra as that of the most commonly used MIR and the recently developed Ge/monolayer/Si methods, but MTR provides much convenience for non-contact and non-destructive measurements of derivatized molecular monolayers on commercial silicon wafers. Moreover the non-destructive and quality-controlled sample monitored by MTR can be used for further device fabrications. The combination of transmissions and reflections gives rise to high enough absorbance and SNR for both p- and s-polarizations. This allows to reliably determine the dichroic ratios of some vibrational modes for the adsorbed species, hitherto accessible only to the MIR spectroscopy, and thereby their relative orientations to the surface, which will be discussed in detail in the next section. While MIR will continue to remain as the choice for *in situ* studies of such interfaces, MTR represents a powerful alternative for *ex situ* investigations of surface-adsorbate interactions.

2.5 Orientation Analyses with IR Spectroscopy

Surface orientation is an important factor to evaluate the order degree of molecules adsorbed on a surface. It is a multi-parameter function of head group size, surface density, surface tension, chain branching, strength of molecular interactions, temperature, kinetics of deposition, and surroundings. Apart from the fundamental interest, characterization of the molecular orientation in adsorbed monolayers has gained considerable attention in connection with elaboration of biomimetic membranes, biological sensors, electronic devices, organic solar cells, etc.[44]. FTIR spectroscopy is one of the very few methods which provide information about the orientation of all parts of adsorbed species simultaneously.

Generally speaking, laborious calculations are necessary for precise orientation analyses with IR spectra. Principally, infrared spectroscopy is a technique based on the vibrations of atoms in a molecule. As shown in Figure 2.22, if the transitional dipole moment (TDM) of a given mode is distributed anisotropically on the surface, the macroscopic result is the selective absorption of linearly polarized radiation propagating in different directions: both parallel components along X (k_x) and Z (k_z) directions are detected by p-polarization (E_p^x and E_p^z), while only the parallel component along Y direction (k_y) is measured by s-polarization (E_s). Because the anisotropic absorption index is the function of orientation parameters as shown in Equation (2.37)-Equation (2.39), the molecular orientation can be calculated by measuring the absorbance of a vibration under linearly polarized radiation. With the optical simulation, it is evident that the orientation may be determined from (1) the absolute polarized absorbance ("spectra fitting"), (2) the ratios of polarized absorbance of two different modes in the same molecule ("relative intensity"), and (3) the absorbance ratios of one vibration mode at p-polarization versus s-polarization ["DR (dichroic ratio) fitting"]. Choosing a suitable approach for orientation analyses is determined by the IR method used for detection. In the following we will discuss the application of each orientation analysis approach combined with those measurements for silicon-based monolayers.

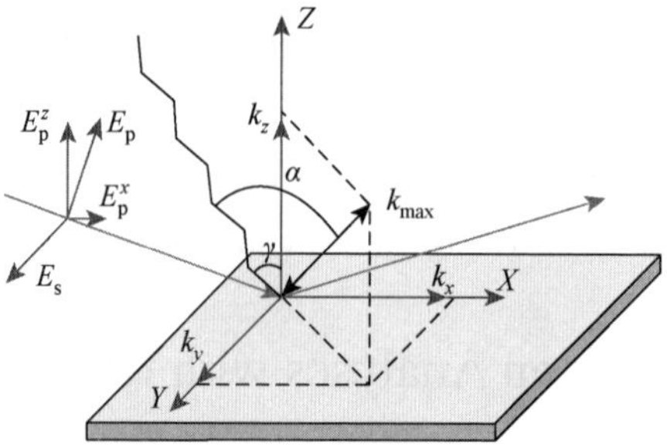

Figure 2.22 Relationship between molecular orientation and absorption index.γ is the tilt angle of the alkyl chain against the surface normal (Z axis) and α is the angle between the transition dipole moment (TDM) and the molecular axis. k_{max} is the absorption index along the TDM, k_z is the perpendicular component of absorption index to the surface, and k_x and k_y are the parallel components to the surface plane. For the all-trans alkyl chain, the TDMs of both symmetric and asymmetric stretching vibrations of methylene are perpendicular to the carbon backbone (α=90°)

For the uniaxial symmetry of alkyl chain on the surface, the imaginary part of the complex refractive index can be determined as follows[40]:

$$k_x = k_y = k_{max}[s(\sin^2\alpha)/2 + (1-s)/3] \tag{2.37}$$

$$k_z = k_{max}[s\cos^2\alpha + (1-s)/3] \tag{2.38}$$

$$s \equiv \langle p_2 \cos\gamma \rangle = \frac{1}{2}\langle 3\cos^2\gamma - 1 \rangle \tag{2.39}$$

The first and second methods are suitable for the measurements that spectra can be only obtained at one polarization. The difference is that the "spectra fitting" is suitable for any measurement, while the "relative intensity" method is usually used to analyze the spectra with a pure normal or parallel component, such as p-polarized spectra on a metallic surface and s-polarized spectra on a transparent surface, because the selective absorption of mixed components in the same TDM depending on the angle of incidence requires rigorous spectral simulations. For example, in the IR-RAS measurement with p-polarization on silicon, both perpendicular and parallel components contribute to the absorbance, and so only "spectra fitting" is available[2]. Although the polarized transmission spectroscopy at an inclined angle of incidence has better sensitivity than IR-RAS for detection of a monolayer on silicon, there are few reports applying transmission on the orientation analyses. At the Brewster angle, p-polarized light transmits through the wafer while s-polarized light is mostly reflected, and it is difficult to get good spectra for both polarizations simultaneously, preventing the use of "DR fitting" for orientation analyses. Additionally, because both perpendicular and parallel components on silicon can be detected by p-polarized transmission, theoretically, only "spectra fitting" can be used. To use the "relative intensity" method, s-polarized spectra including the pure parallel component is used for analyses[37]. By changing the angle of incidence with p-polarization at 30° and 74°, which are sensitive mostly to parallel and perpendicular components respectively, semi-quantitative analyses with the transmission measurement are reported[4, 5]. Theoretically, this method can be used for orientation analyses of a monolayer on a transparent substrate with IR-RAS, such as on the quartz surface at 73° angle of incidence[25]. However, on the silicon surface, no such a report can be found due to the low sensitivity of IR-RAS (see discussion in Section 2.1). The third method "DR fitting", hitherto, is accessible only to the multiple transmission or reflection approaches such as MTR and MIR, which can measure both p- and s-polarized spectra with high sensitivity. Regardless which approach is used, optical simulation is necessary for orientation analyses. The detailed procedure of each approach will be discussed below.

The first approach ("spectra fitting") involves producing simulated spectra to fit the experimental ones in an iterative process, which had been discussed in details by Parikh and Allara[37]. Hoffmann et al. first applied this method on analyses of alkylsiloxane monolayer on Si by fitting the p-polarized spectra from IR-RAS at 80° angle of incidence[2, 13, 17]. Because the simulated spectra are the sum of the absolute absorbance of each peak, a small deviation in the assumed absorption index and film thickness will result in a large deviation of orientation[2]. In the method of Parikh and Allara[37], k_{max} was obtained from the isotropic optical constant k_{iso} (k_{max}=3.0k_{iso}) assuming the oscillator strengths do not change appreciably upon chemisorption of the molecules, and the value of k_{iso} was determined from the transmission spectrum of the molecules in a KBr pellet. However, in fact k_{max} is the function of molecular packing and orientation, and will change when the molecule is chemisorbed on a surface. Additionally, it has been reported that the transmission absorption of KBr pellet was affected by the particle size and shape and therefore by the grinding time. The better way to determine anisotropic optical constants is to fit the spectra with molecules deposited on a metallic surface measured by IR-RAS or on a transparent substrate via transmission as described by Buffeteau, assuming the deposited layer can be reproduced on different substrates[26]. Although usually the film thickness can be obtained from the ellipsometry, there are still measurement deviations in nanometer scale. The necessity of optical parameters and laborious calculations in the "spectra fitting" limits its application.

$$I_1^{\mathrm{SAM}} \propto \left[\vec{E}_z \overrightarrow{\mathrm{TDM}_1}\right]^2 \propto \overrightarrow{\mathrm{TDM}_{1z}} \propto I_1^{\mathrm{iso}} \cos^2(\varphi_1 + \gamma) \tag{2.40}$$

$$\frac{\cos(\varphi_1 + \gamma)}{\cos(\varphi_2 - \gamma)} = \frac{I_1^{\mathrm{SAM}} I_2^{\mathrm{iso}}}{I_2^{\mathrm{SAM}} I_1^{\mathrm{iso}}} \tag{2.41}$$

If there are two or more different modes in a molecule with different oriented TDMs and the peak intensities are strong enough for quantitative analyses, the orientation of the molecule can be deduced from the ratios of absorbances of these peaks in the same polarized spectra with the help of isotropic spectral intensities, which is called as the"relative intensity"method. As shown in Figure 2.23, two oriented modes of $\overrightarrow{\mathrm{TDM}_1}$ and $\overrightarrow{\mathrm{TDM}_2}$ with respective angles of φ_1 and φ_2 between TDM and the molecular axis can be decomposed into two components to the surface, normal ($\overrightarrow{\mathrm{TDM}_z}$) and parallel ($\overrightarrow{\mathrm{TDM}_x}$). $\overrightarrow{\mathrm{TDM}}$ is proportional to the peak intensity in isotropic spectra (transmission spectra of molecules in KBr pellet), and $\overrightarrow{\mathrm{TDM}_z}$ is proportional to the absorbance of SAMs, if only the normal component of SAMs can be detected. The relationship between the peak intensity of a vibrational mode in polarized (I_1^{SAM}) and isotropic (I_1^{iso}) spectra is shown in Equation (2.40) with $\overrightarrow{\mathrm{TDM}_1}$ as an example. To avoid using the absolute value of TDM like in "spectra fitting", the ratio of normal components of two modes can be related with the tilt angle of γ as shown in Equation (2.41). If the two modes of in-plane (φ_1=0°) and out-of-plane (φ_2=90°) are known, the equation will be simplified as $\tan\gamma = I_2^{\mathrm{SAM}} I_1^{\mathrm{iso}} / I_1^{\mathrm{SAM}} I_2^{\mathrm{iso}}$. With this method, the film thickness is cancelled out, and there is no need to extract the absorption index from the isotropic spectra. But like in the "spectra fitting", it is assumed that $k_{\max}$ of TDM does not change with the chemisorption of the molecule on a surface. This assumption will introduce some deviations for orientation analyses. The only way to overcome this problem is "DR fitting", if both p- and s-polarized spectra with high sensitivity are available.

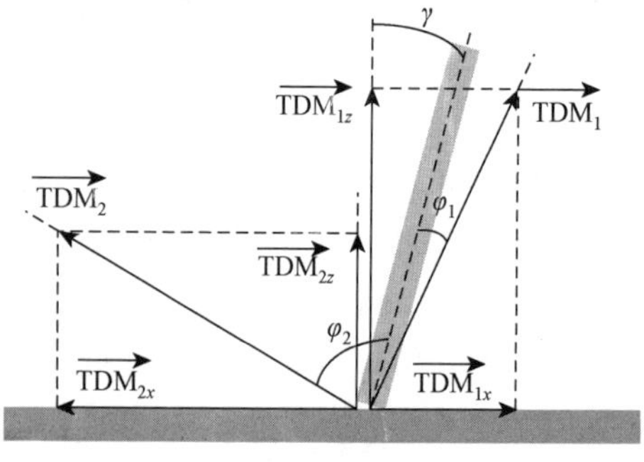

Figure 2.23 Decomposition of two oriented modes of ($\overrightarrow{\mathrm{TDM}_1}$) and ($\overrightarrow{\mathrm{TDM}_2}$) from a molecule (represented with a shadowed dash line) to the surface parallel (x axis) and the surface normal (z axis) directions

The third method ("DR fitting") can give more accurate results by fitting the DR values, which is defined as the ratio of the peak intensity of the absorption band in the s-polarized spectrum, (A_s), to that in the p-polarized spectrum (A_p). The advantage of this method can decrease the deviation effect of optical constants and film thickness on the orientation analyses, compared with the other two methods. This method is widely used in the MIR measurement, which is well-known to give high quality spectra of both p- and s-polarizations. The dependence of DR on k_{max} and that on film thickness are shown in Figure 2.24 (a) and Figure 2.24 (b) respectively. The values of k_{max} from

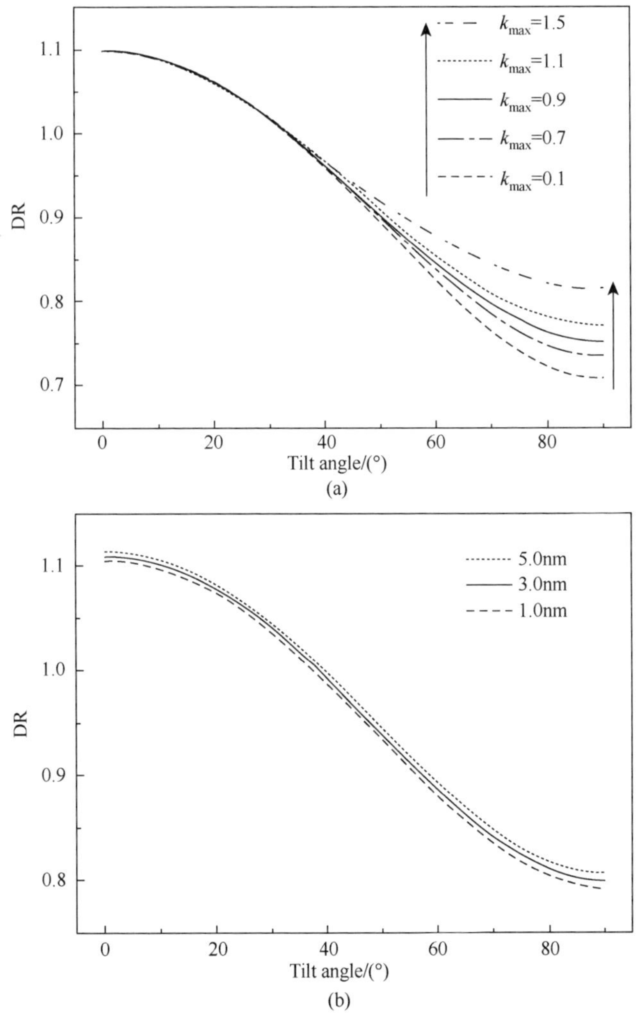

Figure 2.24 DR/tilt-angle curves affected by (a) absorption index k_{max} (0.1, 0.7, 0.9, 1.1, 1.5) and (b) film thickness (1 nm, 3 nm, 5 nm) for MIR measurements, calculated with n=1.5 at 2918 cm^{-1}

0.1-1.5 almost cover all the organic molecules from weak to strong IR absorptions. Usually for the well-ordered monolayer, the tilt angle is lower than 30°, close to the surface normal, and there is nearly no deviation from k_{max} [Figure 2.24 (a)]. The most common approach for the film thickness measurement between 1.0 nm to 5.0 nm is the ellipsometry, which possesses high measurement deviations, whereas only results in a change of the tilt angle of 2°. Obviously, "DR fitting" is the best method for orientation analyses among the three approaches, which does not need laborious calculations and has less deviation.

Because it is well-known that the long chain silane molecules can form highly organized compact monolayers on the silicon surfaces, we use the MTR spectra of this kind of molecules in Section 2.4 (Figure 2.19) to calibrate MTR-IR for orientation analyses. The calculated dependence of DR on the tilt angle for CH_2 at 2918 cm^{-1} with our MTR configuration is shown in Figure 2.25. From these "DR fitting" curves, the tilt angle of alkyl chains can be obtained after determining their DR value in spectra. The typical absorption (k_{max}) and refractive indices (n) of octadecyl ν_{as} (CH_2) are 0.9[2] and 1.5[45-47] respectively. The experimental DR value for ν_{as} (CH_2) of OTS at 2918 cm^{-1} in Figure 2.19 is 1.42 and it corresponds to a tilt angle of 15° in Figure 2.25 (a). The DR value for ν_{as} (CH_2) of DTS is about 0.73, much lower than the value of OTS, and it corresponds to an average tilt angle of about 43°. But this value is physically less meaningful because of the disordered gauche conformation of the alkyl chains in the DTS monolayers.

Alkanephosphates and phosphonates bind strongly onto the surfaces of a variety of metal oxides, including titania, zirconia, niobia, alumina and tantala, to form self-assembled monolayers, which has been investigated widely. However, most IR spectroscopic data for alkanephosphate SAMs adsorbed on metal oxides has been restricted to measurements on powder or nanoparticulate samples. The powder spectra do not reveal information concerning molecular orientation, due to the random distribution of the transition dipole moments with respect to the incident IR radiation. Here, we give the MTR-IR results for the kinetics of phosphate adsorption on the microscopic flat TiO_2 film coated on a silicon surface.

Figure 2.26 (see the Color Inset, p.4) shows a series of s- and p-polarized MTR-IR spectra measured for different adsorption times of n-hexadecylphosphate on TiO_2. Several peaks associated with the C—H stretching modes of the alkyl chain moieties are clearly distinguishable even at very short adsorption time (30 sec, bottom traces in Figure 2.26, see the Color Inset, p.4). Their intensities increase with the incubation time in the phosphate solution, illustrating the increase of surface coverage. Furthermore, the positions of the anti-symmetric and symmetric methylene stretching bands show a

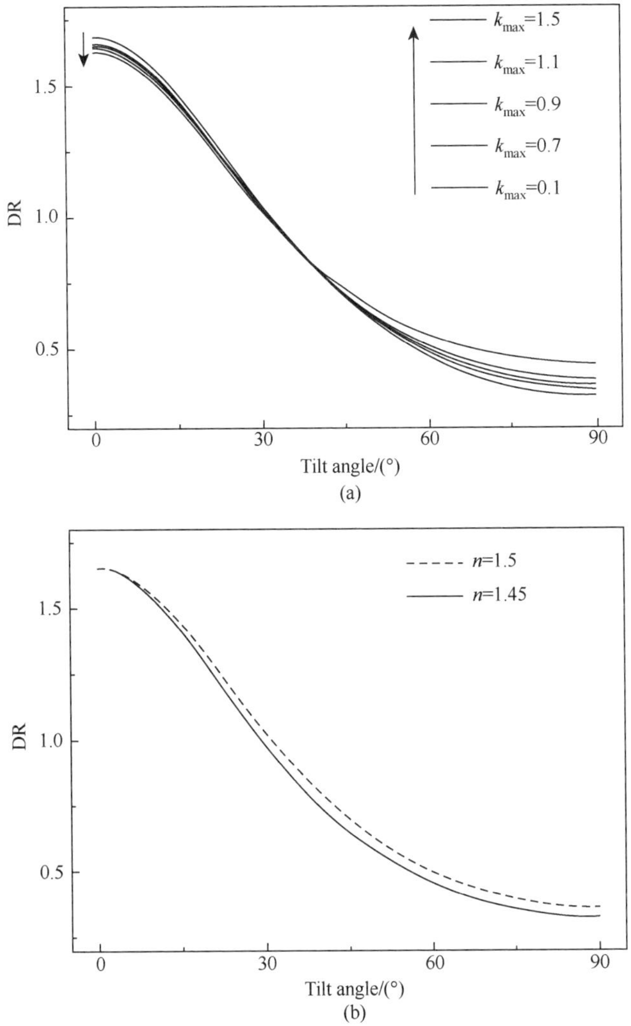

Figure 2.25 The effect of (a) absorption index k_{max} (0.1, 0.7, 0.9, 1.1, 1.5) calculated with refractive index of n (1.5) and (b) refractive index n (1.45 and 1.5) calculated with absorption index k of 0.9 on DR/tilt-angle curves for MTR measurements for a 2.6 nm layer at 2918 cm^{-1}. In calculation, the incident angle is 74°, d_{up}=0.3 mm and L_{Si}=25 mm. A beam divergence of 7.5° and a diameter of the focused spot of 6 mm were considered in the calculations

clear shift to lower wavenumbers at 2918 cm^{-1} and 2848 cm^{-1} with increasing the incubation time, indicating the phase transitions of alkyl chains from a disordered and liquid-like to a crystalline and well-ordered structure[48, 49]. A rapid increase in the thickness of the monolayer corresponding to an increased surface coverage, as measured by variable-angle spectroscopic ellipsometry (VASE) (Figure 2.27, inset in the right bottom), is also observable over the same time scale. The static water contact angles also showed a similar rapid increase with increasing the adsorption time and the full monolayer showed static contact angles close to 110° and 40° with water and hexadecane, respectively, values typical of densely packed alkyl SAMs[50].

A very significant observation from Figure 2.26 (see the Color Inset, p.4) is the distinctly different intensities of all the stretching modes in the s- and p-polarized spectra. The symmetric stretch of the terminal CH_3 group at around 2880 cm^{-1} for the full monolayer is visible only in the p-polarized spectrum and not in the s-polarized spectrum (top spectrum in Figure 2.26, see the Color Inset, p.4). This is entirely consistent with the direction of the transition dipole moment of this vibrational mode. Assuming the monolayer is an organized, tilted structure of the alkyl chains, this vibrational mode will lie nearly orthogonal to the s-polarization direction and is therefore too weak to be seen. A similar effect is also evident for the methyl asymmetric stretching mode at 2950 cm^{-1}, which is composed of two different components. While both the in-plane and out-of-plane components are clearly visible in the p-polarized spectra, only the in-plane component is visible in the s-polarized spectra. It is difficult to quantify their absolute orientations on the surface due to the

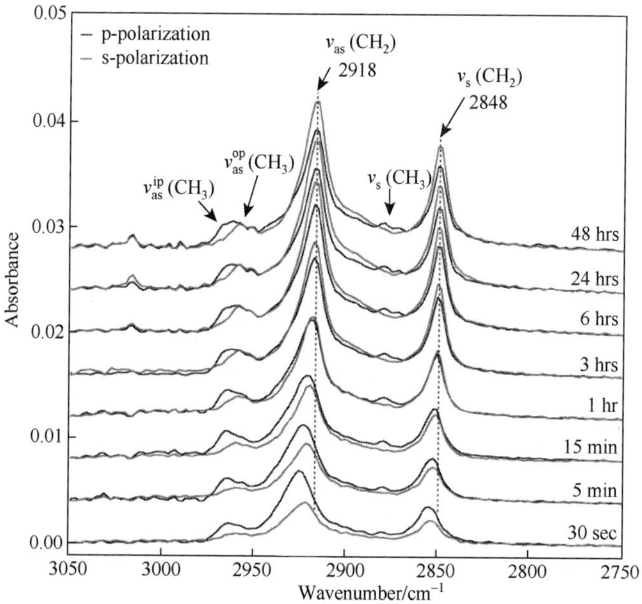

Figure 2.26 A series of p- and s-polarized IR spectra measured at different adsorption times of *n*-hexadecylphosphate onto TiO_2-coated, double-side-polished silicon, measured by the multiple transmission-reflection (MTR) infrared method

low intensity and overlapping of these bands. Nevertheless these delicate observations clearly demonstrate the sensitivity of this methodology for studying orientations of the different vibrational modes of a monolayer on a surface.

Intensity differences could also be observed in the strong methylene stretching bands at 2848 cm^{-1} and 2918 cm^{-1}. Both peaks appear to be relatively more intense in the p-polarized spectra than in the s-polarized ones during the initial periods of adsorption (5 sec, 5 min, and 15 min). Both peaks increase more rapidly with increasing the adsorption time in s-polarized spectra than in p-polarized ones. In fact, with the adsorption time longer than 1 h, they become more intense in s-polarized spectra than in p-polarized ones. A more quantitative way to analyze these differences is to plot the dichroic ratio (DR), defined as the intensity ratio of a vibrational mode in the two polarizations (A_s/A_p). The DR of the methylene anti-symmetric stretching band over the adsorption time is shown in Figure 2.27. A smooth curve over the entire length of the adsorption time can be seen. A transition can be observed from a disordered alkyl chain structure, characterized by a lower DR (and relatively higher methylene stretching band frequencies in lower traces of Figure 2.26) to a highly ordered alkyl chain assembly with a DR close to 1.27 (and relatively lower methylene stretching frequencies in top traces of Figure 2.26).

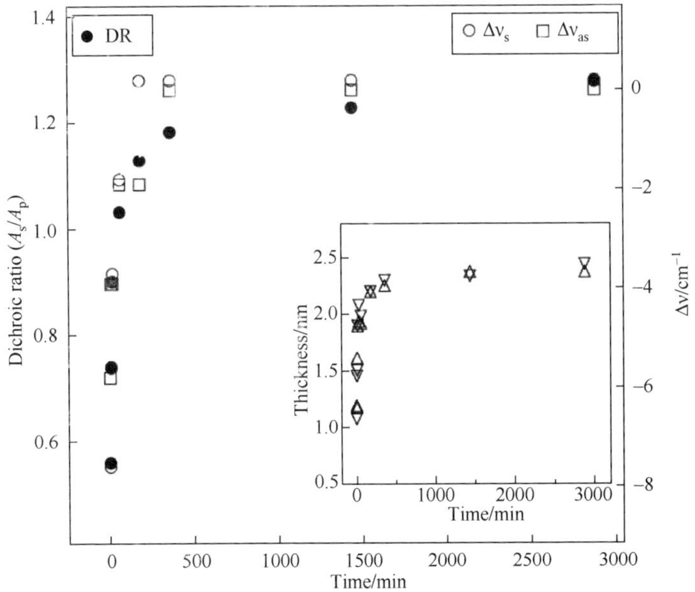

Figure 2.27 Changes in the dichroic ratios and the position of the methylene stretching modes ($\Delta\nu=\nu_{full}-\nu_t$) as a function of adsorption time. Corresponding changes in the adlayer thickness, measured by VASE on both sides of the substrate are shown in the inset

It is interesting to note from Figure 2.26 that both the methylene stretching band positions and the adlayer thickness reach values close to that of a full monolayer quite rapidly, whereas the DR shows a more gradual change especially during the intermediate adsorption period. This is likely due to the sensitivity of DR to the more subtle changes in the monolayer ordering than other parameters. The dichroic ratio could be related to the tilt angle of the alkyl chains by calculating the intensities of these modes using reported values for the absorption indices of the alkyl chain vibrations[2]. Assuming a uniaxial orientation of the alkyl chains, this results in a value of tilt angle for the alkyl chains of 21°±3° from the surface normal. This agrees well with the observed thickness, as determined by ellipsometry. This value is lower than that observed earlier[51] probably due to a slightly different coverage resulting from the different solvents used for adsorption.

2.6 Summary

We have demonstrated that the sensitivity of the GMBR measurement is increased by placing a mirror behind the silicon wafer in a grazing angle configuration. The absorbance of GMBR is 2-3 times than that of the p-polarized single transmission at Brewster angle. The coincidence of maximum absorbance and absorption depth for p-polarization facilitates the choice of the optimum angle of incidence at 78.1°.

With one more mirror on the other side of the silicon wafer, the MTR measurement can obtain higher quality infrared spectra of molecular monolayers chemisorbed on silicon. The MTR method is based on allowing multiple transmissions and reflections to occur on a double-side polished silicon sample between two parallel gold mirrors within a distance of 2 mm. Optimal conditions for spectral acquisition such as the number of transmissions and reflections, incident angle, and mirror distance are chosen from theoretical calculations. Experimental results confirm that the MTR method can reach high sensitivity and obtain high quality spectra, the same as the most commonly used MIR and the Ge-ATR (Ge/monolayer/Si) methods, but MTR provides much convenience for non-contact and non-destructive measurements of molecular monolayers on commercial silicon wafers. Moreover the non-destructive and quality-controlled sample monitored by MTR can be used for further device fabrications. The combination of multiple transmissions and reflections gives rise to high enough absorbance and SNR for both p- and s-polarizations. This allows to reliably determine the dichroic ratios of some vibrational modes for the adsorbed species, hitherto accessible only to the MIR spectroscopy, and thereby their relative orientations to the surface. While MIR will continue to remain as the choice for *in situ* studies of such interfaces, MTR represents a powerful alternative for *ex situ* investigations of surface-adsorbate interactions.

In Table 2.1, the calculated absorbance and absorption depth from different geometric configurations are compared. It is obvious that the three measurements of MTR, Ge-ATR (p-polarization), and MIR have the sensitivity in the same order of magnitude for detection of monolayers on silicon. While only MTR and MIR can give high quality spectra for both p- and s-polarizations, and hence the "DR fitting" method can be used for orientation analyses in these two approaches.

Table 2.1 Comparison of calculated absorbance and absorption depth from different infrared measurement setups for detection of 1 nm monolayer (n=1.5+0.5i) on silicon at 2918 cm^{-1}

Measurement (Incidence angle)	Absorbance		Absorption depth		Sample requirement
	p-polarization	s-polarization	p-polarization	s-polarization	
IR-RAS(80°)	-1.4×10^{-3}	-7.7×10^{-5}	-1.7×10^{-4}	-1.4×10^{-4}	No restriction
Transmission (74°)	9.7×10^{-4}	8.3×10^{-4}	2.2×10^{-3}	5.4×10^{-4}	Double-side-polished
MIR (N times) (45°)	$1.2\times10^{-3}\times N$	$1.1\times10^{-3}\times N$	$2.6\times10^{-3}\times N$	$2.5\times10^{-3}\times N$	ATR crystal
Ge-ATR*(65°)	5.2×10^{-2}	1.9×10^{-3}	1.1×10^{-1}	4.4×10^{-3}	No restriction
GMBR(80°)	4.0×10^{-3}	9.1×10^{-4}	7.7×10^{-3}	1.9×10^{-3}	Double-side-polished
MTR† (N=6)(74°)	1.1×10^{-2}	6.2×10^{-3}	1.8×10^{-2}	1.4×10^{-2}	Double-side-polished

* The results for the Ge-ATR measurement are calculated assuming there is no air gap between the Ge-ATR crystal and the silicon sample. In fact, the experimental absorbance is about 1/3 of the calculated data.

† In the MTR experiment, because the mirror distance is 2 mm, part of the light cannot enter inside the two parallel mirrors, which will decrease the absorption depth. In the calculation, we assigned a beam divergence of ±7.5° for simulation.

References

[1] Chabal Y J. Surface infrared spectroscopy. *Surface Science Reports*, 1988, 8 (5): 211-357.

[2] Hoffmann H, Mayer U, Krischanitz A. Structure of alkylsiloxane monolayers on silicon surfaces investigated by external reflection infrared spectroscopy. *Langmuir*, 1995, 11: 1304-1312.

[3] Faber E J, de Smet L C P M, Olthuis W, et al. Si—C linked organic monolayers on crystalline silicon surfaces as

alternative gate insulators. *ChemPhysChem*, 2005, 6: 2153-2166.

[4] Webb L J, Rivillon S, Michalak D J, et al. Transmission infrared spectroscopy of methyl- and ethyl-terminated silicon (111) surfaces. *The Journal of Physical Chemistry B*, 2006, 110: 7349-7356.

[5] Michalak D J, Rivillon S, Chabal Y J, et al. Infrared spectroscopic investigation of the reaction of hydrogen-terminated, (111)-oriented, silicon surfaces with liquid methanol. *The Journal of Physical Chemistry B*, 2006, 110: 20426-20434.

[6] Caudano Y, Thiry P A, Chabal Y J. Investigation of the bending vibrations of vicinal H/Si (111) surfaces by infrared spectroscopy. *Surface Science*, 2002, 502: 91-95.

[7] Langner A, Panarello A, Rivillon S, et al. Controlled silicon surface functionalization by alkene hydrosilylation. *Journal of the American Chemical Society*, 2005, 127: 12798-12799.

[8] Harrick N J. *Internal Reflection Spectroscopy*. New York: Wiley, 1967.

[9] Haller G L, Rice R W. Study of adsorption on single crystals by internal reflectance spectroscopy. *The Journal of Physical Chemistry*, 1970, 74: 4386-4393.

[10] Ubara H, Imura T, Hiraki A. Formation of Si—H bonds on the surface of microcrystalline silicon covered with SiO_x by HF treatment. *Solid State Communications*, 1984, 50: 673.

[11] Hamers R J, Hovis J S, Lee S, et al. Formation of ordered, anisotropic organic monolayers on the Si (001) surface. *The Journal of Physical Chemistry B*, 1997, 101: 1489-1492.

[12] Picard F, Buffeteau T, Desbat B, et al. Quantitative orientation measurements in thin lipid films by attenuated total reflection infrared spectroscopy. *Biophysical Journal*, 1999, 76: 539-551.

[13] Lummerstorfer T, Hoffmann H. IR reflection spectra of monolayer films sandwiched between two high refractive index materials. *Langmuir*, 2004, 20: 6542-6545.

[14] Rowell N L, Tay L, Lockwood D J, et al. Organic monolayers detected by single reflection attenuated total reflection infrared spectroscopy. *Journal of Vacuum Science & Technology A*, 2006, 24: 668-672.

[15] Anariba F, Viswanathan U, Bocian D F, et al. Determination of the structure and orientation of organic molecules tethered to flat graphitic carbon by ATR-FT-IR and Raman spectroscopy. *Analytical Chemistry*, 2006, 78: 3104-3112.

[16] Milosevic M, Milosevic V, Berets S L. Grazing angle attenuated total reflection spectroscopy: fields at the interface and source of the enhancement. *Applied Spectroscopy*, 2007, 61: 530-536.

[17] Lummerstorfer T, Kattner J, Hoffmann H. Monolayers at solid-solid interfaces probed with infrared spectroscopy. *Analytical and Bioanalytical Chemistry*, 2007, 388: 55-64.

[18] Liu H B, Xiao S J, Chen Y Q, et al. Grazing angle mirror-backed reflection (gmbr) for infrared analysis of monolayers on silicon. *The Journal of Physical Chemistry B*, 2006, 110: 17702-17705.

[19] Liu H B, Venkataraman N V, Bauert T E, et al. Multiple transmission-reflection infrared spectroscopy for high-sensitivity measurement of molecular monolayers on silicon surfaces. *The Journal of Physical Chemistry A*, 2008, 112: 12372-12377.

[20] Liu H B, Venkataraman N V, Bauert T E, et al. Structural evolution of self-assembled alkanephosphate monolayers on TiO_2. *ChemPhysChem*, 2008, 9: 1979-1981.

[21] Lenfant S, Krzeminski C, Delerue C, et al. Molecular rectifying diodes from self-assembly on silicon. *Nano Letters*,

2003, 3: 741-746.

[22] Rakshit T, Liang G C, Ghosh A W, et al. Silicon-based molecular electronics. *Nano Letters*, 2004, 4: 1803-1807.

[23] Richter C A, Hacker C A, Richter L J. Electrical and spectroscopic characterization of metal/monolayer/Si devices. *The Journal of Physical Chemistry B*, 2005, 109: 21836-21841.

[24] Onclin S, Ravoo B J, Reinhoudt D N. Engineering silicon oxide surfaces using self-assembled monolayers. *Angewandte Chemie International Edition*, 2005, 44: 6282-6304.

[25] Lasseter T L, Clare B H, Abbott N L, et al. Covalently modified silicon and diamond surfaces: Resistance to nonspecific protein adsorption and optimization for biosensing. *Journal of the American Chemical Society*, 2004, 126: 10220-10221.

[26] Voicu R, Boukherroub R, Bartzoka V, et al. Formation, characterization, and chemistry of undecanoic acid-terminated silicon surfaces: Patterning and immobilization of DNA. *Langmuir*, 2004, 20: 11713-11720.

[27] Chernyshova I V, Rao K H. A new approach to the IR spectroscopic study of molecular orientation and packing in adsorbed monolayers. Orientation and packing of long-chain primary amines and alcohols on quartz. *The Journal of Physical Chemistry B*, 2001, 105: 810-820.

[28] Buffeteau T, Blaudez D, Pere E, et al. Optical constant determination in the infrared of uniaxially oriented monolayers from transmittance and reflectance measurements. *The Journal of Physical Chemistry B*, 1999, 103: 5020-5027.

[29] Tolstoy V P, Chernyshova I V, Skryshevsky V A. *Handbook of Infrared Spectroscopy of Ultrathin Films*. New York: Wiley, 2003.

[30] Vašíček A. *Optics of Thin Films*. Amsterdam: North-Holland Pub. Co. , 1960.

[31] Yamamoto K, Ishida H. Optical theory applied to infrared spectroscopy. *Vibrational spectroscopy*, 1994, 8: 1-36.

[32] Katsidis C C, Siapkas D I. General transfer-matrix method for optical multilayer systems with coherent, partially coherent, and incoherent interference. *Applied Optics*, 2002, 41 (19): 3978-3987.

[33] Brendel R, Hezel R. Infrared observation of thermally activated oxide reduction within Al/SiO_x/Si tunnel diodes. *Journal of Applied Physics*, 1992, 71: 4377-4381.

[34] Jun Y, Zhu X Y. FTIR spectroscopy of buried interfaces in molecular junctions. *Journal of the American Chemical Society*, 2004, 126: 13224-13225.

[35] Guo D J, Xiao S J, Xia B, et al. Reaction of porous silicon with both end-functionalized organic compounds bearing α-bromo and ω-carboxy groups for immobilization of biomolecules. *The Journal of Physical Chemistry B*, 2005, 109: 20620-20628.

[36] Boukherroub R, Petit A, Loupy A, et al. Microwave-assisted chemical functionalization of hydrogen-terminated porous silicon surfaces. *The Journal of Physical Chemistry B*, 2003, 107: 13459-13462.

[37] Parikh A N, Allara D L. Quantitative determination of molecular structure in multilayered thin films of biaxial and lower symmetry from photon spectroscopies. I. Reflection infrared vibrational spectroscopy. *The Journal of Chemical Physics*, 1992, 96: 927-945.

[38] Higashi G S, Chabal Y J, Trucks G W, et al. Ideal hydrogen termination of the Si surface. *Applied Physics Letters*, 1990, 56: 656-658.

[39] Tolstoy V P, Bogdnnova L P, Aleskovski V B. *Dokladi Akademii Nauk USSR, Phys. Khimia*, 1986, 291: 913.

[40] Chernyshova I V, Rao K H. A new approach to the IR spectroscopic study of molecular orientation and packing in adsorbed monolayers. Orientation and packing of long-chain primary amines and alcohols on quartz. *The Journal of Physical Chemistry B*, 2001, 105: 810-820.

[41] Greenler R G. Reflection method for obtaining the infrared spectrum of a thin layer on a metal surface. *The Journal of Physical Chemistry*, 1969, 50: 1963-1968.

[42] Barth J, Johnson R L, Cardona M. *Handbook of Optical Constants of Solids II*. New York: Academic Press, 1991.

[43] Kaplan S G, Hanssen L M. Silicon as a standard material for infrared reflectance and transmittance from 2 to 5 μm. *Infrared Physics & Technology*, 2002, 43: 389-396.

[44] Krapchetov D A, Ma H, Jen A K Y, et al. High-sensitivity transmission IR spectroscopy for the chemical identification and structural analysis of conjugated molecules on gallium arsenide surfaces. *Langmuir*, 2006, 22: 9491-9494.

[45] Allara D L, Nuzzo R G. Spontaneously organized molecular assemblies. 1. Formation, dynamics, and physical properties of *n*-alkanoic acids adsorbed from solution on an oxidized aluminum surface. *Langmuir*, 1985, 1: 45-52.

[46] Maoz R, Sagiv J. On the formation and structure of self-assembling monolayers. I. A comparative atr-wettability study of Langmuir-Blodgett and adsorbed films on flat substrates and glass microbeads. *Journal of Colloid Interface Science*, 1984, 100: 465-496.

[47] Tillman N, Ulman A, Schildkraut J S, et al. Incorporation of phenoxy groups in self-assembled monolayers of trichlorosilane derivatives. Effects on film thickness, wettability, and molecular orientation. *Journal of the American Chemical Society*, 1988, 110: 6136-6144.

[48] Snyder R G, Strauss H L, Elliger C A. Carbon-hydrogen stretching modes and the structure of *n*-alkyl chains. 1. Long, disordered chains. *The Journal of Physical Chemistry*, 1982, 86: 5145-5150.

[49] Macphail R A, Strauss H L, Snyder R G, et al. Carbon-hydrogen stretching modes and the structure of *n*-alkyl chains. 2. Long, all-trans chains. *The Journal of Physical Chemistry*, 1984, 88: 334-341.

[50] Bain C D, Troughton E B, Tao Y T, et al. Formation of monolayer films by the spontaneous assembly of organic thiols from solution onto gold. *Journal of the American Chemical Society*, 1989, 111: 321-335.

[51] Foster T T, Alexander M R, Leggett G J, et al. Friction force microscopy of alkylphosphonic acid and carboxylic acids adsorbed on the native oxide of aluminum. *Langmuir*, 2006, 22: 9254-9259.

Chapter 3 Molecular Interactions at Model Cell Membranes Investigated Using Nonlinear Optical Spectroscopy

Zhan Chen

3.1 Introduction

Cell membrane separates the interior of a cell and its outside environment. It has many important biological functions such as selective ion transportation, cell signaling, and cell adhesion[1]. It is a physical barrier which protects a cell. Therefore in order to ensure a drug molecule or a drug delivery vehicle to reach its target inside a cell, it is necessary for these molecules to overcome this barrier. The main structure of cell membrane is a lipid bilayer, with many other different molecules such as proteins imbedded[1].

It is important to study molecular interactions between cell membranes and other molecules or materials to elucidate the underlined interaction mechanisms. However, such a task is difficult due to the lack of an appropriate analytical method which can sensitively probe a single lipid bilayer *in situ* in contact with a liquid environment in real time at the molecular level. Extensive research has been applied to investigate such interactions, but detailed understanding on molecular interactions between cell membranes and other molecules is still lacking.

Since the real cell membranes can be quite complicated, solid substrate supported lipid bilayers have been widely used as simple models for cell membranes[2]. It has been shown that such lipid bilayers hold many characteristics as real cell membranes and can be used to effectively model the molecular behavior of the real cell membranes. In this chapter, we will report our progress in studying molecular interactions between model cell membranes and a variety of molecules and materials using a nonlinear optical laser method, sum frequency generation (SFG) vibrational spectroscopy at the molecular level *in situ*. The membrane interaction materials studied include small drug molecules, nanoparticles, and biological molecules such as peptides and proteins.

Antibiotics have been widely used to treat infectious diseases but recently, it was reported that many bacteria can develop drug resistance against traditional antibiotics[3]. Alternatives, such as antimicrobial oligomer[4, 5], were developed to effectively disrupt cell membranes of bacteria, which are difficult for bacteria to develop drug resistance. Here we will report the molecular mechanism for such oligomers to interact with model cell membranes. In addition, we will also investigate molecular interactions between model cell membranes and chlorpromazine (CPZ), a drug widely used to treat certain psychiatric disorders, as well as amantadine, an anti-viral and anti-parkinsonian drug. It was found that these two drugs interact with model cell membranes differently, and many factors influence such interactions.

Nanomaterials have been extensively studied as drug delivery vehicles and bio-imaging agents[6-8]. However, the detailed molecular interactions between such nanomaterials and cell membranes are not known. Here we report the interactions between Au nanoparticles and a model mammalian cell membrane. This book focused on the effect of nanoparticle size on the interaction with cell membranes.

Membrane associated peptides and proteins are important for many biological processes. Here we will focus on the studies of interactions between antimicrobial peptides (AMPs) and model cell membranes to understand AMPs' antimicrobial activity and selectivity. Peptides with different secondary structures, such as α-helical, 3_{10}-helical, and β-sheet were studied. In addition, we will review our recent results on the orientation determinations of membrane associated proteins, such as G proteins and G protein-coupled receptor kinase (GRK).

In these studies, we developed sum frequency generation (SFG) vibrational spectroscopy into a powerful technique to elucidate the molecular interactions between the above discussed molecules and model cell membranes *in situ*. SFG is a second order nonlinear optical laser spectroscopy, which can provide vibrational spectra of surfaces and interfaces[9-15]. In the next section, we will introduce this technique in more detail. The studies in this chapter present an innovative methodology to probe interactions with model cell membranes. The knowledge obtained here will facilitate our understanding on (1) drug-cell membrane and nanomaterial-cell interactions, aiding in the design of drugs and nanomaterials with improved properties; (2) activity and selectivity of AMPs, aiding in the development of antimicrobial materials to overcome bacteria drug resistance; and (3) orientation of membrane associated proteins, elucidating the structure-function relationship of membrane proteins.

3.2 Sum Frequency Generation Vibrational Spectroscopy

In a typical SFG experiment, two laser beams, a visible beam with a frequency ω_1 and a mid-IR beam with a tunable frequency ω_2, overlap temporally and spatially on a surface or interface, generating a signal at the sum frequency of the two input laser beams (Figure 3.1). The surface sensitivity of SFG comes from its selection rule. As a second order nonlinear optical method, its selection rule indicates that the sum frequency signal can only be generated from a medium with no inversion symmetry under the electric dipole approximation[9-15]. Most bulk materials possess inversion symmetry, therefore no SFG signal can be observed. While at surfaces and interfaces, due to the inversion symmetry broken, strong SFG signals can be detected. Both experiments and simulations indicated that SFG is sub-monolayer surface specific. The SFG signal intensity vs. the input IR frequency provides a vibrational spectrum of the surface or interface. According to the SFG signal peak centers, we can tell which functional groups are present on a surface or at an interface (e.g., a methyl, methylene, hydroxyl, or carbonyl group, etc.). Using different polarization combinations of the input and output laser beams, orientations of various surface molecules or functional groups can be deduced[9-15].

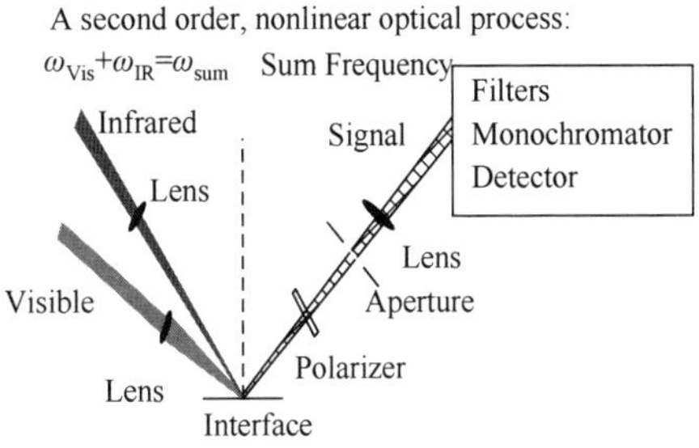

Figure 3.1 SFG experimental geometry [Reproduced with permission from *Progress in Polymer Science*, 34: 1376-1402 (2010). Copyright 2010 Elsevier]

Various SFG spectrometers have been developed in the last two decades or so. The two main types of SFG spectrometers include frequency tunable pico-second SFG system and broadband IR beam femto-second SFG system[16]. Details of such spectrometers have been extensively reported before and will not be repeated here. We are using pico-second frequency tunable SFG systems in our lab[17-20] to perform the experiments reported in this chapter. Each SFG system is composed of four components: the pumping laser, the nonlinear optical system, the sample stage, and the detection system. The output laser beam from a pumping laser is a 1064 nm near IR beam with 20 ps pulsewidth. The 532 visible input beam for the SFG experiment is generated by frequency doubling the 1064 nm beam, and the frequency tunable IR beam was generated by the optical parametric generation/amplification and difference frequency generation stages in the nonlinear optical component of the SFG spectrometer. The visible and frequency tunable IR beams overlap on the sample surface/interface, generating SFG signal. More details about the SFG experimental geometry and procedure can be found in our previous publications[17-20].

3.3 Interactions Between Model Cell Membranes and Small Drug Molecules

We have performed in-depth studies on molecular interactions between model cell membranes and drug molecules, including antimicrobial oligomers[21, 22], chlorpromazine (CPZ)[23], amantadine (AMA)[24], as well as other molecules[25, 26]. As we discussed above, we used solid substrate supported lipid bilayers as model cell membranes. The details about preparing such model cell membranes have been extensively published[21, 22] and will not be repeated here. The SFG experimental geometry of studying molecular interactions at the model cell membranes is shown in Figure 3.2.

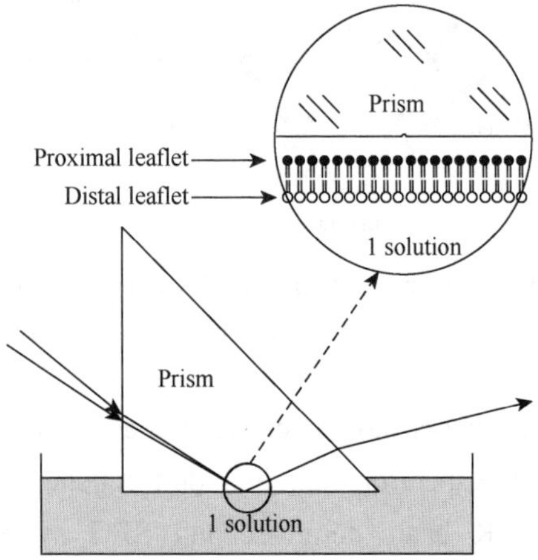

(a)

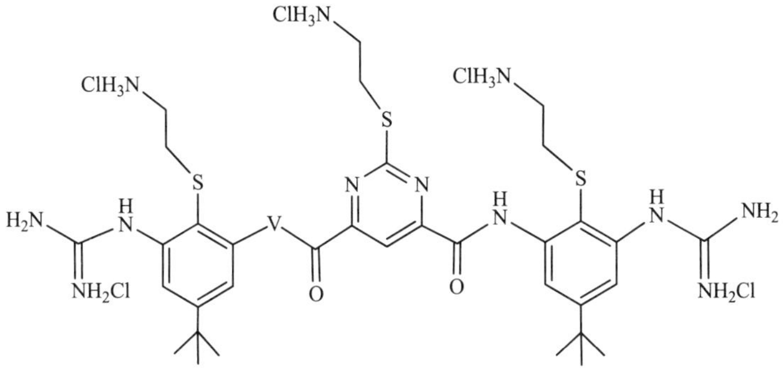

(b)

Figure 3.2 (a) Schematic of SFG experimental geometry used to study lipid bilayer/solution interfaces. (b) Molecular formula of oligomer 1 [Reproduced with permission from *Journal of the American Chemical Society*, 128: 2711-2714 (2006). Copyright 2006 American Chemical Society]

First, we will present our results on molecular interactions between antimicrobial oligomer 1 and model cell membranes[21]. The molecular formula of oligomer 1 is also shown in Figure 3.2. These antimicrobial oligomers were designed to disrupt bacteria cell membrane to kill bacteria, and therefore can overcome the possible drug resistance developed by the bacteria. To probe the interactions between each leaflet of the lipid bilayer and the oligomer, lipid bilayers with one leaflet deuterated and one leaflet hydrogenated were used in the study. More specifically, hydrogenated and deuterated 1, 2-dipalmitoyl-*sn*-glycero-3-phosphoglycerol (DPPG and d-DPPG) were used for the proximal (or inner) and distal (or outer) leaflets respectively so that the structure of both leaflets can be monitored with SFG at the C—H and C—D stretching frequency ranges simultaneously. SFG spectra were collected from the supported lipid bilayer in contact with oligomer 1 solutions with varied concentrations. If oligomer 1 does not interact with the lipid bilayer, both leaflets will not change order and no change in SFG signals would be detected. If oligomer 1 can disrupt the lipid bilayer, SFG signal intensities should decrease. It was found that the outer leaflet can be disrupted (shown by SFG intensity decrease) even at a very low oligomer 1 concentration of 0.1 μg/mL (Figure 3.3). However, the inner leaflet cannot be disrupted at this low concentration, but can be disrupted at higher concentrations (Figure 3.3). In order to kill bacteria, both inner and outer leaflets of the cell membrane need to be disrupted. Therefore the oligomer 1 concentration needs to be high enough to kill bacteria—indeed the concentration tested in this book to disrupt both leaflets (~0.8 μg/mL) is well correlated to the minimum inhibition concentration (MIC) obtained in bacteria testing. This book clearly shows that oligomer 1 strongly interacts with the lipid bilayer. When oligomer 1 interacts with a lipid bilayer, it interacts with the first leaflet, then with the second leaflet.

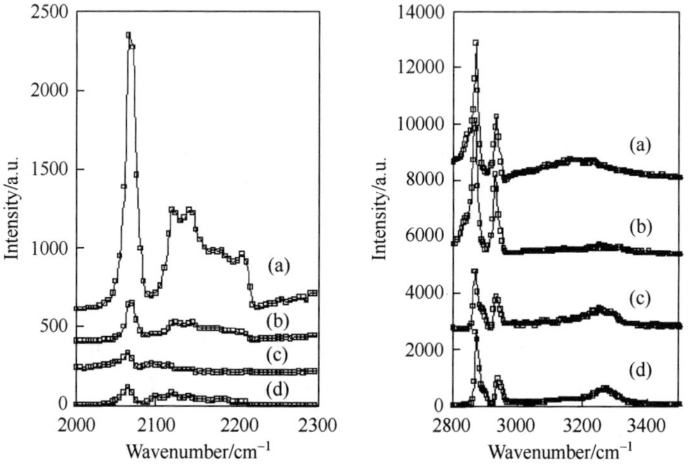

Figure 3.3 SFG spectra collected from the DPPG/d-DPPG lipid bilayer before (a) and after contacting oligomer 1 at three different solution concentrations, (b) 0.1 μg/mL, (c) 0.8 μg/mL, and (d) 4.0 μg/mL in the C—D (left) and C—H (right) stretching frequency regions. [Reproduced with permission from *Journal of the American Chemical Society*, 128: 2711-2714 (2006). Copyright 2006 American Chemical Society]

SFG spectra were also collected from the C—H stretching modes of the oligomer 1 molecules while interacting with a DPPG lipid bilayer with both leaflets deuterated (d-DPPG/d-DPPG bilayer) using different polarization combinations of the input and output laser beams[21]. In addition, the C=O signals of the oligomer 1 were also detected. Based on both C—H and C=O stretching signals, it can be shown that the plane of the oligomer 1 is perpendicular to the bilayer surface, like a knife cutting into a piece of cheese. We believe that this is the first time to elucidate such detailed interaction mechanisms between a drug molecule and model cell membranes. Later, several oligomers with similar structures were studied using SFG to understand the effect of oligomer structure (e.g., hydrophobicity of the functional groups and charge distribution on the molecule) on the interaction mechanism with the model cell membranes[22].

In addition to the antimicrobial oligomers, we also study molecular interactions between model cell membranes and two small drug molecules, chlorpromazine (CPZ)[23] and amantadine (AMA)[24]. The structures of the two drug molecules are shown in Figure 3.4. CPZ is widely used to treat certain psychiatric disorders. It is amphipathic and can bind to and partition into cell membranes. Two different kinds of lipid bilayers, a zwitterionic PC bilayer (1, 2-Distearoyl-*sn*- glycerol-3-phosphocholine or DSPC) and a negatively charged PG (1, 2-dipalmitoyl-sn-glycero-3-phospho-(1′-rac-glycerol) or DPPG) lipid bilayer were used in this book to understand how CPZ interacts with different lipids in the real cell membrane. For the PC bilayer, when the CPZ concentration is larger than 0.2 mM, it induces lipid molecules to flip-flop. The higher the CPZ solution concentration was, the faster the flip-flop rate was observed. Differently, for the PG bilayer, even at 0.1 mM, CPZ immediately disordered the outer leaflet and then gradually reduced the ordering of the inner leaflet. It is interesting to see that CPZ interacts with different lipid bilayers differently[23].

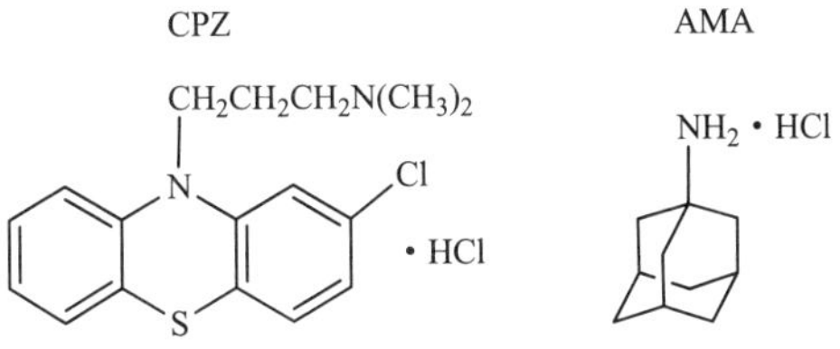

Figure 3.4 Molecular formulae of CPZ and AMA

AMA is an anti-viral and anti-parkinsonian drug, which was also used for cell surface modification. Different from CPZ, AMA does not affect the flip-flop process of the PC lipids, unless at very high solution concentrations[24]. AMA was found to adsorb to the PC bilayer—The lower the AMA solution concentration was, the slower the adsorption process was observed[24]. For PG bilayer, amantadine only interacts with the outer leaflet. AMA can be well fitted within the inter-lipid regions of PGs, binding tightly to the surrounding negatively charged phosphate groups of lipid molecules (like a glue). A schematic to show the interaction mechanisms of AMA and lipid bilayers is shown in Figure 3.5 (see the Color Inset, p.5). From the above studies, we can see that CPZ and AMA interact with lipid bilayers differently. Likely this is due to their structural difference: Although both AMA and CPZ are positively charged, AMA's hydrophobic ring is much smaller than CPZ's.

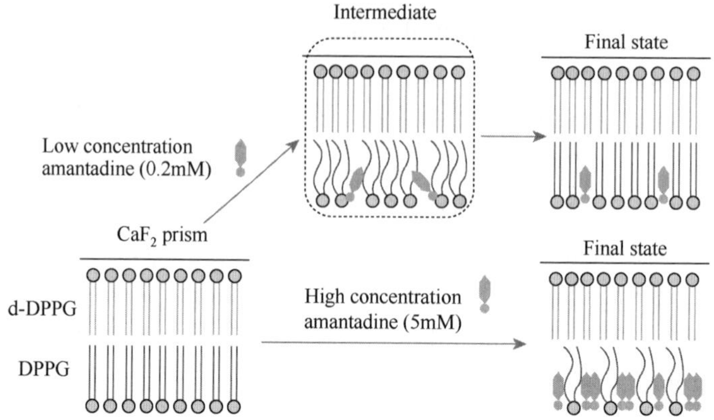

Figure 3.5　Schematic to show AMA interaction mechanism with lipid bilayers [Reproduced with permission from *Langmuir*, 30: 8491-8499 (2014). Copyright 2014 American Chemical Society]

3.4　Interactions Between Model Cell Membranes and Nanomaterials

Nanomaterials have been extensively studied as drug delivery vehicles and bio-imaging agents[6-8]. For both applications, it is necessary for nanomaterials to get into cells. Nanomaterials can enter into cells by endocytosis, or direct penetration through cell membranes. Here we studied molecular interactions between Au nanoparticles (Au NPs) and model cell membranes. Especially we focused on the investigations of the Au NP size effect[27].

SFG signals collected from a d-DSPC/DSPC bilayer in contact with water do not exhibit any time-dependent changes, indicating that the bilayer is stable and no flip-flop occurs. After certain amount of Au NP stock solution was added to the subphase of the lipid bilayer, SFG signals collected from both leaflets of the bilayer decreased, due to the interactions between the lipid bilayer and Au NPs. SFG signal intensity from each leaflet was monitored as a function of time to follow the kinetics of the membrane-Au NP interaction. Figure 3.6 (see the Color Inset, p.5) shows the SFG spectra from each leaflet before and after in contact with the Au NP solution, and the time-dependent SFG signal decreases[27]. The Au NPs studied are 20 nm Au NPs with a solution concentration of 4.00×10^{-7}g/mL. The SFG signal decrease from both leaflets may be caused by lipid removal from the substrate, lipid disruption, and lipid flip-flop induced by Au NPs. Interestingly, the SFG signal decay rates for the deuterated and hydrogenated leaflets are similar, showing very likely Au NPs induced lipid flip-flop in the bilayer. To confirm that Au NPs can induce lipid flip-flop, two additional experiments were conducted. First, a bilayer was constructed using a surface immobilized self-assembled monolayer (SAM) prepared with octadecyltrichlorosilane (OTS) as an inner leaflet and a DSPC leaflet as an outer layer. Since the inner layer is chemically immobilized on the surface, no flip-flop should occur. Indeed, SFG signal decreases from both lipid leaflets were much slower. Second, we carried out ATR-FTIR experiments. Our ATR-FTIR data on PC lipid bilayer in contact with Au NP solution showed no lipid signal decrease as a function of time. If lipid removal or bilayer disruption happened, ATR-FTIR signal from the lipids should decrease. However, ATR-FTIR signals from a d-DSPC/DSPC bilayer with or without flip-flop should be identical. Therefore, both the SAM and ATR-FTIR experiments showed that the lipid bilayer was not removed or seriously damaged/disrupted[27]. We therefore conclude that the decrease of the SFG signal was caused by flip-flop induced by the lipid bilayer-Au NP interaction.

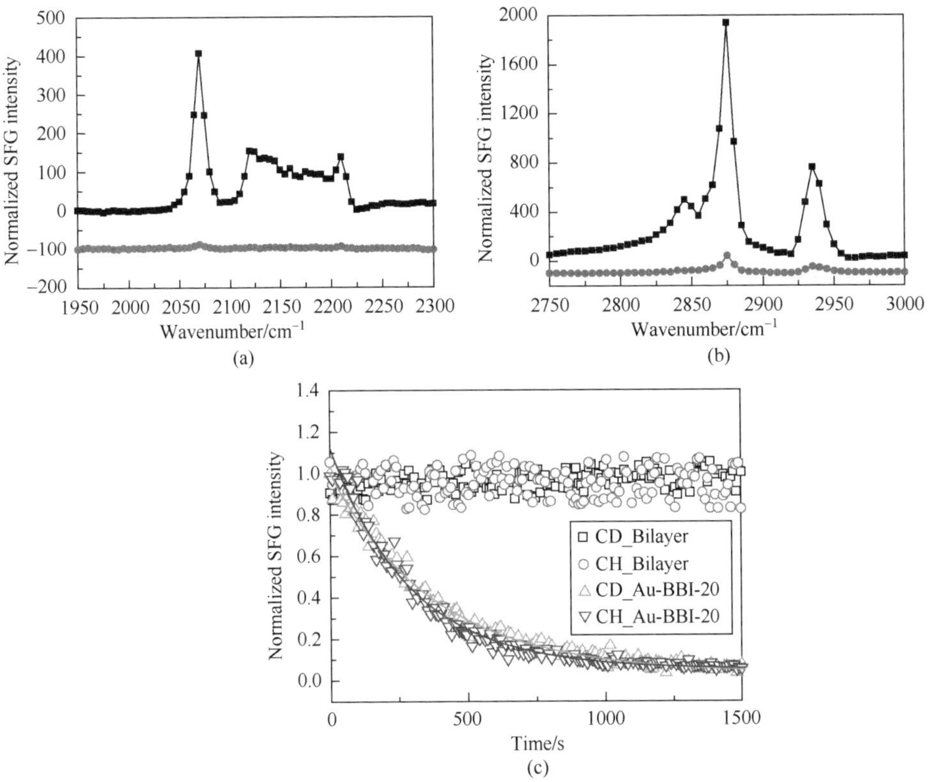

Figure 3.6 SFG spectra in the (a) C—D and (b) C—H stretching frequency ranges before (black square) and after (red circle) the introduction of Au-BBI-20 for d-DSPC/DSPC lipid bilayer respectively. (c) Time dependent SFG signals in C—D (green) and C—H (blue) stretching frequency ranges for Au-BBI-20 with fittings [Reproduced from *Physical Chemistry Chemical Physics*, 17: 9873-9884 (2015) with permission from the *Physical Chemistry Chemical Physics* Owner Societies]

The flip-flop rates of the lipid bilayers were then investigated using solutions of Au NPs with different sizes at different concentrations. Au NPs with diameters of 5, 20, 50, and 100 nm were used in the study[27]. We compared the Au NP size effect using the same particle number, the same overall surface area, and the same mass of Au NPs of different sizes in the solution in contact with the lipid bilayer. It was found that when compared at the same NP number and the same NP surface area, the larger the Au NP was, the faster the flip-flop of the bilayer induced by the Au NP happened, evidenced by the fast SFG signal decreases. However, at the same Au NP mass, the SFG signal decreased at the same speed, indicating that the lipid flip-flop rate was independent on the NP size (Figure 3.7, see the Color Inset, p.5). Therefore we believe that the lipid flip-flop rate depends on the Au NP mass in the solution in contact with the bilayer, not the Au NP size. As shown in Figure 3.8, regardless of the Au NP size, the lipid flip-flop rate monotonically increased as the Au NP mass increased in the solution[27].

It is worth mentioning that in the real situation when Au NPs are interacting with live cells, situations can be more complicated. Cell membranes contain many different components in addition to lipids and the lipid compositions can be very complicated. Also, the Au NPs may enter into cells through endocytosis. Therefore in the future it is necessary to use more complex model cell membranes in the study and the systematic comparisons between the model cell membrane study and live cell study should be performed.

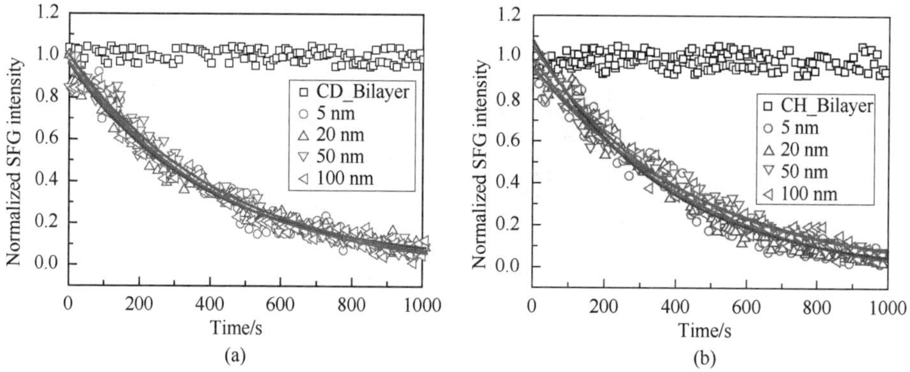

Figure 3.7　Time-dependent SFG signal intensity changes in (a) C—D and (b) C—H stretching frequency ranges when Au NPs (5, 20, 50, and 100 nm) interacting with the d-DSPC/DSPC bilayer at the same mass concentration of 4.00×10^{-7} g/mL. The dots are experimental data and the lines are fitting results [Reproduced from *Physical Chemistry Chemical Physics*, 17: 9873-9884 (2015) with permission from the *Physical Chemistry Chemical Physics* Owner Societies]

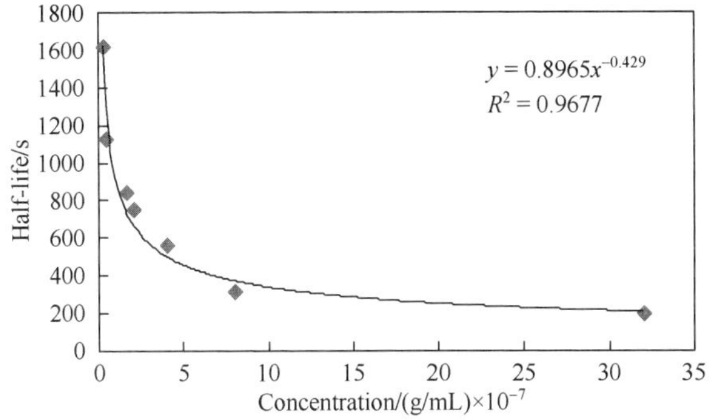

Figure 3.8 The observed flip-flop half-lives of the d-DSPC/DSPC bilayer lipid translocation induced by Au NPs of 5 nm, 20 nm, 50 nm and 100 nm at different mass concentrations [Reproduced from *Physical Chemistry Chemical Physics*, 17: 9873-9884 (2015) with permission from the *Physical Chemistry Chemical Physics* Owner Societies]

3.5 Interactions Between Model Cell Membranes and Peptides

Antimicrobial peptides (AMPs) have been extensively examined in the recent years because they have the potential to overcome the drug resistance developed by bacteria[28-32]. It has been shown that AMPs kill bacteria by disrupting bacteria cell membranes. Understanding the detailed interaction mechanisms between AMPs and bacteria cell membranes will lead to the design and development of AMPs with improved performance.

We have examined the molecular interactions between various AMPs and model cell membranes using SFG vibrational spectroscopy, supplemented with other analytical methods such as attenuated total reflectance (ATR)-FTIR[33-43]. A variety of AMPs with different secondary structures such as α-helical (magainine 2[34], MSI-78[35], cecropin P1[36], mellitin[37], LL-37[38], ovisprin-1[39]), 3_{10}-helical (alamethicin)[40-42], and β-sheet (tachyplesin I)[43] structures have been investigated. In such studies, we used zwitterionic lipid bilayers as models for mammalian cell membranes and negatively charged lipid bilayers as models for bacterial cell membranes.

3.5.1 Magainin 2

Magainin 2 (G-I-G-K-F-L-H-S-A-K-K-F-G-K-A-F-V-G-E-I-M-N-S-X) is a widely studied AMP, which was isolated from African clawed frog *Xenopus laevis*[44-46]. We studied molecular interactions between Magainin 2 and the bacterial cell membrane model 1-palmitoyl-2-oleoyl-*sn*-glycero-3-[phospho-*rac*- (1-glycerol)] (POPG)/POPG lipid bilayer as well as the mammalian cell membrane model 1-palmitoyl-2-oleoyl-*sn*-glycero-3-phosphocholine (POPC)/POPC lipid bilayer. Figure 3.9 shows the SFG ssp and ppp amide I spectra collected from the interface between the POPG/POPG bilayer and the Magainin 2 solution with a concentration of 800 nM. The amide I signal is dominated by the contribution from the backbone C=O stretches, which can be used to study the secondary structure and orientation of peptide molecules. Clearly, even at this low concentration, substantial Magainin 2 molecules segregate to the cell membrane, contributing to strong SFG signals. SFG amide I spectra shown in Figure 3.9 are centered at ~1650 cm^{-1}, indicating that Magainin 2 adopts an α-helical structure when interacting with the POPG/POPG bilayer. Using SFG spectra collected using different polarization combinations of the input and output beams, the orientation of interfacial molecules can be deduced. We have extensively studied orientation determination of α-helical secondary structure at interfaces; the detailed data analysis has been published[47]. From the spectral fitting results, it was deduced that Magainin 2 tilts at about 22 degrees vs. the surface normal in a POPG bilayer. Similar results can be obtained from the polarized ATR-FTIR study (Figure 3.10), validating the SFG data analysis method[34].

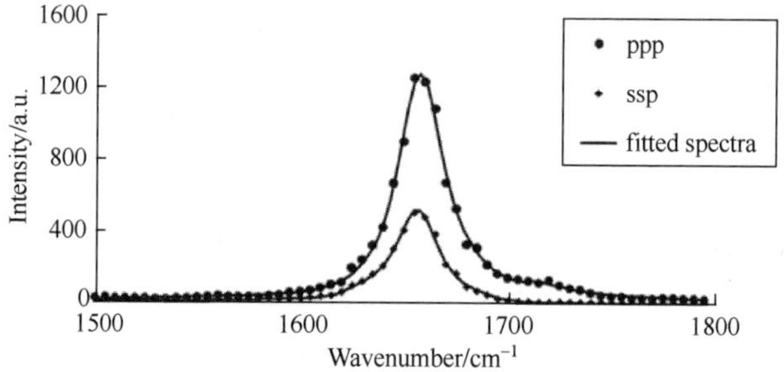

Figure 3.9 SFG ssp and ppp spectra of the C=O stretching frequency region collected from a POPG/POPG bilayer in contact with an 800 nM Magainin 2 solution [Reproduced with permission from *The Journal of Physical Chemistry B*, 113: 12358-12363 (2009). Copyright 2009 American Chemical Society]

We then studied molecular interactions between Magainin 2 and POPC/POPC lipid bilayer. No SFG signal could be observed from the interface between a POPC/POPC bilayer and a Magainin 2 solution with a concentration of 800 nM. This indicates that Magainin 2 has a lower affinity to a model mammalian cell membrane compared to a model bacterial cell membrane (e.g., a POPG/POPG bilayer). SFG ssp and ppp spectra can be collected when the Magainin 2 concentration in the subphase increased to 2000 nM. The collected SFG spectra are still centered at 1650 cm^{-1}, showing that Magainin 2 adopts an α-helical structure associated with a POPC/POPC bilayer. The spectral fitting results indicated that Magainin 2 more or less lies down on the POPC/POPC bilayer (75 degrees vs. the surface normal). At this concentration, Magainin 2 cannot insert into the POPC/POPC bilayer. The different orientations of Magainin 2 when associated with different model cell membranes demonstrate the antimicrobial selectivity of this peptide. In this situation, no ATR-FTIR signal can be detected from Magainin 2 molecules associated with the POPC/POPC bilayer. This indicates that SFG is a much more sensitive technique with a much lower detection limit compared to ATR-FTIR.

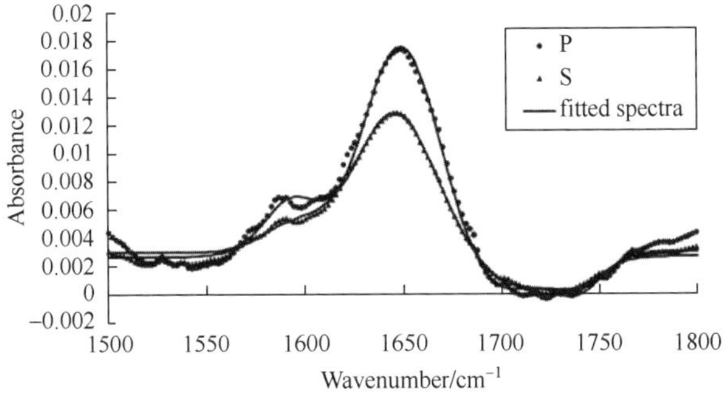

Figure 3.10 Polarized ATR-FTIR spectra of the C=O stretching frequency region collected from a POPG/POPG bilayer in contact with an 800 nM Magainin 2 solution [Reproduced with permission from *The Journal of Physical Chemistry B*, 113: 12358-12363 (2009). Copyright 2009 American Chemical Society]

3.5.2 MSI-78

SFG has also been used to investigate another antimicrobial peptide, MSI-78[35], which is an analogue of Magainin 2 but has much better activity and selectivity[48, 49]. The sequence of the MSI-78 is G-I-G-K-F-L-K-K-A-K-K-F-G-K-A-F-V-K-I-L-K-K-NH_2. It was found by SFG that the MSI-78 orientation when associated with a DPPG/DPPG bilayer depends on the concentration of the MSI-78 solution in contact with the lipid bilayer[35]. At a low solution concentration of 400 or 500 nM, the peptide tilts on the lipid bilayer towards the surface (Figure 3.11, see the Color Inset, p.6). At a higher concentration of 600 or 800 nM, the peptide tilts more towards the surface normal, inserting into the bilayer (Figure 3.11, see the Color Inset, p.6). At an even higher concentration of 2000 nM, the polarized SFG data cannot be interpreted by a single peptide orientation (delta orientation distribution). We believe that at such a high concentration, MSI-78 peptides form toroidal pores inside the lipid bilayer. The peptides adopt a multiple-orientation distribution (Figure 3.11, see the Color Inset, p.6). SFG has also been applied to study interactions between MSI-78 and a zwitterionic lipid bilayer 1, 2-dipalmitoyl-*sn*-glycero-3-phosphocholine (DPPC)/DPPC. No SFG signals from MSI-78 can be observed even at a concentration of 12000 nM. This demonstrates the excellent selectivity of MSI-78 against the model bacteria cell membrane vs. the model mammalian cell membrane.

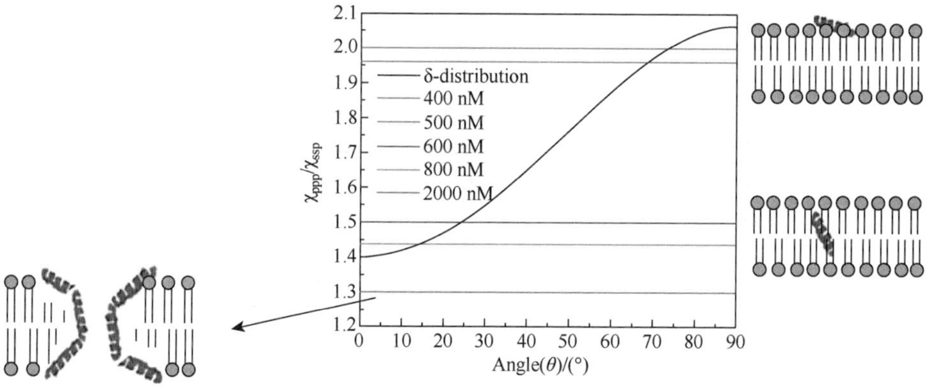

Figure 3.11　Relation between the SFG measured χ_{ppp}/χ_{ssp} ratio and the orientation angle of MSI-78 molecules in a DPPG/d-DPPG bilayer. Schematics of the membrane orientations of MSI-78 at different solution concentrations were also depicted. [Reproduced with permission from *Langmuir*, 27: 7760-7767 (2011). Copyright 2011 American Chemical Society]

3.5.3　Melittin

To quantify a multiple-orientation distribution of peptide molecules associated with a lipid bilayer, more measured parameters are needed. We combined SFG and ATR-FTIR spectroscopic techniques to determine the complex orientation distribution of melittin associated with a DPPG/DPPG bilayer[37]. As we extensively discussed previously[50], ATR-FTIR and SFG measure different orientation parameters. If an interfacial α-helical structure tilts at an angle of θ vs. the surface normal, ATR-FTIR measures $\langle\cos^2\theta\rangle$, while SFG measures $\langle\cos\theta\rangle$ and $\langle\cos^3\theta\rangle$ ("$\langle\rangle$" means average).

Melittin is naturally found in bee venom toxin. It is composed of 26 amino acid residues and is highly positively-charged in physiological conditions[51]. SFG and ATR-FTIR spectra were successfully collected from melittin associated with a DPPG/DPPG bilayer. The measured SFG ssp/ppp intensity ratio cannot lead to a single delta angle distribution. We first assume that melittin molecules can adopt two different orientations (θ_1 and θ_2, dual-delta orientation distribution). With this assumption and after fitting and analyzing the ATR-FTIR and SFG data, we found that θ_1=13°, θ_2=100°, and population N_1=0.274 and N_2=0.726. This shows that about a quarter of the mellitin molecules more or less stand up in the lipid bilayer, while three quarters lie down, slightly tilting towards another direction.

We then deduced the complex orientation function using a trial function based on the maximum entropy theory with the form of $G(\theta)=\mathrm{e}^{-(\lambda_0+\lambda_1\cos\theta+\lambda_2\cos^2\theta+\lambda_3\cos^3\theta)}$ (without using the dual-delta distribution assumption). The deduced orientation distribution is shown in Figure 3.12. It successfully reproduced the results obtained with the dual-delta distribution assumption. The deduced complex orientation distribution contains two peaks, with centers at 6° and 100° respectively, similar to those obtained under the dual-delta distribution assumption. The populations for the two orientations are 0.263 and 0.737 respectively, also similar to the above presented results under the dual-delta distribution assumption. A schematic showing the melittin complex orientation distribution in the DPPG/DPPG lipid bilayer is displayed in Figure 3.13.

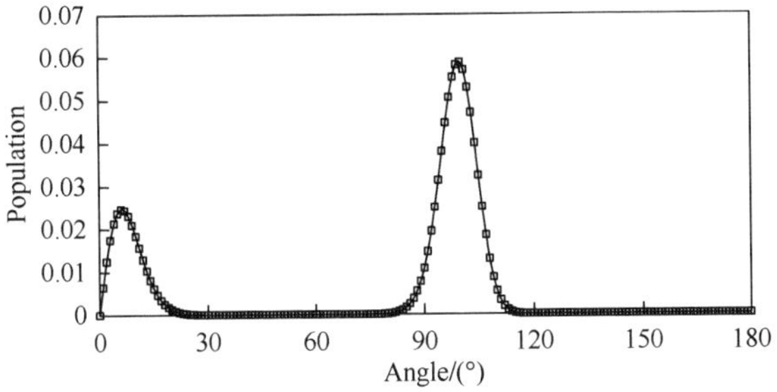

Figure 3.12 Complex melittin orientation distribution derived based on the maximum entropy theory and ATR-FTIR/SFG data [Reproduced with permission from *Journal of the American Chemical Socitey*, 129: 1420-1427 (2007). Copyright 2007 American Chemical Society]

ATR-FTIR and SFG provide independent measurements. The combined use of ATR-FTIR and SFG can be applied to determine a more complex orientation distribution than the use of ATR-FTIR or SFG alone. We have presented previously with the combination of higher order nonlinear optical spectroscopic methods, more parameters can be deduced to study complicated molecules at interfaces[50].

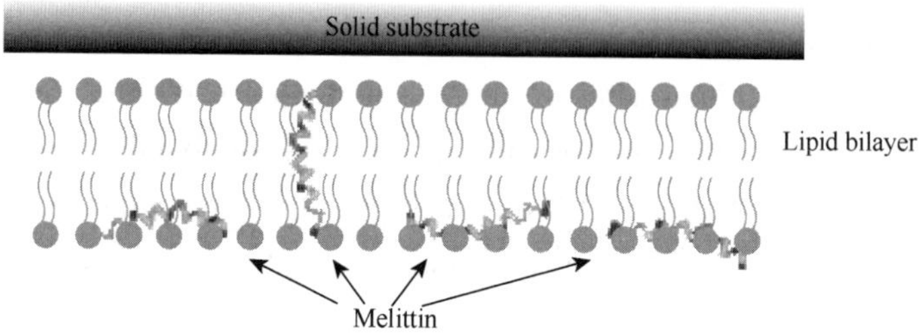

Figure 3.13 Schematic of the two orientations of melittin associated with a lipid bilayer [Reproduced with permission from *Journal of the American Chemical Society*, 129: 1420-1427 (2007). Copyright 2007 American Chemical Society]

3.5.4 Cecropin P1

Here we are using solid supported lipid bilayers as model cell membranes. In the above presented research, normally we used CaF_2 windows or prisms as solid supports. We have demonstrated that the interaction between a lipid bilayer and a CaF_2 solid substrate is minimal and can be ignored by comparing the structure of a lipid bilayer built on a CaF_2 surface to that on a polymer (poly L-lactic acid or PLLA) cushion on a CaF_2 substrate. The bilayer structure has no noticeable difference observed on a surface with or without a polymer cushion[36]. In addition, we studied the molecular interactions between an AMP, cecropin P1, and the lipid bilayers constructed on the above two surfaces. Same results were obtained, further showing that the substrate- lipid bilayer interaction can be ignored, and a supported lipid bilayer is a good model for cell membrane[36]. It was found that cecropin P1 adopts an α-helical structure and its membrane orientation is dependent on the peptide solution concentration.

3.5.5 LL-37

In addition to the linear α-helical peptide, we also developed a systematic data analysis method to study membrane orientations of bent α-helices. Such a method was successfully applied to study AMP LL-37 associated with different lipid bilayers, including lipid bilayers with cholesterol. It was found that the lipid compositions greatly influence the LL-37-cell membrane interaction (Figure 3.14, see the Color Inset, p.6)[38].

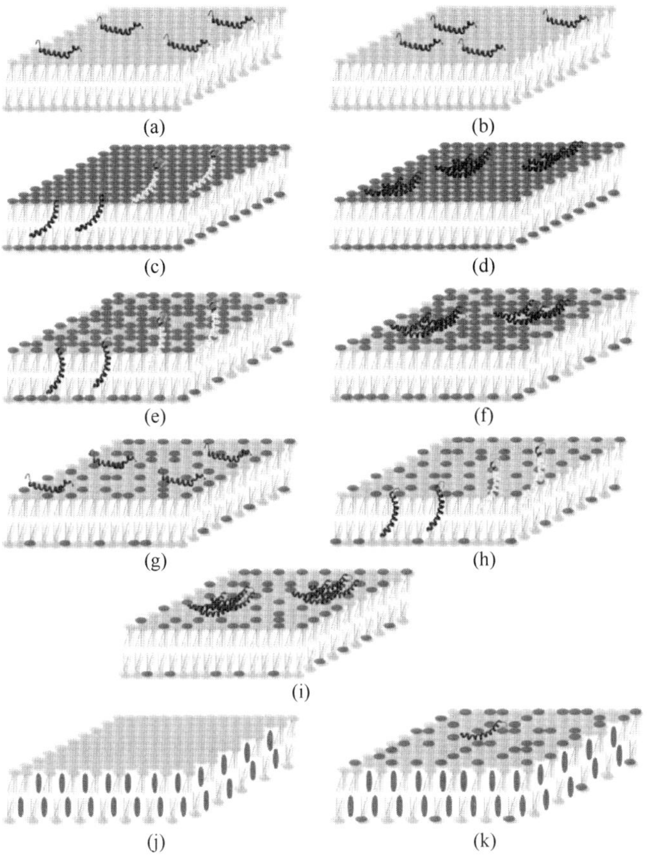

Figure 3.14 Schematics showing interactions between LL-37 and different lipid bilayers. (a) POPC bilayer at a low concentration and (b) a high concentration; (c) POPG bilayer at a low concentration and (d) a high concentration; (e) 3∶7 POPC∶POPG lipid bilayer at a low concentration and (f) a high concentration; (g) 7∶3 POPC∶POPG lipid bilayer at a low concentration and (h) a high concentration and (i) even higher concentrations; (j) 1∶1 POPC∶CHO lipid bilayer at a high concentration; (k) 0.3∶0.7∶1 POPG∶POPC∶CHO lipid bilayer at a high concentration [Reproduced from *Scientific Reports*, 3: 1854 (2013)]

3.5.6 Alamethicin

As we presented above, we have extensively investigated membrane orientation of peptides with α-helical structures. We have also studied interfacial orientations of peptides with other secondary structures. For example, we have studied alamethicin associated with different lipid bilayers[40-42]. It was found that when associated with a fluid phase lipid bilayer such as a 1, 2-dimyritoyl-*sn*-glycero-3-phosphocholine (DMPC) bilayer, alamethicin adopts a mixed α-helical and 3_{10}-helical structure. An α-helix has 3.6 amino acids per turn (or 18 amino acids every 5 turns), while a 3_{10}-helix has 3 amino acids per turn. The orientation determination methodology for a 3_{10}-helix has been developed in detail previously[47]. SFG results showed that the alamethicin 3_{10}-helix at the C-terminus tilts at 43° vs. the surface normal, while the α-helix at the N-terminus tilts at 63° vs. the surface normal. ATR-FTIR data are well correlated to the SFG results, showing that the orientation distributions for both helices are narrow. We therefore believe that both the α-helix and the 3_{10}-helix have a delta distribution[40]. When alamethicin interacts with another fluid phase lipid bilayer POPC, similar results were obtained. For example, at alamethicin solution pH of 6.7, the α-helical and 3_{10}-helical structures were determined to orient at 72° and 50° vs. the surface normal, respectively[41]. Interestingly, when the solution pH was raised to 11.9, the alamethicin orientation was found to change to 56.5° and 45° for the α-helical and the 3_{10}-helical structures. These orientation changes are due to the local pH change near the lipid bilayer, which changed the membrane potential. This book provides direct evidence that peptides can change membrane orientation when the membrane potential is varied.

We also studied the molecular interactions between alamthcin and various gel phase lipid bilayers such as DPPC, DSPC, and DPPG bilayers[40]. Different from the results we presented above, when alamethicin molecules interact with a gel phase lipid bilayer, they exhibit different behaviors: they aggregate and/or lie down on the lipid bilayer surface. Figure 3.15 (see the Color Inset, p.7) displays a schematic to show that alamethicin interacts with different lipid bilayers differently.

To further understand the different interactions between alamethicin and lipid bilayers with different phases, we used the same lipid bilayer at different temperatures in the study[42]. At different temperatures, a lipid bilayer can have different phases. In this study, we found that the alamethicn solution concentration plays an important role in alamethcin-cell membrane interactions. At a low alamethicin solution concentration of 0.84 μM, alamethicin lies down on both fluid (DMPC bilayer at a high temperature: 30℃) and gel phase (DMPC bilayer at a low temperature: 10℃) lipid bilayers. At a medium concentration of 10.80 μM, alamethicin inserts into the DMPC bilayer at 30℃ (fluid phase). When the lipid temperature decreased to 10℃, the alamethicin changed its orientation and lay down on the gel phase lipid bilayer. At a high solution concentration of

21.60 μM, alamethicin can insert into both fluid and gel phase lipid bilayers. Schematics showing different interactions of alamethicin with a DMPC bilayer at different temperatures (different phases) are presented in Figure 3.16 (see the Color Inset, p.7).

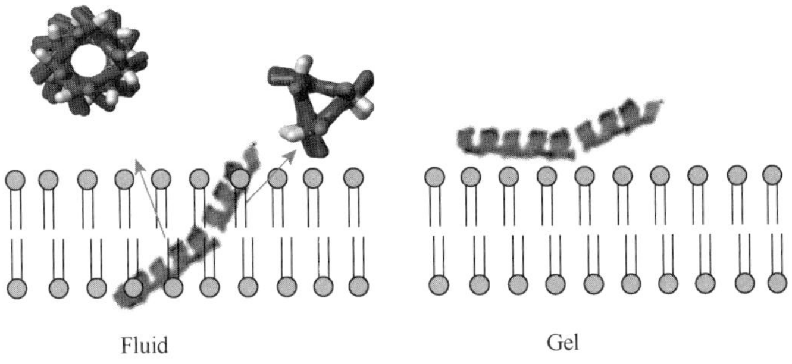

Figure 3.15 Schematic showing when alamethicin interacts with different lipid bilayers, it exhibits different behaviors. It can insert into a fluid phase lipid bilayer, but aggregates/lies down on a gel phase lipid bilayer [Reproduced with permission from *The Journal of Physical Chemistry B*, 114: 3334-3340 (2010). Copyright 2010 American Chemical Society]

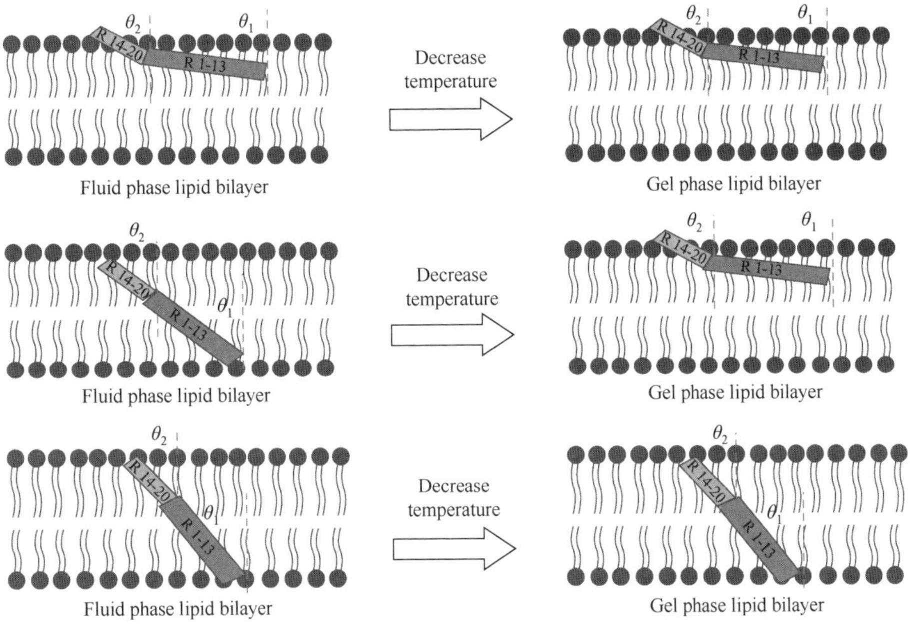

Figure 3.16 Schematics showing when alamethicin interacts with different lipid bilayers, it exhibits different behaviors at different solution concentrations. Alamethicin solution concentrations are 0.84 (top), 10.80 (middle), and 21.60 (bottom) μM respectively [Reproduced with permission from *The Journal of Physical Chemistry C*, 117: 17039-17049 (2013). Copyright 2013 American Chemical Society]

3.5.7 Tachyplesin I

In addition to the peptides with helical structures, we also studied β-sheet peptide such as tachyplesin I. Besides the tilt angle, to determine the orientation of a β-sheet, it is necessary to deduce the twist angle as well. We showed by combining SFG and ATR-FTIR, it is feasible to determine both the tilt and twist angles of a β-sheet at the interface[43]. Importantly, we demonstrated the feasibility to detect chiral SFG signals from β-sheet structure at the interface, using tachyplesin I as an example[52]. The chiral signal can be used along with the non-chiral SFG signal to study β-sheet orientation.

3.6 Interactions of Membrane Associated Proteins

3.6.1 Gβγ

Secondary structures are building blocks for proteins. After presenting our research on membrane related peptides, here I want to discuss the studies on membrane proteins using SFG. The first protein we investigated was Gβγ. Heterotrimeric guanine nucleotide-binding proteins (G proteins) are important membrane proteins, which play crucial roles in signal transduction[53, 54], and relay the extracellular signals sensed by G protein-coupled receptors (GPCRs) to down stream effectors. G proteins have three subunits (Gα, Gβ, and Gγ), with Gβ and Gγ forming a tightly associated dimer. The membrane orientation of Gβγ is important but has not been determined due to the lack of appropriate analytical techniques.

SFG ssp and ppp spectra were collected from the Gβγ molecules associated with a POPG/POPG bilayer (Figure 3.17, see the Color Inset, p.7)[55]. The SFG signals were centered at 1650 cm^{-1}, contributed by the α-helical components in Gβγ. Using the SFG spectral fitting results, the calculated hyperpolarizability of the helical domain, and by assuming the twist angle of the membrane associated Gβγ is fixed, we can deduce the membrane orientation of Gβγ by determining the tilt angle θ. The deduced orientation of Gβγ is also shown in Figure 3.17. We also collected SFG spectra from Gβγ associated with different lipid bilayers and found that the lipid composition in the bilayer influences the Gβγ membrane orientation[55].

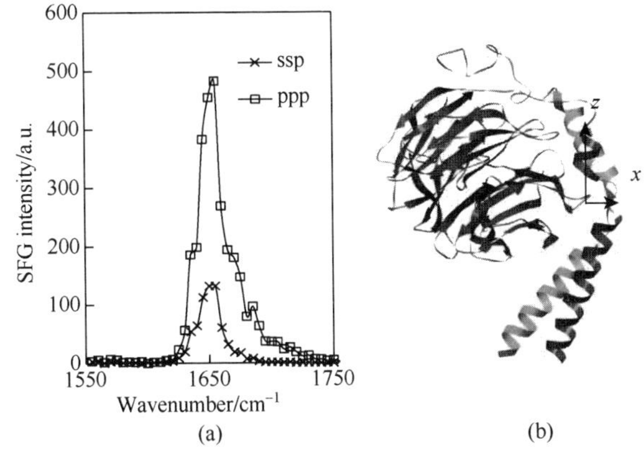

Figure 3.17 (a) SFG amide I spectra of interfacial Gβγ adsorbed onto a POPC/POPC bilayer; (b) Gβγ orientation deduced based on the SFG intensity ratio [Reproduced with permission from *Journal of the American Chemical Society*, 129: 12658-12659 (2007). Copyright 2007 American Chemical Society]

3.6.2 Gβγ-GRK2 complex

To study the orientation of more complicated proteins at interfaces, we developed a computer software package which can be used to read any protein structure in the protein data bank. Then this program will locate the α-helical structures in the protein and calculate the SFG second order nonlinear optical susceptibility components of the overall α-helical structure as a function of protein orientation. Two angles, the title angle and the twist angle, can be used to define the protein orientation.

SFG has been applied to investigate the formation of the complex between Gβγ and G protein-coupled receptor (GPCR) kinase 2 (GRK2) at a lipid bilayer and the membrane orientation of the complex. GPCRs are integral membrane proteins which play important roles in many biological processes including signaling. GRK2 interacts with the cell membrane, Gβγ, and GPCRs, and is playing a role in the assembly and organization of signaling complexes at GPCRs[56]. The orientation of Gβγ-GRK2 is important and was determined using SFG, with the help of the computer program discussed above[57]. Since Gβγ-GRK2 is a large protein, with the SFG measurements alone, we could not uniquely determine the membrane orientation of this protein. However, with other knowledge (e.g., the physical constraints of the membrane orientation), the possible orientation regions (possible combinations of tilt and twist angles) were determined[57].

3.6.3 Gαβγ complex

As we discussed above when we studied membrane orientations of peptides, the combined SFG and ATR-FTIR studies provide more independently measured parameters, which can be used to determine more complex orientation distributions. Here with the help of ATR-FTIR measurements, SFG can be used to more accurately deduce the protein membrane orientation. As for SFG data analysis, here we also developed a software package for ATR-FTIR data analysis. The software package can calculate the ATR-FTIR dichroic ratio as a function of protein orientation. The comparison of the calculated dichroic ratio and the experimentally measured data can be used to determine the possible protein orientation, represented by a heat map[58]. Similarly, according to SFG measurements, a heat map can also be obtained[57]. Since SFG and ATR-FTIR provide independent measurements, the final orientation should be the overlapping regions of the two heat maps generated by SFG and ATR-FTIR measurements. Using this method, we studied the membrane orientation of a Gαβγ complex. A unique membrane orientation of this complex has been determined, as shown in Figure 3.18 (see the Color Inset, p.8).

3.6.4 GRK5

As discussed above, GPCRs are involved in many biological processes and are also targeted by many drugs. GRKs play important roles in GPCR regulations. There are three GRK subfamilies, including GRK1, GRK2, and GRK4[59]. GRK 4 subfamily members include GRK4, GRK5, and GRK6, which bind to negatively charged cell membranes. It is believed that GRK4 family proteins (e. g., GRK5) bind to negatively charged lipid bilayers by positively charged amino acid region close to the N-terminus. But it is also believed that the C-terminal amphipathic helix may play a role in membrane binding. Therefore here we want to study whether the GRK5 N-terminal region, C-terminal helix, or both play roles in membrane binding. To investigate the influence of the C-terminal helix in membrane binding, here we studied a full length GRK5 (residue 1 to 590) and a truncated GRK5 with C-terminal region deleted, $GRK5_{1\text{-}531}$ (with residues from 1 to 531)[60].

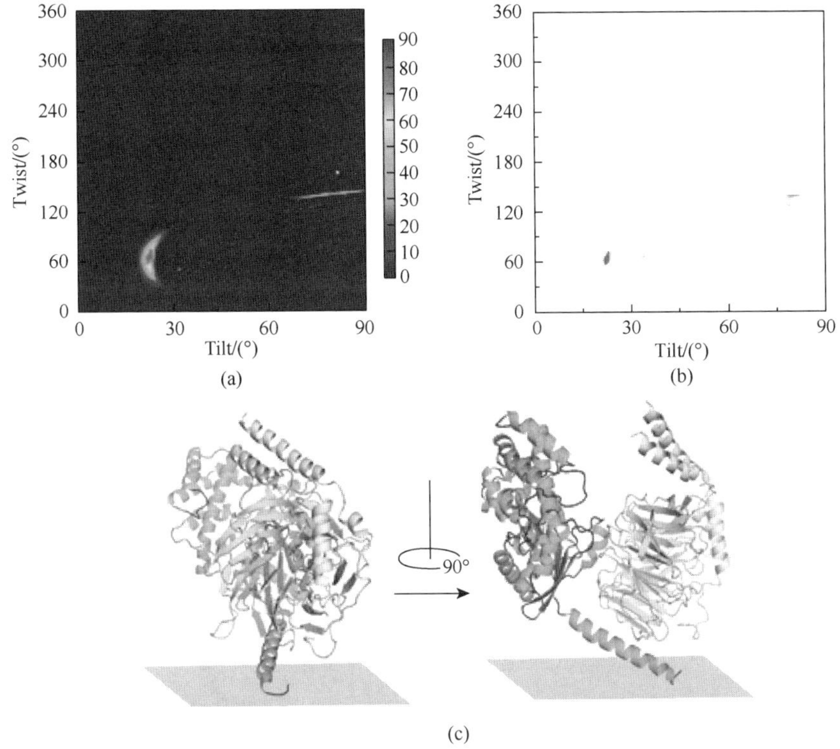

Figure 3.18 (a) Heat map to show the possible membrane orientation regions of Gαβγ. (b) Most possible orientation regions of Gαβγ. Only regions with a score >70 are shown. (c) Representative orientation of Gαβγ [Reproduced with permission from *Journal of the American Chemical Society*, 135: 5044-5051 (2013). Copyright 2013 American Chemical Society]

SFG spectra were collected from a mixed lipid bilayer with a POPC : POPG ratio of 9 : 1 in contact with a GRK5 solution. No SFG signal was observed (Figure 3.19)[60]. This may be because GRK5 does not associate with the lipid bilayer, or the bilayer associated GRK5 adopts a random orientation. Strong ATR-FTIR signal can be collected from the lipid bilayer/GRK5 solution interface (Figure 3.19), showing that GRK5 does associate with the lipid bilayer, but with a random orientation[60]. We then replaced the 9 : 1 POPC : POPG mixed bilayer with a 1 : 1 POPC : POPG mixed bilayer. Weak SFG signals can then be detected from GRK5 associated with this bilayer, but the signal/noise ratio is not sufficiently high for detailed data analysis (Figure 3.19)[60]. After we replaced the lipid bilayer with a POPG bilayer, strong SFG signal from GRK5 can be obtained [Figure 3.19 (see the Color Inset, p.8)]. This shows that it requires enough negatively charged lipid compositions to order the associated GRK5 molecules. After the data analysis of the SFG and ATR-FTIR spectra, the POPG bilayer associated GRK5 orientation can be deduced, as shown in Figure 3.20 (see the Color Inset, p.9). The two representative protein orientations are depicted in the same figure[60].

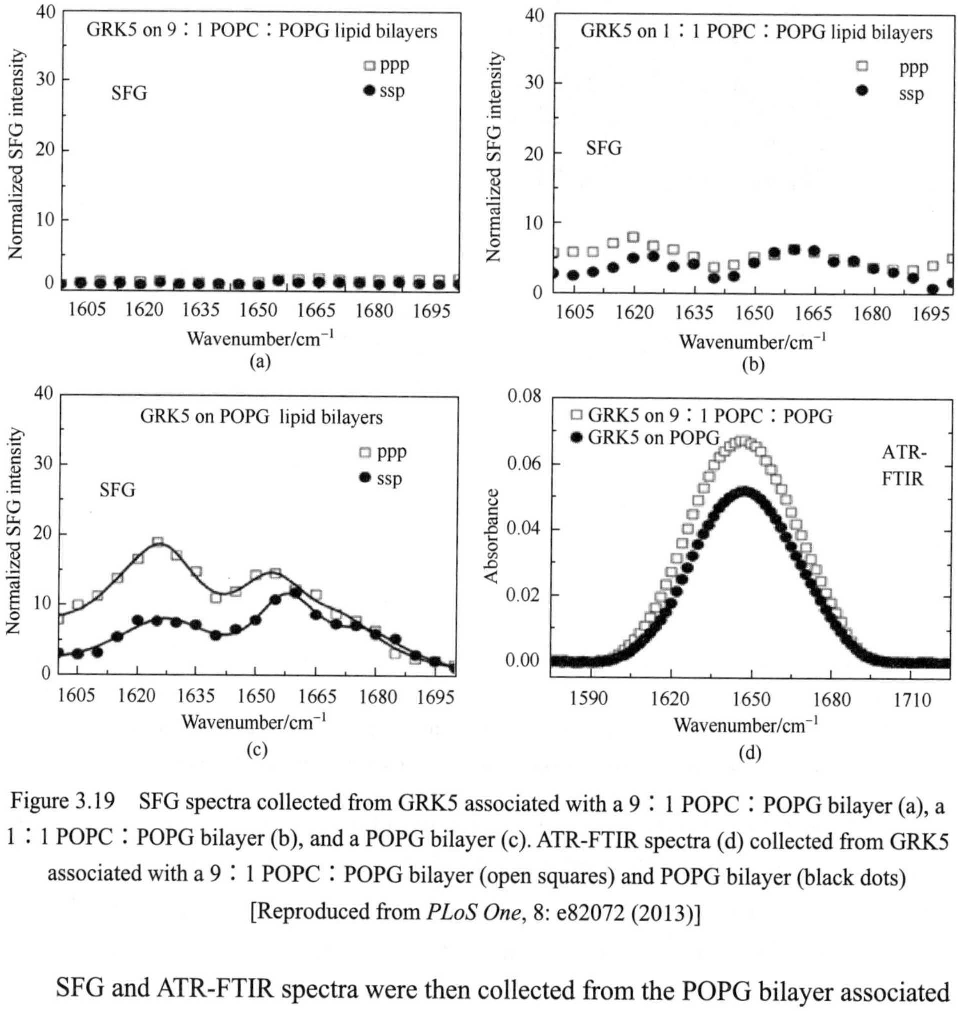

Figure 3.19 SFG spectra collected from GRK5 associated with a 9 : 1 POPC : POPG bilayer (a), a 1 : 1 POPC : POPG bilayer (b), and a POPG bilayer (c). ATR-FTIR spectra (d) collected from GRK5 associated with a 9 : 1 POPC : POPG bilayer (open squares) and POPG bilayer (black dots) [Reproduced from *PLoS One*, 8: e82072 (2013)]

SFG and ATR-FTIR spectra were then collected from the POPG bilayer associated $GRK5_{1\text{-}531}$. From the fitting parameters of these spectra, the membrane orientation of $GRK5_{1\text{-}531}$ can be deduced, as also shown in Figure 3.20 (see the Color Inset, p.9). Two representative protein orientations were displayed in Figure 3.20 (see the Color Inset, p.9) as well. Clearly, the full length GRK5 and the C-terminal region deleted $GRK5_{1\text{-}531}$ adopt similar orientations, probably showing that the C-terminal region does not play substantial roles in determining GRK5 membrane orientation[60].

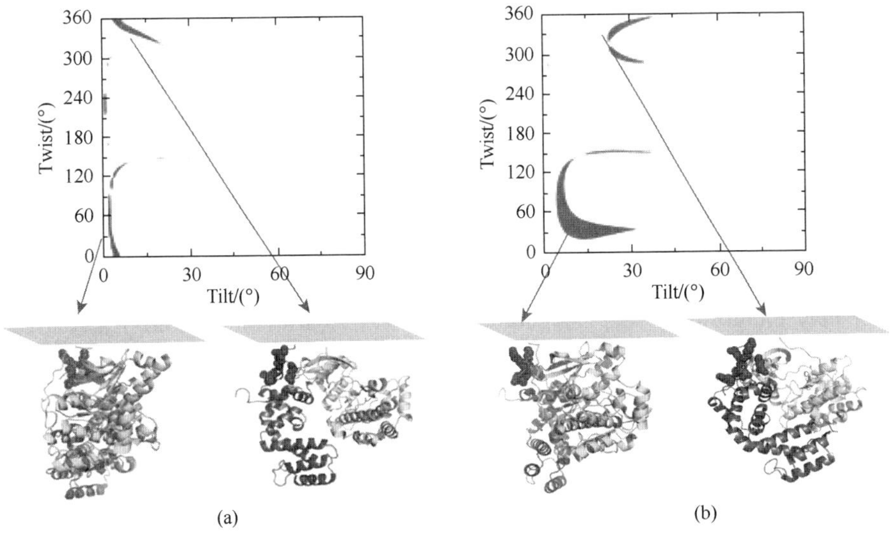

Figure 3.20 Deduced membrane orientations for GRK5 (a) and truncated $GRK5_{1\text{-}531}$ (b). The orientations are similar, showing that the deleted C-terminal region does not play a major role in GRK5 membrane binding [Reproduced from *PLoS One*, 8: e82072 (2013)]

3.7 Conclusions and Outlooks

This chapter discussed the use of a nonlinear optical vibrational spectroscopy, SFG, to study molecular interactions between model cell membranes and a variety of molecules and materials including drug molecules, nanoparticles, peptides, and proteins. SFG has many advantages in such studies. Not like many traditional surface sensitive techniques, SFG does not require high vacuum to operate. Therefore it can study solid/liquid interfaces *in situ* in real time. As a result, SFG can be used to study lipid bilayers in contact with various solutions *in situ*. SFG has an excellent surface/interface specificity—molecules in bulk media would not interfere with the studies on molecules on surfaces and interfaces. SFG provides vibrational spectra, therefore provide molecular level information. For lipid bilayers, it can probe the structure of each leaflet of the bilayer and the interacting molecules at the same time. SFG has a superb detection limit, and only requires very small amount of sample.

As we presented in this chapter, SFG can be used to elucidate detailed interactions between small drug molecules and various model cell membranes. We clearly demonstrated that the drug concentration and lipid composition both influence the interactions between the drugs and model cell membranes. We therefore can use SFG to study the activity and selectivity of various drugs on various cell types. Nanomaterials have been extensively researched for drug delivery and bio-imaging. We showed in this book that Au NPs can induce lipid flip-flop. SFG can be used to study many different nanoparticles and nanomaterials to elucidate their interactions with cell membranes in the future.

We have extensively examined molecular interactions between peptides and model cell membranes. We have developed systematic methodologies to deduce interfacial peptide orientation using SFG. Peptides with different secondary structures including α-helix, β-sheet, and 3_{10}-helix have been investigated. Here we only focused on the studies of AMPs, but we have investigated different peptides. For example, we examined cell penetrating peptides using SFG[61]. To further understand the detailed structure of peptides at interface, it is necessary to use isotope labeled peptides. The isotope labeled peptide unit can be studied one by one to get more detailed structural information. We demonstrated the feasibility to use SFG to study isotope labeled peptides[39]. In the future, such an approach needs to be applied to develop SFG into a powerful tool to elucidate detailed structure of interfacial biological molecules.

The current model cell membrane is quite simple, mostly only containing one or two lipid compositions. We showed that the lipid bilayers prepared using the lipid mixture extracted from *E. coil* interacts with various AMPs differently compared to the simple POPG bilayer[62]. In the future, it is necessary to use more complicated model to better represent the cell membranes. Such models should contain more types of lipids, cholesterol, and proteins as well.

SFG has been shown to be powerful to determine protein membrane orientation. As we discussed above, we have developed computer software packages to calculate SFG and ATR-FTIR signals as a function of protein orientation. Such calculated data can be compared to the experimentally measured data to deduce protein orientation. We presented our results on studying G proteins in this chapter. In the future, SFG should be applied to study more membrane related proteins, including ion channels.

SFG study on interactions at model cell membranes is still at its early stage. We believe that as better lasers, optics, and detections are quickly developed, SFG will become a major analytical tool to study surfaces and interfaces in the future.

3.8 Acknowledgement

This work is supported by the University of Michigan. I want to thank all the students, postdoctoral fellows, colleagues, and collaborators for their great contributions to the related research.

References

[1] Voet D, Voet J G. *Biochemistry*. 4th ed. New York: John Wiley and Sons, 2011.

[2] Tamm L K, Mcconnell H M. Supported phospholipid bilayers. *Biophysical Journal*, 1985, 47 (1): 105-113.

[3] Stewart P S, Costerton J W. Antibiotic resistance of bacteria in biofilms. *Lancet*, 2001, 358(9276): 135-138.

[4] Tew G N, Liu D H, Chen B, et al. De novo design of biomimetic antimicrobial polymers. *Proceedings of the National Academy of Sciences of the United States of America*, 2002, 99 (8): 5110-5114.

[5] Rennie J, Arnt L, Tang H Z, et al. Simple oligomers as antimicrobialpeptide mimics. *Journal of Industrial Microbiology Biotechnology*, 2005, 32(7): 296-300.

[6] Yao J, Yang M, Duan Y. Chemistry, biology, and medicine of fluorescent nanomaterials and related systems: New insights into biosensing, bioimaging, genomics, diagnostics, and therapy. *Chemical Reviews*, 2014, 114(12): 6130-6178.

[7] de Jong W H, Borm P J A. Drug delivery and nanoparticles: Applications and hazards. *International Journal of Nanomedicine*, 2008, 3(2): 133-149.

[8] Hubbell J A, Chilkoti A. Chemistry nanomaterials for drug delivery. *Science*, 2012, 337 (6092): 303-305.

[9] ShenY R. *The Principles of Nonlinear Optics*. New York: Wiley, 1984.

[10] ShenY R. Surface-properties probed by 2nd-harmonic and sum-frequency generation. *Nature*, 1989, 337(6207): 519-525.

[11] Shen Y R. Phase-sensitive sum-frequency spectroscopy. *Annual Review of Physical Chemistry*, 2013, 64: 129-150.

[12] Bain C D. Sum-frequency vibrational spectroscopy of the solid/liquid interface. *Journal of the Chemical Society, Faraday Transactions*, 1995, 91(9): 1281-1296.

[13] Eisenthal K B. Liquid interfaces probed by second-harmonic and sum-frequency spectroscopy. *Chemical Reviews*, 1996, 96 (4): 1343-1360.

[14] Chen Z, Shen Y R, Somorjai G A. Studies of polymer surfaces by sum frequency generation vibrational spectroscopy. *Annual Review of Physical Chemistry*, 2002, 53: 437-465.

[15] Moore F G, Richmond G L. Integration or segregation: How do molecules behave at oil/water interfaces? *Accounts of Chemical Research*, 2008, 41: 739-748.

[16] Smith J P, Hinson-Smith V. SFG coming of age. *Analytical Chemistry*, 2004, 76 (15): 287A-290A.

[17] Chen C Y, Even M A, Wang J, et al. Sum frequency generation (SFG) vibrational spectroscopy studies on molecular conformation of liquid polymers poly (ethylene glycol) (PEG) and poly (propylene glycol) (PPG) at

different interfaces. *Macromolecules*, 2002, 35 (24): 9130-9135.

[18] Wang J, Buck S M, Chen Z. Sum frequency generation (SFG) vibrational spectroscopy studies on protein adsorption. *The Journal of Physical Chemistry B*, 2002, 106(44): 11666-11672.

[19] Wang J, Woodcock S E, Buck S M, et al. Different surface-restructuring behaviors of polymethacrylates detected by SFG in water. *Journal of the American Chemical Society*, 2001, 123 (38): 9470-9471.

[20] Wang J, Chen C Y, Buck S M, et al. Molecular chemical structure on poly (methyl methacrylate) (PMMA) surface studied by sum frequency generation (SFG) vibrational spectroscopy. *The Journal of Physical Chemistry B*, 2001, 105 (48): 12118-12125.

[21] Chen X Y, Tang H Z, Even M A, et al. Observing a molecular knife at work. *Journal of the American Chemical Society*, 2006, 128 (8): 2711-2714.

[22] Avery C W, Som A, Xu Y, et al. Dependence of antimicrobial selectivity and potency on oligomer structure investigated using substrate supported lipid bilayers and sum frequency generation vibrational spectroscopy. *Analytical Chemistry*, 2009, 81 (20): 8365-8372.

[23] Wu F, Yang P, Zhang C, et al. Investigation of drug-model cell membrane interactions using sum frequency generation vibrational spectroscopy: A case study of chlorpromazine. *The Journal of Physical chemistry C*, 2014, 118 (31): 17538-17548.

[24] Wu F, Yang P, Zhang C, et al. Molecular interactions between amantadine and model cell membranes. *Langmuir*, 2014, 30: 8491-8499.

[25] Zhang C, Wu F G, Hu P P, et al. Interaction of polyethylenimine with model cell membranes studied by linear and nonlinear spectroscopic techniques. *Journal of Physical Chemistry C*, 2014, 118 (23): 12195-12205.

[26] Avery C W, Palermo E F, McLaughlin A, et al. Investigations of the interactions between synthetic antimicrobial polymers and substrate-supported lipid bilayers using sum frequency generation vibrational spectroscopy. *Analytical Chemistry*, 2011, 83 (4): 1342-1349.

[27] Hu P P, Zhang X X, Zhang C, et al. Molecular interactions between gold nanoparticles and model cell membranes. *Physical Chemistry Chemical Physics*, 2015, 17 (15): 9873-9884.

[28] McPhee J B, Hancock R E W. Function and therapeutic potential of host defence peptides. *Journal of Peptide Science*, 2005, 11 (11): 677-687.

[29] Zasloff M. Antimicrobial peptides of multicellular organisms. *Nature*, 2002, 415 (6870): 389-395.

[30] Ding J L, Ho B. Antimicrobial peptides: Resistant-proof antibiotics of the new millennium. *Drug Development Research*, 2004, 62 (4): 317-335.

[31] Brogden K A. Antimicrobial peptides: Pore formers or metabolic inhibitors in bacteria? *Nature Reviews Microbiology*, 2005, 3 (3): 238-250.

[32] Yeaman M R, Yount N Y. Mechanisms of antimicrobial peptide action and resistance. *Pharmacological Reviews*, 2003, 55 (1): 27-55.

[33] Chen X Y, Chen Z. SFG studies on interactions between antimicrobial peptides and supported lipid bilayers. *Biochimica et Biophysica Acta*, 2006, 1758 (9): 1257-1273.

[34] Nguyen K T, Le Clair S, Ye S, et al. Molecular interaction between magainin 2 and model membranes *in situ*. *The Journal of Physical Chemistry B*, 2009, 113 (36): 12358-12363.

[35] Yang P, Ramamoorthy A, Chen Z. Membrane orientation of MSI-78 measured by sum frequency generation vibrational spectroscopy. *Langmuir*, 2011, 27 (12): 7760-7767.

[36] Wang T, Li D W, Lu X L, et al. Single lipid bilayers constructed on polymer cushion studied by sum frequency generation vibrational spectroscopy. *The Journal of Physical Chemistry C*, 2011, 115 (15): 7613-7620.

[37] Chen X Y, Wang J, Boughton A P, et al. Multiple orientation of melittin inside a single lipid bilayer determined by combined vibrational spectroscopic studies. *Journal of the American Chemical Society*, 2007, 129 (5): 1420-1427.

[38] Ding B, Soblosky L, Nguyen K, et al. Physiologically-relevant modes of membrane interactions by the human antimicrobial peptide, LL-37, revealed by SFG experiments. *Scientific Reports*, 2013, 3: 1854.

[39] Ding B, Laaser J E, Liu Y, et al. Site-specific orientation of an α-helical peptide ovispirin-1 from isotope-labeled SFG spectroscopy. *The Journal of Physical Chemistry B*, 2013, 117(47): 14625-14634.

[40] Ye S J, Nguyen K T, Chen Z. Interactions of alamethicin with model cell membranes investigated using sum frequency generation vibrational spectroscopy in real time *in situ*. *The Journal of Physical Chemistry B*, 2010, 114 (9): 3334-3340.

[41] Ye S J, Li H C, Wei F, et al. Observing a model ion channel gating action in model cell membranes in real time *in situ*: Membrane potential change induced alamethicin orientation change. *Journal of the American Chemical Society*, 2012, 134 (14): 6237-6243.

[42] Yang P, Wu F G, Chen Z. Lipid fluid-gel phase transition induced alamethicin orientational change probed by sum frequency generation vibrational spectroscopy. *The Journal of Physical Chemistry C*, 2013, 117 (33): 17039-17049.

[43] Nguyen K T, King J T, Chen Z. Orientation determination of interfacial β-sheet structures *in situ*. *The Journal of Physical Chemistry B*, 2010, 114 (25): 8291-8300.

[44] Berkowitz B A, Bevins C L, Zasloff M A. Magainins: A new family of membrane-active host defense peptides. *Biochemical Pharmacology*, 1990, 39 (4): 625-629.

[45] Zasloff M, Martin B, Chen H C. Antimicrobial activity of synthetic magainin peptides and several analogues. *Proceedings of the National Academy of Sciences of the United States of America*, 1988, 85 (3): 910-913.

[46] Maloy W L, Kari U P. Structure-activity studies on magainins and other host defense peptides. *Biopolymers*, 1995, 37 (2): 105-122.

[47] Nguyen K J, Le Clair S, Ye S, et al. Orientation determination of protein helical secondary structure using linear and nonlinear vibrational spectroscopy. *The Journal of Physical Chemistry B*, 2009, 113 (36): 12169-12180.

[48] Hallock K J, Lee D K, Ramamoorthy A. MSI-78, an analogue of the magainin antimicrobial peptides, disrupts lipid bilayer structure via positive curvature strain. *Biophysical Journal*, 2003, 84 (5): 3052-3060.

[49] Giacometti A, Ghiselli R, Cirioni O, et al. Therapeutic efficacy of the magainin analogue MSI-78 in different intra-abdominal sepsis rat models. *Journal of Antimicrobial Chemotherapy*, 2004, 54 (3): 654-660.

[50] Wang J, Paszti Z, Clarke M L, et al. Deduction of structural information of interfacial proteins by combined vibrational spectroscopic methods. *The Journal of Physical Chemistry B*, 2007, 111 (21): 6088-6095.

[51] Dempsey C E. The actions of melittin on membranes. *Biochimica Biophysica Acta*, 1990, 1031 (2): 143-161.

[52] Wang J, Chen X Y, Clarke M L, et al. Detection of chiral SFG vibrational spectra of proteins and peptides at interfaces. *Proceedings of the National Academy of Sciences of the United States of America*, 2005, 102 (14): 4978-4983.

[53] Neves S R, Ram P T, Iyengar R. G protein pathways. *Science*, 2002, 296(5573): 1636-1639.

[54] Cabrera-Vera T M, Vanhauwe J, Thomas T O, et al. Insights into G protein structure, function, and regulation. *Endocrine Reviews*, 2003, 24(6): 765-781.

[55] Chen X Y, Boughton A P, Tesmer J J G, et al. *In situ* investigation of heterotrimeric G protein beta gamma subunit binding and orientation on membrane bilayers. *Journal of the American Chemical Society*, 2007, 129 (42): 12658-12659.

[56] Dupre D J, Robitaille M, Rebois R V, et al. The role of $G\beta\gamma$ subunits in the organization, assembly, and function of GPCR signaling complexes. *Annual Review of Pharmacology and Toxicological*, 2009, 49: 31-56.

[57] Boughton A P, Yang P, Tesmer V M, et al. Heterotrimeric G protein $\beta1\gamma2$ subunits change orientation upon complex formation with G protein-coupled receptor kinase 2 (GRK2) on a model membrane. *Proceedings of the National Academy of Sciences of the United States of America*, 2011, 108 (37): E667-E673.

[58] Yang P, Boughton A, Homan K T, et al. Membrane orientation of $G\alpha_i\beta_1\gamma_2$ and $G\beta_1\gamma_2$ determined via combined vibrational spectroscopic studies. *Journal of the American Chemical Society*, 2013, 135 (13): 5044-5051.

[59] Mushegian A, Gurevich V V, Gurevich E V. The origin and evolution of G protein-coupled receptor kinases. *PLoS One*, 2012, 7 (3): e33806.

[60] Yang P, Glukhova A, Tesmer J G, et al. Membrane orientation and binding determinants of G protein-coupled receptor kinase 5 as assessed by combined vibrational spectroscopic studies. *PLoS One*, 2013, 8 (11): e82072.

[61] Ding B, Chen Z. Molecular interactions between cell penetrating peptide pep-1 and model cell membranes. *The Journal of Physical Chemistry B*, 2012, 116 (8): 2545-2552.

[62] Soblosky L, Ramamoorthy A, Chen Z. Membrane interaction of antimicrobial peptides using *E. coli* lipid extract as model bacterial cell membranes and SFG spectroscopy. *Chemistry and Physics of Lipids*, 2015, 187: 20-33.

Chapter 4 Microparticle Based Biochips' Preliminary Subsections

Xiangwei Zhao, Junjie Yuan, Zhongze Gu

4.1 Introduction

After the sequencing of human genome, there's a great demand for the fast and multiplex detection of proteins or nucleic acids. In comparison with traditional technologies, biochip technology provides the cutting-edge solutions, such as the DNA/protein microarray. However, the fact is that the strength of microarray technology is limited in practices due to its intrinsic technique shortcomings such as deficiencies in data reproducibility and poor flexibility. A flora of novel chip technology based on microparticles which is more flexible and practical grows up. In this chapter, the recent development of microparticle based biochips will be reviewed under the scope of bioelectronics, from both the material aspect and the instrumentation aspect.

4.2 Plain Microparticles

4.2.1 Protein microarrays

Microarrays are matrices with spots of different chemical compounds on a surface, the number of which ranges from a few dozen to up to several millions. DNA or protein microarray is fabricated with lithography or robotic spotter by immobilizing proteins or oligonucleotides onto a planar carrier surface such as glass and silicon to form microscopic spots. The detection scheme is illustrated in Figure 4.1 (a) (see the Color Inset, p.9). The signal of the analyte is detected by the detecting label, which is an enzyme, fluorescence dye or radioactive isotope. Although microarrays are marked by their high-density screening (multiplicity up to hundreds of thousands), they're still deficient in data and array reproducibility and the reduced reaction kinetics, high cost and poor flexibility are also their drawbacks[1, 2].The main reason is that the fixed planar arrangement of probes is diffusion-limited and the sensitivity is restricted by the low surface to volume ratio of the planar substrate[3]. One solution is to use the active flow in microfluidic chips to accelerate the reaction on the microparticle carriers[4-7]. And another alternative is to utilize the suspension array to release them into the bulk sample solution in order to realize approximately homogenous reaction [Figure 4.1 (b), see the Color Inset, p.9][8]. In this case, new encoding strategy should be employed so that different binding events in the same solution can be distinguished. For example, in the xMAP technology of Luminex Corp., microparticles are dyed with different fluorescence color and intensity in order to track the reactions[9].

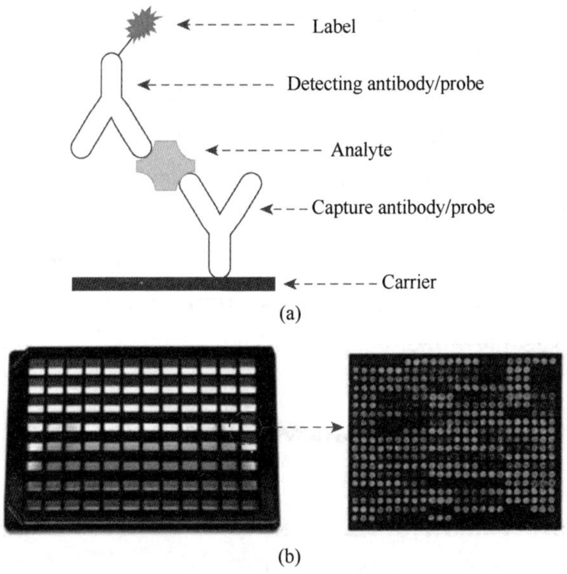

Figure 4.1 (a) Scheme of a typical solid-phase immunoassay. Capture antibody immobilized on a solid carrier, analyte, and labeled detecting antibody form sandwich immunocomplex. The signal of the analyte is detected by the detecting label, which is an enzyme, fluorescence dye or radioactive isotope. (b) 96 well glass bottom plates for multi-patient diagnostics. Each well contains high density arrayed features for multiplex assay, and the minimum spot and space size are 65 μm and 50 μm, respectively (Arrayit Corporation)

4.2.2 Magnetic microparticles

Magnetic microparticles usually consist of a superparamagnetic core surrounded by a polymer shell and can be manipulated using electromagnetic fields.

Magnetic microparticles with immobilized reagents can be conveniently located inside microfluidic devices without the need for microfabricated structures for physical retention in the device[10]. And furthermore, they can be used to vigorously mix a sample during washing and binding steps to overcome diffusion limitations[11-13]. Fuh et al. utilized anti-CRP labeled magnetic microparticles to react with the CRP samples in a well, and then anti-CRP labeled fluorescent microparticles were added to perform magnetic sandwich immunoassay [Figure 4.2 (a), see the Color Inset, p.10][14]. A magnetic field was applied to operate the micropartocles. For example, they were reconcerated after removing the unreacted species. This approach made it a nearly homogeneous condition, so all of these reactions were highly effective. Moreover, a new format for magnetic particle based immunoassays is developed and illustrated in Figure 4.2 (b) (see the Color Inset, p.10). In this approach, fluids were electrostatically controlled as discrete droplets on an array of insulated electrodes. Both noncompetitive and competitive immunoassays using the magnetic particles were realized within the platform. This new format allowed the realization the particle separation and resuspension, which is capable of removing greater than 90% of unbound reagents in one step.

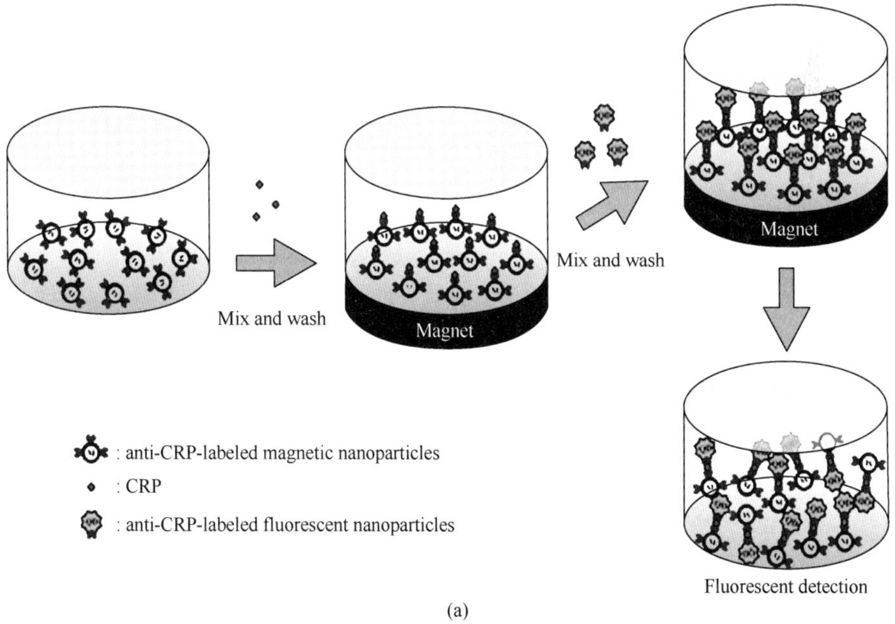

Figure 4.2 (a) Schematic view of sandwich reaction steps in a magnetic immunoassay; (b) Immunoassays and digital microfluidics

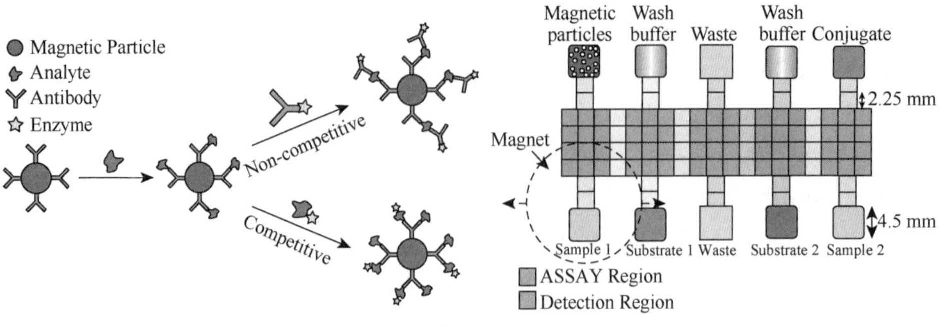

(b)

Figure 4.2 (Continued)

4.2.3 Charged microparticles

Electronic interactions could also be used to operate microparticles inside microfluidic devices. For instance, Gijs et al. developed an approach to use positively charged patterns inside a microchannel to self-assemble streptavidin-coated, negatively charged polystyrene microparticles for sandwich immunoassays [Figure 4.3 (a), see the Color Inset, p.10][15]. And charge switchable beads[16] are also demonstrated to automatically perform a series of essential processes for the fast but sensitive detection

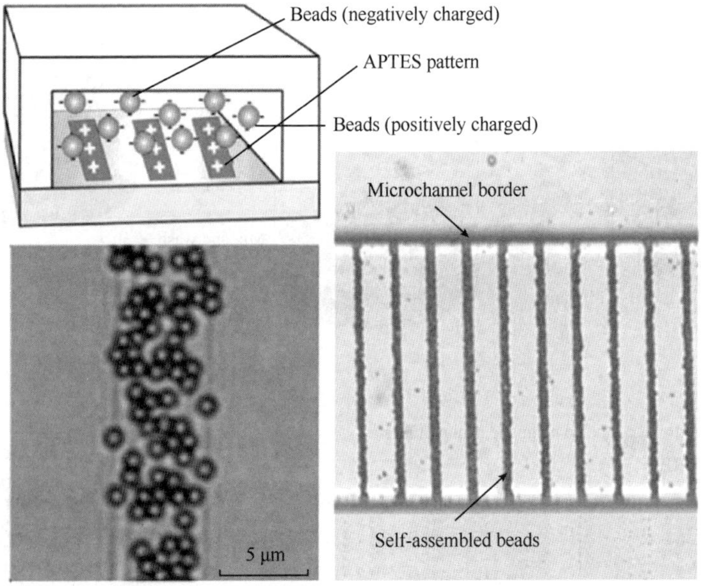

(a)

Figure 4.3 (a) Scheme of whole detection processes of murine NoV by microfluidic chip module; (b) Shape switchable microfluidic chip module consisting of charge switchable microbeads

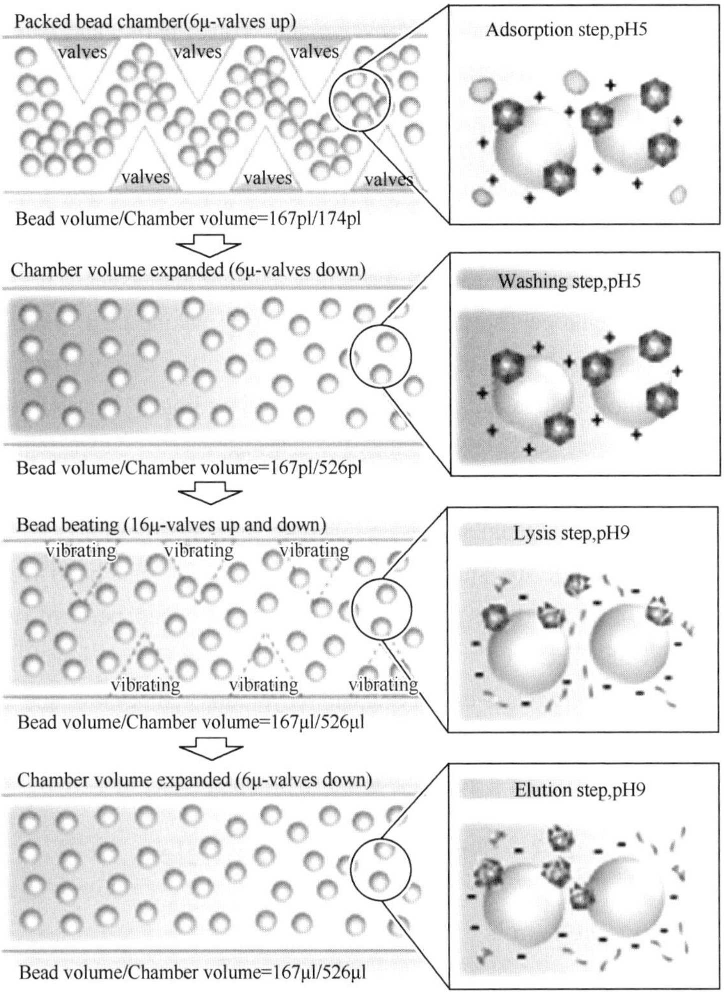

(b)

Figure 4.3 (Continued)

of norovirus in oysters, including sequentially concentration, lysing murine NoV, and amplifying RNA of murine NoV[Figure 4.3(b), see the Color Inset, p.11]. The sample solution was first loaded into a shape switchable sample preparation chamber consisting of charge switchable microbeads. Three different charge switchable glass beads were used in the assay. At pH of 5-7, murine NoV (negatively charged) was adsorbed on the microbeads (positively charged) by electrostatic force. Several pneumatic microvalves were equipped to change the volume of the sample preparation chamber and make the microbeads vibrated to lyse Murine NoV by mechanical forces. After that, the extracted RNA was transferred to the detection chamber by pneumatic microvalve and syringe pump, and was successfully amplified using Nucleic Acid Sequence Based Amplification (NASBA).

4.2.4 Hydrogel microparticles

Hydrogels are permeable to solvents as they comprise a porous structure, and can be functionalized with reagents like capture molecules. Analyte molecules can be bind to the chemically modified hydrogels specifically and efficiently, making them promising carriers for immobilized reagents. The pore size of the hydrogel plays an important factor in the diffusion of molecules through the porous structure of a hydrogel. Agarose gels with pore sizes of 45-620 nm have been developed for gel electrophoresis[17, 18]. Gels with larger pores can be produced by cryogelation. This technique utilizes the phase separation of water from dissolved reagents at subzero temperatures. In this state, the water molecules form ice crystals to work as porogens while the surrounding reagents crosslink and build a macroporous structure. Macroporous gel particles produced from polyvinyl alcohol performed pore sizes of up to 150 μm and functionalized with ligands for sorption of Cu (II) ions[19]. However, a disadvantage of some hydrogels is that drying and rehydration can be slow, and the processes may change the morphology or pore size of the hydrogel. Hence, devices containing such hydrogel compounds need to be stored in a humid environment after fabrication.

4.3 Encoded Microparticles

4.3.1 Encoding the labels

1. Förster resonance energy transfer encoding

Homogeneous reaction in solution phase simulates the state of nature interaction like antibody-antigen binding and DNA hybridization *in vivo*, comparing to the binding events happening on the 2D solid carrier and liquid interface. The process reacts rapidly, operates conveniently and is potentially compatible with *in vivo* or real-time applications. Distance-dependent fluorescence quenching by Förster resonance energy transfer (FRET) are usually utilized to probe homogenous binding events. And by the virtue of fast-developing nanomaterials, nanogold, quantum dots (QDs), nanowire and single wall carbon nanotubes are studied and applied in FRET to increase the transfer efficiency[20-24]. Fan and Song et al. designed a FRET system based on gold nanoparticles as superquencher and multiple fluorescence dyes as encoding labels of aptamers, which are oligonucleotides that can bind specifically with analytes like ligands[25]. A multiplex assay is illustrated in Figure 4.4 (a) (see the Color Inset, p.11), at first, three aptamers, including anti-adenosine aptamer, potassium- specific Gquartet and anti-cocaine aptamer, are encoded by different fluorescence dyes at the 5′ end respectively. And gold nanoparticles are modified with multiple 3′-thiolated oligonucleotides, the sequences of which are complementary to those aptamers. After they form complex nanostructures by hybridization, the dyes are in close proximity to the gold nanoparticles which produces FRET. The aptamers will dissipate from the complex and bind with the analytes while in the presence of adenosine (A), potassium, and cocaine because of the higher affinity between them, and therefore fluorescence will be detected in a signal-on mode.

Meanwhile, by this means or using grapheme oxide as a selective quencher for charged single strand DNA, multiplex DNA detection with sensitivity in nM and pM can also be realized [Figure 4.4(b), see the Color Inset, p.11][26, 27]. Apart from the fast

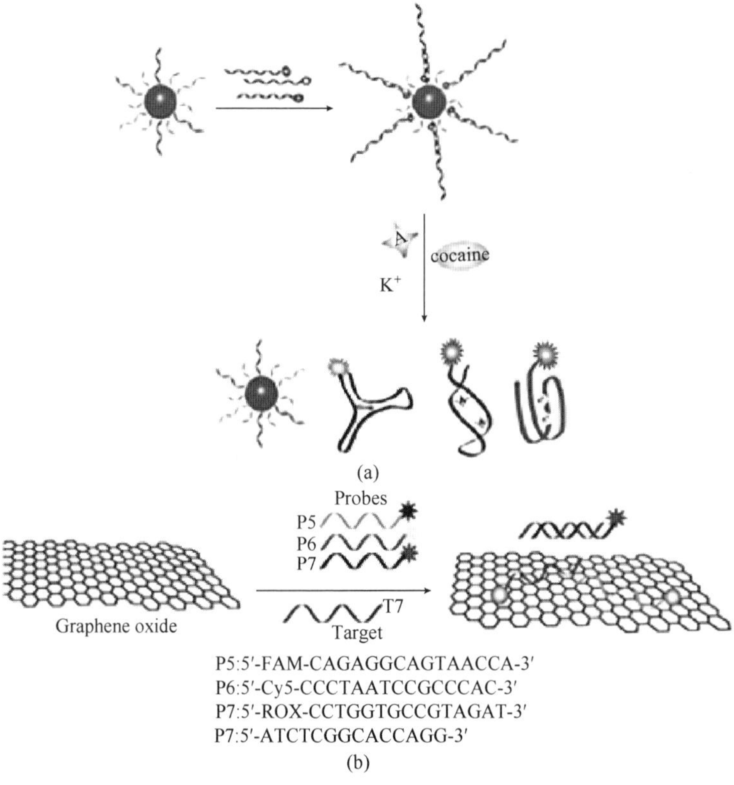

Figure 4.4 (a) Encoded aptamers for the multiplex signal-on detection of adenosine (A), potassium, and cocaine; (b) scheme for multicolor DNA analysis based on grapheme oxide as a selective quencher for charged single strand DNA

binding kinetics, perhaps the greatest strength of this encoding method and multiplex detection is the manner of mix-and-detection without washing steps for the signal-on mode, which saves mounts of time and holds a high throughput potential. Although, this encoding strategy can be used to multiplex protein detection, multiplicity is poor resulting from the limited number of optional aptamers and non-overlapping fluorescence dyes. In addition, multiple light wavelengths are needed to excite and detect the encoding fluorescence dyes.

2. DNA nanobarcodes

Lately, Luo's group have synthesized the dendrimer-like DNA (DL-DNA) nanostructures (DNA nanobarcodes) with diameter less than 30 nm[28, 29], which is assembly through hybridization of DNA strands from Y-shaped DNA (Y-DNA) as building unit. The unit has two ends labeled with different fluorescence dyes and one sticky end used for conjugation with probe molecule [Figure 4.5 (a), see the Color Inset, p.12]. Both dye number and dye type can be precisely controlled to generate multicolor fluorescence intensity-encoded nanobarcodes in the method. As Figure 4.5 (b) and Figure 4.5(c) (see the Color Inset, p.12) demonstrated, biotinylated probe oligonucleotides complementary to the target DNA sequences were bound to avidincoated polystyrene beads. The mixture of beads coated with different probe sequences were then reacted with the target strands. Fluorescent nanobarcodes attached to oligonucleotides complementary to part of each target sequence are then hybridized to the target sequences attached to the bead. Each bead bounded with targets has a particular combination of fluorescent dyes. And their intensities, the fluorescence spectrum of which indicates the DNA sequence on its surface. The label encoding strategy was applied in a 4-plex assay for the detection of DNA sequences from the pathogenic organisms *B. anthracis*, Francisella tularensis, Ebola virus and SARS coronavirus with detection limit low to 620 attomol. And the beads were successfully decoded by using fluorescence microscope or two-color flow-cytometry. The authors also showed their usage in blotting-based detection (southern, northern and western) through gel imaging system. Compared with conventional multicolor label encoding methods, the reaction kinetics of this way is promoted by using beads as carriers. In addition, the coding capacity (C) of nanobarcode is determined and enlarged by the labeled branch number (P) and the color number (L), and calculated by the following formula: $C= (P+L-1)!/[P!\ (L-1)!]$, where P is determined by the generation number (n) of DL-DNA ($P=3\times2^n$)[29]. Theoretically, 325 different nanobarcodes can be produced by three fluorescent colors and a third generation (G3) DL-DNA. However, in fact, factors such as FRET, choices of dyes, equipment sensitivity and signal to noise ratios, and detection methods may decrease the number of codes that can be distinguished. Not only will increasing the generation number (n) enlarge the capacity exponentially, but the decoding readout also becomes more expensive and complex.

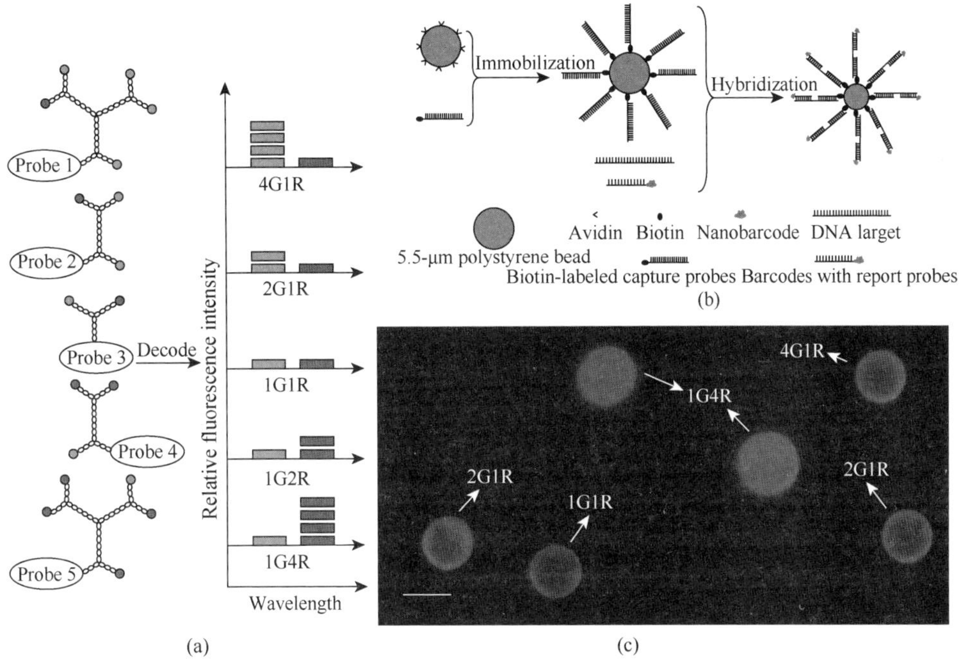

Figure 4.5 (a) Schematic illustration of barcoding of the DNA dendrimer fluorescent labels by fluorescence intensity ratio; (b) DNA nanobarcode, DNA target and polystyrene microbeads form sandwich structure in multiplex detection; (c) four targets were detected using DNA nanobarcodes and microbeads. Scale bars, 5 μm

3. Nanostring encoding

Nanostring, as another similar label encoding strategy, which based on combination of fluorescence dyes was proposed by NanoString Technologies Inc[30, 31]. In this strategy, unique pairs of reporter probes and capture are designed and prepared to detect target mRNAs at first. As shown in Figure 4.6 (a) (see the Color Inset, p.12), tripartite structures are formed and comprised of a target mRNA bound to its specific reporter and capture probes after a homogenous hybridization reaction with total RNA in solution. The single strand DNA backbone is annealed to a unique pool of seven dye-coupled RNA segments (nanostring) corresponding to a single code of the target mRNA. After affinity purification, unhybridized reporter and capture probes are removed, and the remaining complexes are washed across a streptavidin coated surface and captured by biotinylated oligonucleotides annealed to the 3′ repeats. Then, each complex in the solution are oriented in the same direction when applied an electric field. The complexes are then immobilized in an elongated state [Figure 4.6 (b), see the Color Inset, p.12] by biotinylated oligonucleotides annealed to the 5′ repeats and imaged [Figure 4.6 (c), see the Color Inset, p.12] with a fluorescence microscope. All target molecule of interest could be identified by the color code. The level of expression is measured by counting the number of codes for each mRNA.

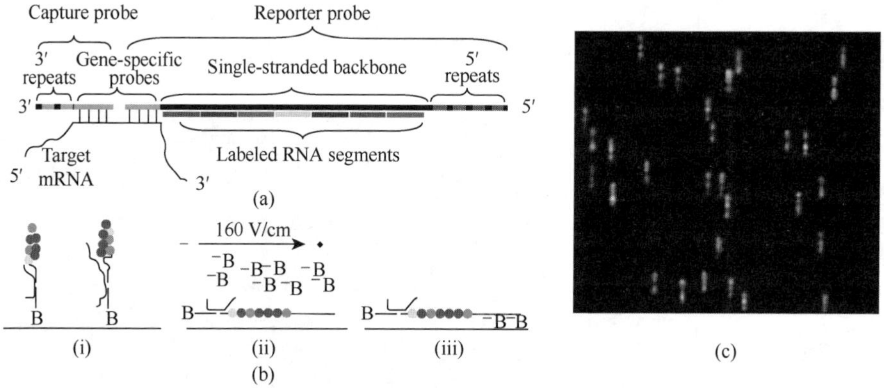

Figure 4.6 Illustration of nanostring encoded assays. (a) Complex structures formed after hybridization of a target mRNA, its specific reporter and capture probes; (b) biotinylated 3′ and 5′ repeats are used to fix the starched tripartite structure on a streptavidin coated slide; (c) the nanostring codes are imaged with a fluorescence microscope equipped with high resolution and magnification lens

The number of effective codes is determined by both the length of the DNA backbone and the minimum spot size that can be resolved under current imaging conditions. The authors used seven backbone positions [Figure 4.6 (a), see the Color Inset, p.12] and four colors to generate nanostring codes to minimize spectral overlap of fluorescence dyes during imaging. The encoding capacity of the strategy is therefore 4^7=16384 codes, which is a relatively large encoding capacity suitable for big scale gene expression assays. Their high sensitivity (0.1-0.5 fM detection limit) up to real-time PCR is also demonstrated. However, the main drawbacks are the complex assay process and the long detection time (up to 6 h) due to slow and high resolution codes scanning, which may lower its analysis throughput at a large extent.

4. Surface enhanced Raman spectroscopy encoding

The weakness of using fluorescence dyes in molecule labeling or encoding is obvious, such as photobleaching, spectral overlapping, narrow excitation with broad emission profiles and multiple excitation requiring in multiplexed assays. Hence, alternative labels are in pursuit all along, among which are Raman dyes or tags. They can be excited at any wavelength and also have quite stable unique spectral fingerprints containing high information content with narrow peak widths (20 cm^{-1}). Despite the low intensity of Raman spectra since the typical Raman cross section is in the order of 10^{-30} cm^2/molecule, it can be overcome and increased million-to trillion-fold when Raman active molecules is in very close proximity to roughened noble metals like gold or silver nanoparticles[32, 33]. For example, commercialized Nanoplex™ biotag of Nanoplex technology, latterly acquired by Oxonica limited, is composed of Raman active molecule tagged gold nanoparticles encapsulated with silica shells [Figure 4.7 (a), see the Color Inset, p.13][34]. Irradiation of these tags with monochromatic light yields the SERS spectrum of the reporter. Simply employs a different reporter molecule can achieve different encode tags. It is possible to create many distinct and simultaneously quantifiable tags results from narrow spectral features and large spectral window of SERS. In a typical multiplex assay, the magnetic beads, analytes and biotags immobilized with capture probes [Figure 4.7 (b), see the Color Inset, p.13] form two-particle sandwich complexes, which are then concentrated and then detected at a specific location within the reaction vessel in the presence of the sample matrix and excess free detection particles. This assay permits no-washing, fast homogenous reaction, multiplexed detection and quantification [Figure 4.7 (c)-Figure 4.7 (f), see the Color Inset, p.13] of DNA sequences and proteins. This encoding technology has great potentials in robotic high throughput screening, in miniaturized point-of-care and field applications such as lateral flow immunoassays. In the presence of biological matrices such as serum, plasma and whole blood, the protein detection sensitivity can reach pg/mL according to the company's data. However, the encoding capacity of Nanoplex™ biotag is determined by the level of different spectral precisions required for specific applications and the cost of the decoding instrument is around 10 dollars for low-cost handheld decoding readers.

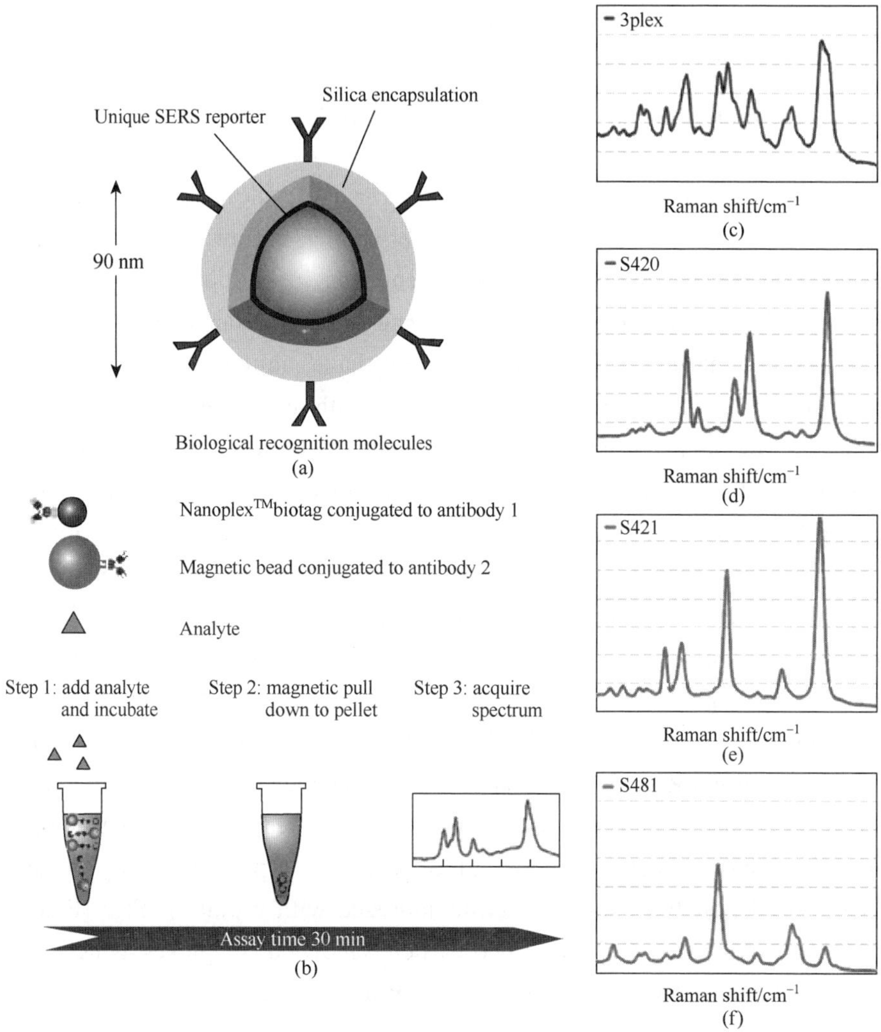

Figure 4.7　(a) Diagram of Nanoplex™ biotags with gold core; (b) multiplex assay procedure using Nanoplex™ biotags as encoded labels and magnetic beads as carriers; (c) composite spectra of three different tags as measured in a multiplexed assay; (d)-(f) spectra of each tag quantitatively deconvoluted by proprietary software (www.oxonica.com)

According to theoretical calculation, their aggregations can signally increase the SERS signal intensity by several orders of magnitude while compared with single noble metal nanoparticles[35, 36]. Mirkin group has applied SERS encoding in ultrasensitive and multiplex detection of DNA[37]. Gold nanoparticle probes are coated with Raman-dye-labeled oligonucleotides to achieve spectroscopic codes for individual targets of interest in this approach. Silver staining catalyzed by gold confirmes the presence of the target and then the amplified Raman signal of the dye. The unoptimized detection limit of this method is 20 femtomolar for DNA, and only single-wavelength laser radiation is needed to scan a highly multiplexed array with numerous target-specific Raman dyes, showing great values for gene express profiling. However, the aforementioned Raman labels still suffer from small Raman scattering cross-sections and degradation under long time laser irradiation.

5. Bio-barcode

To overcome the drawbacks associated with microarray technology such as low reaction kinetics and low sensitivity, Mirkin et al. have utilized bio-barcode assay (BCA)[38-40]. They employs oligonucleotides as bar codes for protein analytes or multiple target DNA. In a typical assay (Figure 4.8, see the Color Inset, p.13), gold nanoparticles are functionalized with bar-code DNA composed of target recognition sequence and universal sequence, and magnetic microparticles functionalized with sequences that capture a different target region. In the presence of target DNA, the gold nanoparticles and the magnetic microparticles form sandwich structures which are magnetically separated from solution. The bar-code DNA strands are then released by ligand-exchange with dithiothreitol (DTT) and detected by using the scanometric approach which is approved by Food and Drug Administration (FDA) of USA. During the scanometric detection, the released strands form sandwich structures again with universal sequence specific probes on other gold nanoparticles and target-specific capture probes on a microarray. The detection limits could be as low as 500 zM (10 strands in solution) with silver staining catalyzed by gold nanoparticles. Since the kinetics of the target binding process can be controlled by adjusting probe concentrations, the method obviates the need for PCR amplification in usually DNA detection regardless of target concentration. Depending upon capture antibody and background signal, the same scheme can also be adapted for multiplex protein detection with limits of detection that are 4-6 orders of magnitude lower than ELISAs. Though unlimited barcodes can be synthesized for virtually any target of interest, for example, there can be 4^{10} codes for a 10-mer oligonucleotide sequence. The ultra-sensitivity resulted from the unique signal amplification mechanism which is detected by scanometry is the main advantage of the BCA mainly. The multiplexed bio-barcode assay can be performed within a 96-well-plate in a facile and high throughput manner significantly. Integrated with microfluidic chips, it can also realize more miniaturized and labor-saving assays[41, 42].

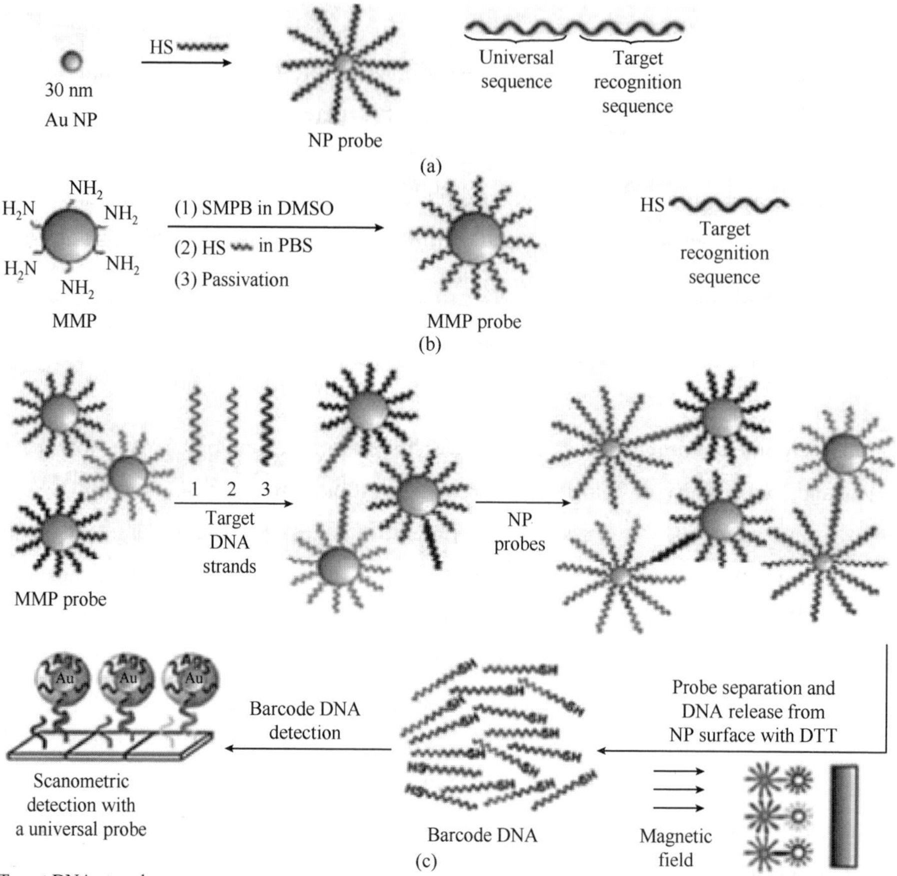

Target DNA strands:

HBV: 5′TTGGCTTTCAGTTAT-ATGGATGTGGTA-3′
VV: 5′AGTTGTAACGGAAGA-TGCAATAGTAATCAG-3′
EV: 5′-GGAGTAAATGTTGGA-GAACAGTATAACAA-3′
HIV: 5′-AGAAGATATTTGGAATAA-GATGACCTGGATGCA-3′

un=TACGAGTTGAGAATC
PEG_{15}=$(\cdot OCH_2CH_2\cdot)_6$

MMP DNA strands:

MMP-HBA: 5′ATAACTGAAAGCCAA-A_{10}-SH-3′
MMP-W: 5′-TCTTCCGTTACAACT-A_{10}-SH-3′
MMP-EV: 5′-TCCAACATTTACTCC-A_{10}-SH-3′
MMP-HIV: 5′-TTATTCCAAATATCTTCT-A_{10}-SH-3′

Barcode DNA strands:

B-HBV: 5′-HS-un-PEG_{10}-TACCACATCATCCAT-3′
B-VV: 5′-HS-un-PEG_{10}-CTGATTACTATTGCA-3′
B-EV: 5′-HS-un-PEG_{10}-TTGTTGATACTGTTC-3′
B-HIV: 5′-HS-un-PEG_{10}-TGACTCCAGGTCATG-3′

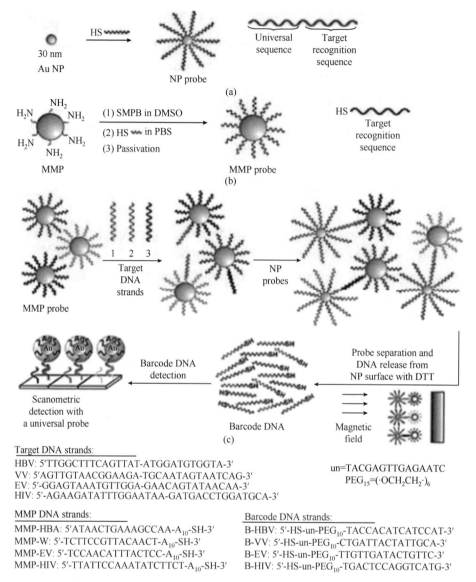

Figure 4.8 The procedure of multiplex protein detection by bio-barcode

6. Electrochemical label encoding

Electroactive labels are commonly used in the detection of bio-binding events on solid electrodes in the same way as fluorescence labels in optical detection since electrochemical detection has obvious merits of simplicity, low-cost, well-integratable, size-scalable. For instance, not only fluorescence but also quantum dots yield well resolved and highly sensitive stripping voltammetric signals. Wang et al. demostrated an electrochemical coding technology for the simultaneous detection of multiple DNA targets which based on QDs tags with diverse redox potentials[43]. Distinct DNA hybridizations hence can be detected by stripping voltammetric signatures of the encoding QDs that can be systematically tuned by changing their compositions. In the technology, a sandwich assay in which target capture strands are attached to magnetic beads was employed (Figure 4.9, see the Color Inset, p.14). It is labeled with oligonucleotide-functionalized QDs for the target, once the target DNA hybridizes to the capture strand. Magnetic field can easily separate and transfer the sandwich system to an electrochemical cell where stripping voltammetric signals of encoding nanoparticles are detected. Although this method is amenable to multiplexing and quantification since the magnitude of the stripping peak corresponds to the concentration of target DNA, the encoding number is also limited as in the fluorescence encoding. By taking advantages of electroactivity and highly sensitive stripping response of the adenine (A) and guanine (G) nucleobases, the Wang group also put forward multiplex protein electrochemical detection[44]. The target proteins are sandwiched by magnetic bead probes and polystyrene microsphere probes coated with oligonucleotide barcodes in a same manner as bio-barcode assays, followed by alkaline release and acidic dipurinization of barcodes, and adsorptive chronopotentiometric stripping measurements of the free nucleobases. It is possible to create a larger number of identifiable oligonucleotide barcodes for electrochemical immunoassays by designing oligomers with different predetermined A/G ratios in this case. In contrast favorably with values obtained by ELISA, the coupling of carrier-loaded amplification of barcodes with the preconcentration feature of electrochemical stripping detection leads to extremely low detection limits down to pg/mL for protein molecules, but still higher than bio-barcode assays mentioned above.

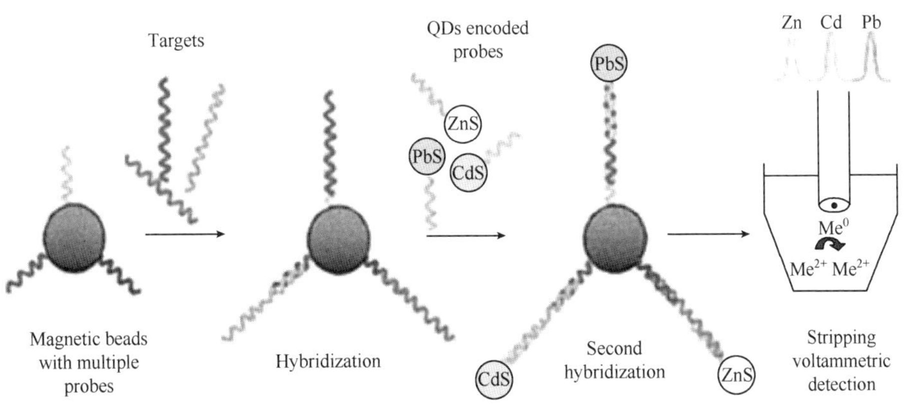

Figure 4.9 Multiplex electrochemical assay by QDs encoded probes

4.3.2 Encoding the carrier

Compared with the label encoding, plenty of techniques have been proposed and extensively studied for encoding the carriers, such as electronic encoding, fluorescence dye/QDs encoding, barcode encoding, graphic encoding[45, 46]. However, it is still not a perfect encoding technology that can fulfill the multiplex and high throughput requirements of biomedical applications until now.

1. Fluorescence color encoded nanoarrays

Despite some drawbacks mentioned above of QDs or fluorescence dyes as labels, fluorescence color is the most widely and well-established employed encoding elements for their convenience both in encoding and decoding. Microspheres are dyed with multiple fluorescence dyes or QDs in different intensities to enlarge the encoding capacity[8, 29, 30, 47]. However, it is hard to control the dyeing progress and the repeatability is very poor. To construct 2D lattices as nanoarrays, Yan et al. used DNA tile, a programmable building block for self-assembly of micro- and nano-architectures based on the simple rules of Watson-Crick base paring[48]. As illustrated in Figure 4.10 (a) (see the Color Inset, p.15), the nanoarray is composed of three basic DNA tiles. A1 and A2 tiles are dyed with "red" Cy5 and "green" Rhodamine Red-X as encoding tiles, while probe tiles B [Figure 4.10 (a) and Figure 4.10 (b), see the Color Inset, p.15] are dyed with "blue" Alexa Fluor 488 for single strand probes, which dangle out of the array plane through base-paring with the anchor strands that are single-stranded extensions of one of the oligos within the DNA tile. They are associated with each other alternatively by the sticky ends of three kind tiles to form 2D lattices. The blue dyed single strand DNA will bind with target molecules when they appear and dissociate from the encoding nanoarrays, enabling a signal-off mode detection mechanism. As shown in Figure 4.10 (c) (see the Color Inset, p.15) and Figure 4.10 (d) (see the Color Inset, p.15), four nanoarray probes were used in bovine serum during an encoded detection. The advantage of this encoding method is that the codes coming from combination of different dye ratios can be precisely designed with a high reproducibility and yield for the base pairing rules, in addition to washing-free due to the signal-off mode and the fast homogenous reaction speed (15-20 min). The encoding capacity is governed by the similar mechanism of DNA nanobarcodes since the maximum encoding number is determined by the number of different intensity levels ("red" and "green") that can be distinguished by the fluorescence microscope detector in image-decoding. If no image processing software is available, a conclusion that the decoding is toilsome due to the stacking or overlap of nanoarrays could get from Figure 4.10 (c) (see the Color Inset, p.15) and Figure 4.10 (d) (see the Color Inset, p.15).

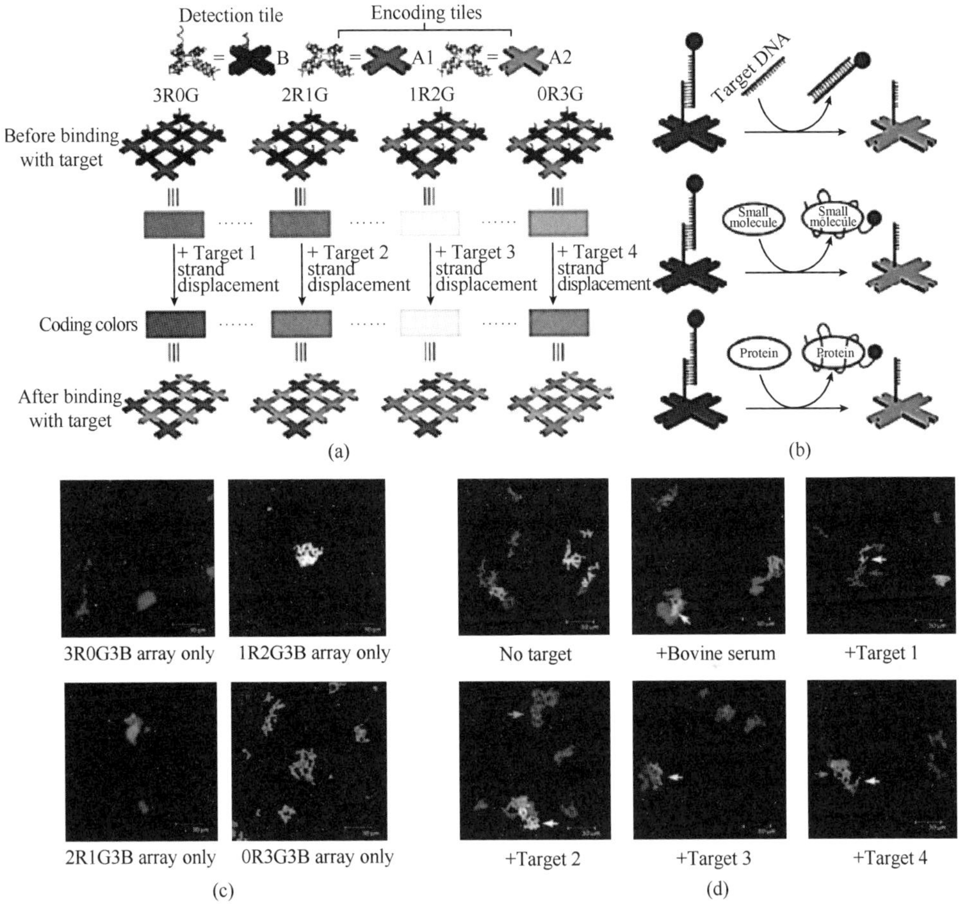

Figure 4.10 (a) Encoding strategy of nanoarray composed of different ratios of red and green fluorescence dyed DNA tiles, and blue fluorescence dyed detection tiles. (b) The signal-off mechanism of the detection tiles using oligos as probes for DNA and RNA, and aptamers for proteins and small molecules. (c) Four fluorescence color encoded probes: probes 1 (3R0G3B) and 2 (2R1G3B) are the complementary sequences of the two virus DNA sequences, severe acute respiratory syndrome virus (SARS) and the human immunodeficiency virus (HIV); probes 3 (1R2G3B) and 4 (0R3G3B) are the aptamer sequences that can specifically bind to human R-thrombin and adenosine triphosphate (ATP), respectively. (d) Detection of oligos (4 μM) in bovine serum using probes in (c). In the present of only bovine serum, greenish yellow (1R2G) encoding color was revealed since a micromolar concentration of thrombin exists in serum. SARS DNA (target 1), HIV DNA (target 2), and the complementary strand to ATP aptamer (target 4) can all be detected without any ambiguity. Target 3 is the complementary strand to the thrombin binding aptamer. The arrows point to the appearance of the encoding colors. Scale bar 30 μm

2. Structure color encoded microcarriers

Structure colors generated due to the reflection of periodic nanostructures of dielectric material called photonic crystals, which can prohibit the propagation of electromagnetic waves in a range of frequencies[49]. The color or reflection wavelength is determined by system refractive index and the structural period of the dielectric system according to the Bragg's law, $m\lambda=2nd\sin\theta$. Therefore, they are tunable and resistant to photo destroy like photo quenching or photo bleaching, which make them a kind of ideal color encoding strategy. The first application as carriers in the encoded DNA and protein detection is explored by Sailor et al[50]. In their work, porous silicon photonic crystal flakes were fabricated by galvanostatic anodic etch of crystalline silicon wafers. The porosity and thickness of the porous silicon were controlled by the current density, the composition of the etchant solution, and the duration of the etch cycle, different colors were generated as reflection peaks in the VISNIR region with a full-width at half-maximum (FWHM) of 11 nm by varying a computer-generated pseudo-sinusoidal current waveform. And it is much smaller than emission spectra of QDs (20 nm) and fluorescence dyes (more than 50 nm). And etching multiple porous layers on a single flake could generate more encoding spectra. For example, a ten-bit binary code is composed of combinations of different reflection intensities and peaks[51, 52]. Therefore, it is possible to generate 4^{10} codes for ultrahigh level of multiplexing. But the anisotropic 1D microparticles still need to be properly oriented to avoid standing or stacking in the decoding process.

Gu et al. have solved this problem by doping pearl pigments, mixed polystyrene microspheres with one kind of low cost 1D photonic crystal and rendering them with uniform photo stable colors[53][Figure 4.11(a) (see the Color Inset, p.15)]. Or use isotropic 3D photonic crystals such as photonic crystal beads. The bottom-up colloidal self-assembly maybe the most cost-effective way to this purpose while compared with top-down microfabrication technology. We have also fabricated colloidal photonic crystal beads (PCBs) by droplet-templated self-assembly of monodispersed nanoparticles [Figure 4.11 (b) (see the Color Inset, p.15) and Figure 4.11 (c) (see the Color Inset, p.15)]. Uniform water-in-oil droplets of colloidal solution were generated with a microfluidic device. In the droplets silica or polymer colloids formed ordered lattices by van der Waals force and capillary force during the dry process[54-56]. Then, the mechanical strength of hence formed structure was enhanced by sticking or sintering of the nanoparticle intercourse, therefore they can bear any collision during the analysis applications[57]. The color and size of the opal photonic beads [Figure 4.11 (d) (see the Color Inset, p.15) and Figure 4.11 (e) (see the Color Inset, p.15)] can be well controlled by the nanoparticle diameter and the microfluidics. The nanoparticle assemblies were inverted with the interstices between them[58] to fabricate inverse structure PCBs [Figure 4.11 (f) (see the Color Inset, p.15) and Figure 4.11 (g) (see the Color Inset, p.15)]. When they are used as microcarriers in encoded detection, no orientation is needed owing to the isotropic morphology. Theoretically, by self-assembling of one size of nanoparticles in the range of 100-500 nm, one reflection spectrum or color in the UV-VIS-NIR range can be simply gained. However, the FWHM of the beads (~25 nm) is relatively larger, which make the encoding number of photonic beads limited less than 20 in the visible range.

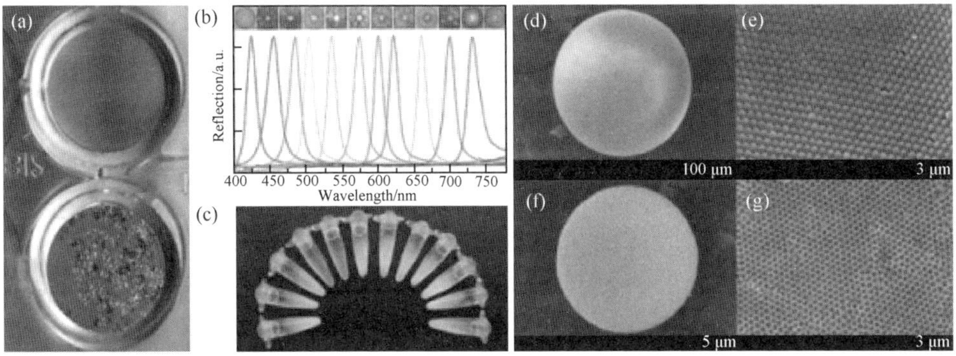

Figure 4.11 (a) 1D photonic crystal doped polystyrene microspheres in a 96-well plate; (b) and (c) reflection spectra of 12 kinds of colloidal photonic crystal beads in a centrifuge tube; (d) and (e) SEM image of opal colloidal photonic crystal bead and its surface structure; (f) and (g) SEM image of inverse opal colloidal photonic crystal bead and its surface structure

For enlarge the encoding capacity, complex encoding method was introduced by combination of QDs and PCBs with short emission spectra and broad exciting[59, 60]. For example, the QD color and intensity are precisely controlled since QDs are immobilized on the surface of PCBs by LbL (layer by layer) assembly [Figure 4.12 (a), see the Color Inset, p.16]. In this case, the encoding number is $k\ (n^m-1)$, in which m and k is the QDs and PCB color number respectively, and n is the intensities of the QDs that could be well distinguished. Therefore, under conservative estimation, the encoding number will be as large as $10\ (2^{10}-1)=10230$ if 2 intensities and 10 PCB and QD colors are used. To enlarge the encoding number, a more convenient way combing carrier encoding and label-encoding is illustrated in Figure 4.12 (b) (see the Color Inset, p.16). The code numbers will be their product (to be published), that is 100 codes for 10 PCBs and QDs, which is equal to the maximum of Luminex 100/200.

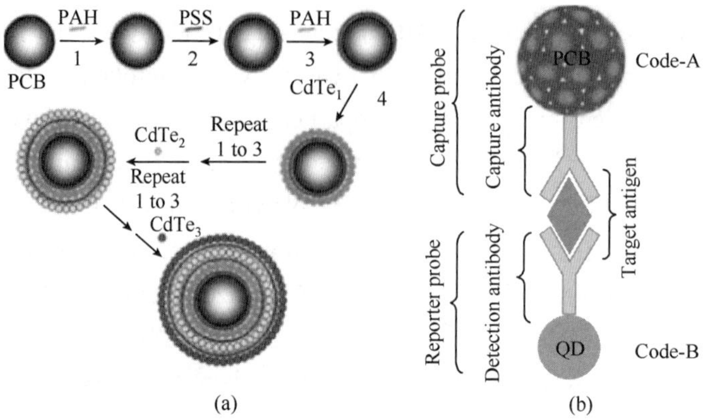

Figure 4.12　(a) Scheme of complex encoding by assembly of QDs on photonic crystal bead (PCB); (b) scheme of combination of label-encoding and carrier-encoding by QDs and PCB, respectively

Another advantage of PCBs as encoded microcarrier is that the high surface to volume ratio (SVR) resulting from the porous and ordered nanostructure contributes a lot to enhanced assay sensitivity. In our research, we demonstrated that the detection limit of assays based on microarray is about 100-fold higher than that on PCBs for DNA detection. And the results of a multiplex detection of four tumor markers indicated that carcinoembryonic antigen (CEA), carcinoma antigen 125 (CA 125) and carcinoma antigen 19-9 (CA 19-9), α-fetoprotein (AFP), were assayed with detection limit of 0.95 ng/mL, 0.99 U/mL, and 2.30 U/mL, 0.68 ng/mL, respectively, comparable with those clinically used high sensitive chemiluminescence ELISA assay Kits[61]. Moreover, the photonic bandgap of PCBs also enhances the signals for the detection[60, 62].

Whereas, the encoded detections aforementioned all use QDs or fluorescence dyes as labels. Typically, as shown in Figure 4.12 (b) (see the Color Inset, p.16), two steps of binding are needed in the multiplex sandwich format immunoassays. Each step is followed by several times of washings in order to remove unreacted entities. Then encoding signals and detection label signals are interrogated and acquired respectively, as shown in Figure 4.13 (a) (see the Color Inset, p.16). While in high throughput screening, this label-involved detection will be costly and time-wasting difficulties and complexities on the atomization. And labeling the probe brings an additional cost in spite of the activity of the probe being affected possibly. Hence, label-free detection is preferred in multiplex and high throughput applications.

For photonic crystals, structure period or system refractive index changing caused by the biomolecular binding will lead to the shift of their refection peak, which makes photonic crystals a kind of promising biosensing material[63, 64]. Our research showed that inverse opal photonic crystal films and beads can be used as encoded carriers for multiplex biosensing. The key point is that if the pore surface is saturated with bound complex, the peak shift caused by binding events will stop. It means that the specific peak shift range determined by the original reflection peak of PCB carrier can be used to encode multiple label-free detection as long as the ranges will not overlap with each other, as illustrated in Figure 4.13 (see the Color Inset, p.16). Three tumor markers, CA125, CA19-9 and CEA using three encoding inverse opal silica PCBs showed their priority over label-involved ones in a multiplex label-free detection. Hydrogel inverse opal PCBs could be utilized for further improvements in encoded detection of DNA or small molecule proteins, onto the skeleton of which recognizing moieties were incorporated to enlarge the peak shift by double response of period and refractive index to binding events. However, still limited to the FWHM of PCBs, the lowest detection limit for DNA is 10^{-9} M and the largest peak shift is less than 20 nm. Larger peak shift preferred for high sensitivity will cut down the encoding number to less than 10 in the visible region for example, but higher multiplex and label-free detection become available since complex encoding withQDs can be

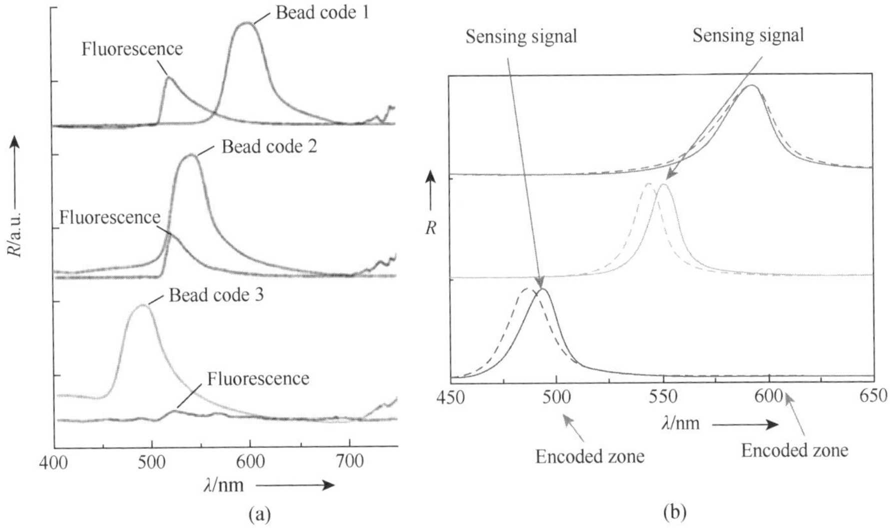

Figure 4.13 (a) Fluorescence and reflection spectra of three encoded photonic beads after multiplex detection FITC-tagged goat anti-human IgG and goat anti-rabbit IgG. Beads with codes 1, 2 and 3 were immobilized with human, mouse, and rabbit IgG, respectively. (b) Reflection spectra of three encoded photonic beads with molecular imprint of bovine Hb (hemoglobin), HRP (horseradish peroxidase), and BSA (bovine serum albumin), respectively. The cyan, green, and red dashed lines and solid lines are the spectra of the imprinted photonic beads before and after multiplex detection of Hb and HRP. The gray areas are the encoding zone of the beads

adopted to counteract it. Whether or not, multiplex and label-free detection based on encoded microparticles lowers the cost and simplifies the assay procedure for highly automated high throughput detections.

3. Bar-coded DNA tiles

Yan et al. proposed to use bar-coded for the detection of RNA without PCR amplifications[65]. Self-assembly of a circular single-stranded M13 viral DNA composed of 7249 bases construct the DNA tiles, and more than 200 short synthetic DNA strands are used as helper [Figure 4.14 (a), see the Color Inset, p.16]. Two features of tiles, the first is the bar-code at the top left corner used to index the tiles and orient them for the AFM imaging, the second is probes dangling out of the tiles and used to detect multiplex RNA targets. The probes are optimized to be immobilized on the edge and then are encoded by their position on the tile and the index of the tile since compared with at the inner positions, probes at the edges of the rectangle tiles have the highest binding efficiency. The strands form a stiff V-shaped structure upon hybridization of the target to a pair of half-probes [Figure 4.14 (a), see the Color Inset, p.16]. This local stiffening is readily sensed mechanically with an AFM cantilever and appears in the image as local high-spot [Figure 4.14 (b), see the Color Inset, p.16]. The probe is identified by retrieving the index of the tiles. As illustrated in Figure 4.14 (b) (see the Color Inset, p.16), the multiplex RNA detection demonstrated its minimum sample volume and ultrahigh sensitivity. Therefore, by virtue of the very simple assay procedure, the method appeals to the clinical applications. However, it's hard to be used for high-level multiplex detection although the bar-code has a high encoding capacity due to the signal readout by AFM takes a long time.

4. Nanodisk barcoding

Combing decoding and detection signal acquisition into one process is another method to simplify the detection process. For example, Mirkin et al. demonstrated that gold or silver nanodisk encoded microcarriers fabricated by "on-wire lithography"[66, 67]. First, nanowires of nickel with pairs of gold segments [Figure 4.15 (a), see the Color Inset, p.17] are synthesized by electroplating and then deposited randomly on a surface and covered with evaporated silica. After nickel is etched away, arrays of gold nanodisk pairs fixed along a silica "backbone" are gained. Then the nanodisk pairs acted as a binary barcodes. The capture probes are immobilized on the gold nanodisks when they are used in binding detection. And owing to the SERS of gold nanodisk pairs, the signal of the Raman labels from the captured targets will be enhanced. And hence, a visible code will be seen under the Raman microscope. As shown in Figure 4.15 (c) (see the Color Inset, p.17), the nanodisk pairs, which can be removed from the string or not to create 0 or 1 digits, generate a 5-bit binary barcode. Then, the encoding capacity of these nano-disk arrays is 2^5=32 codes. By using more disks in the array, the encoding capacity can potentially be increased. And using a number of different Raman-active dyes can also increase the numbers of encoding, and they can be identified by filtering the Raman images according to the unique Raman properties of each dye. In this method, the decoding and target signal acquisition are at the same time like the process on microarray since the codes will be seen only when there are target captured. And a multiplex DNA detection with concentration as low as 100 fM was successfully performed owning to the high sensitivity of SERS. However, the decoding interference due to the stacking or aggregation of the nanodisk array will be unavoidable when larger numbers of codes are used.

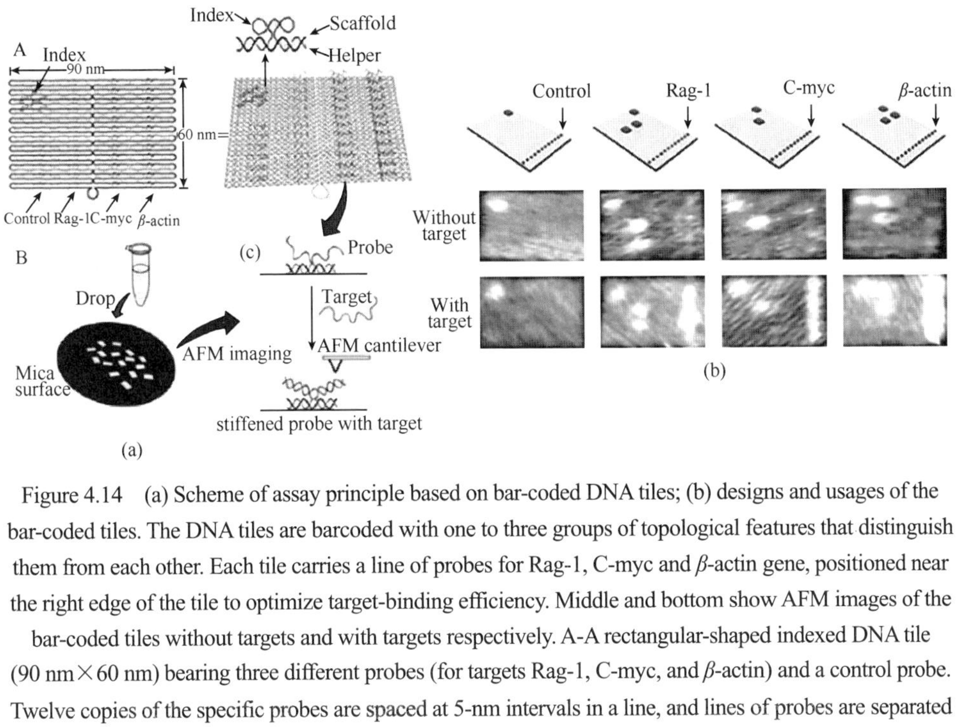

Figure 4.14 (a) Scheme of assay principle based on bar-coded DNA tiles; (b) designs and usages of the bar-coded tiles. The DNA tiles are barcoded with one to three groups of topological features that distinguish them from each other. Each tile carries a line of probes for Rag-1, C-myc and β-actin gene, positioned near the right edge of the tile to optimize target-binding efficiency. Middle and bottom show AFM images of the bar-coded tiles without targets and with targets respectively. A-A rectangular-shaped indexed DNA tile (90 nm×60 nm) bearing three different probes (for targets Rag-1, C-myc, and β-actin) and a control probe. Twelve copies of the specific probes are spaced at 5-nm intervals in a line, and lines of probes are separated by 20 nm. An index spot at the top left corner is used to orient the AFM image of each individual tile. B-Illustration of the operation process: probe tiles hybridized with targets are dropped onto the mica surface for AFM imaging. (c) Probe design and detection mechanism. A probe sequence is immobilized on a pair of neighboring helper strands extended out of the tile plane. Upon target hybridization, the double helix of the stiffer DNA-RNA duplex is readily detected with the AFM cantilever

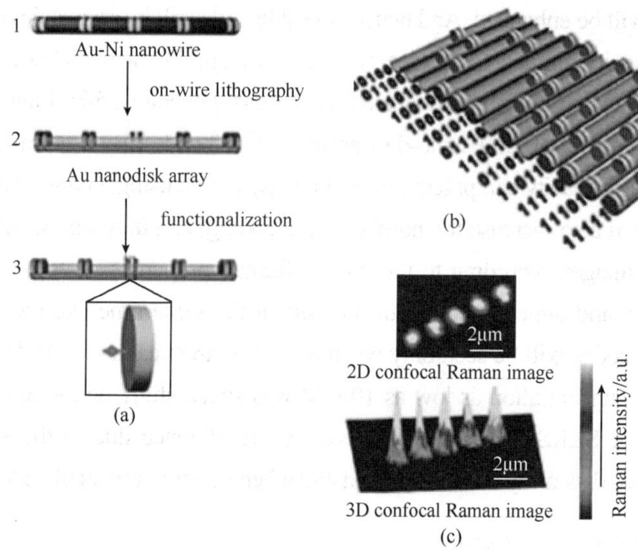

Figure 4.15 (a) Scheme of preparation of gold nanodisk arrays; (b) binary codes arranging sequences of “0” (nanodisk absent) and “1” (nanodisk present); (c) the binary code is read by observing the arrays using confocal Raman spectroscopy

5. Pattern encoded multifunctional microparticles

It is crucial to avoid decoding interference for large numbers of microcarrier codes, such as pattern, barcodes or graphic codes on anisotropic microcarriers like flakes or bar. Doyle et al. transferred the encoding and decoding processes into microfluidic chip in a high throughput way[68]. Figure 4.16 (a) (see the Color Inset, p.17) shows two laminar stream pass through a microfluidic channel together. Red stream contains PEG monomers with oligonucleotide probes attached for the analyte section of the particle and blue stream contains fluorescent labeled PEG monomers for the code section of the particle [Figure 4.16 (b) and Figure 4.16 (c), see the Color Inset, p.17]. The monomers are polymerized into hydrogel microparticles with dotted pattern as code adjacent to the target capturing part by irradiating with patterned high intensity UV light through a mask and lens. In detection, the microparticles are flowed-focus through a channel only slightly larger than the particle width in a microfluidic chip. Finally, all image sequences captured by an inverted fluorescence microscope are analyzed to determine the particle code and quantify targets [Figure 4.16 (f), see the Color Inset, p.17].

In this method, if it is not limited by the resolution of photo-polymerization and the size of the encoding section, the encoding capacity of the pattern can be infinity. The number is 2^{20} as a pattern of 20 dots used here. And multiple probes or probe gradients can be incorporated into the probe section [Figure 4.16 (d) and Figure 4.16 (e), see the Color Inset, p.17], which both multiplying the encoding number and allowing for comparison of several targets on a single particle or broadening the detection range of targets with fixed detection sensitivity. Another feature of this method is that the probes are incorporated into the encoded particles by simply copolymerizing the acrylate modified probes with hydrogel monomers, which makes the assays more convenient as it is applicable both for protein and DNA[68-70]. In addition, since the target label and pattern codes share the same fluorescence dye, the decoding process and target signal acquisition are completed simultaneously by imaging in a flow-focus microfluidic channel, which efficiently avoids the stacking of microcarriers. Owing to the high surface to volume ratio nature of the hydrogel polyme used herein, DNA oligomers of 500 attomoles can be detected without biotin-avidin-aided signal amplification. However, the large size of multifunctional particles [as can be seen from Figure 4.16 (d) and Figure 4.16 (e), see the Color Inset, p.17] will cost sample volume as large as sub-milliliter. It affects the reaction kinetics severely especially in high level multiplex.

Mixing the reaction solution is important in improving the reaction kinetics in microparticle based multiplex assays and enabling the facile separation and washing. Kwon et al. also manufactured pattern encoded microparticles for multiplex detection of DNA in a similar way[71]. However, in their method, magnetic function and structure color encoding also are incorporated. The dots in the encoding pattern were dyed with a single M-Ink material to enlarge the encoding capacity[72]. When applied a magnetic field, monodispersed superparamagnetic colloidal nanocrystal clusters (CNCs) aligned to form the photonic crystal structure, where the M-Ink color comes from. By using different magnetic intensity, which will change the lattice plane distances, the structure color can be easily turned. In Doyle's work, the colored dot pattern [Figure 4.17 (a), see the Color Inset, p.18] was formed color-by-color in a microfluidic channel in magnetic field by modulation the UV light with a digital micromirror device (DMD). For a ten-dot pattern with 8 colors, the encoding number in this method amplified to $8^{10}=2^{30}$. The microparticles can be easily oriented instead of using flow-focus like in Doyle's method while in multiplex detection. The best feature is that, the microparticles can be rotated at a fast rate during the assays to create microscale rotating stirrers (3D reaction) depending on the direction of the external magnetic-field lines [Figure 4.17 (b), see the Color Inset, p.18]. And this will overcome the kinetic problems associated with big flake particles. The oligonucleotide hybridization time was reduced to 10 min while compared with assays in Doyle's method without rotation operation. Also, the particles can be rotated to lie horizontally with respect to the vial surface, enabling solution exchange or code identification. However, due to the big size of the microparticles, the sample volume need is still high. In addition, the target label reading and color code reading are separated here, which means high detection complexity and more time-consuming.

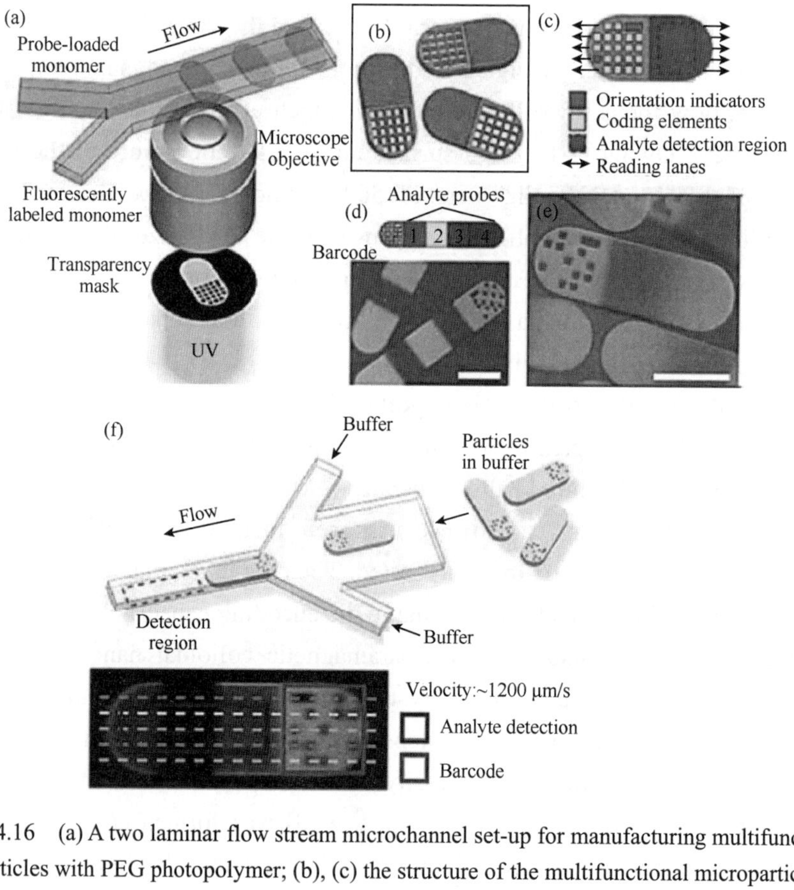

Figure 4.16　(a) A two laminar flow stream microchannel set-up for manufacturing multifunctional microparticles with PEG photopolymer; (b), (c) the structure of the multifunctional microparticles; (d), (e) overlap of DIC and fluorescence image of microparticles with multiprobe and gradient probe; (f) decoding microfluidic set-up for the multifunctional microparticles and a typical image of a microparticle flowing through the microfluidic channel, five dash lines indicate the fluorescence intensity measurement position used for decoding and target signal acquisition. Scale bars in (d) and (e) are both 100 μm

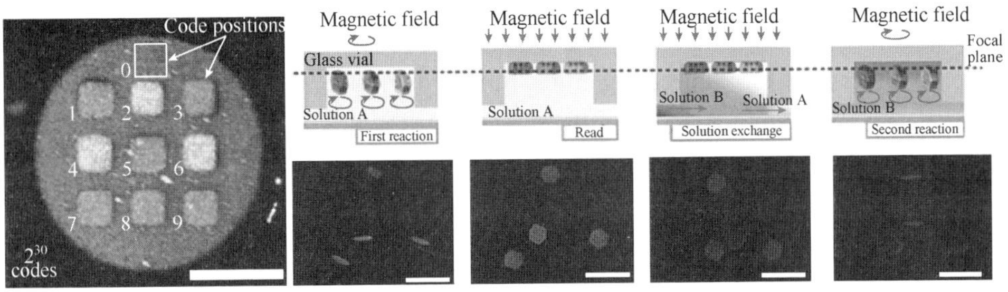

(a) (b)

Figure 4.17 (a) A colored pattern encoded magnetic microparticle; (b) 3D reaction and 2D reading scheme for a multistep reaction: first 3D reaction, 2D reading, solution exchange and second 3D reaction in a different solution. The scale bar is 100 μm in (a) and 500 μm in (b)

4.4 Microparticles and Chips

4.4.1 Materials for chips

Materials largely determine the properties of the microfluidic flow path. The flow rate, pressure, optical properties, wetting, adhesion of biomolecules and the cost of the microfluidic device are all related to the materials used and their fabrication. Original microfluidic devices were fabricated using microelectromechanical systems (MEMS), namely using photolithography and etching in a cleanroom. The first generation of microfluidic flow paths were fabricated in silicon and glass[73]. Since then, a mass of lab-on-a-chip devices have been fabricated in a wide range of materials, using different fabrication techniques[74].

1. Microfluidic flow paths

The microfluidic flow path is the combination of geometry and chemistry of the materials used to define the volumes inside which samples flow from an initial loading zone throughout the microfluidic device. Silicon microfabrication is a reliable technique for microfluidic devices using MEMS and microelectronic infrastructure. However silicon has some disadvantages. For example, it is opaque to ultraviolet and visible light, and impermeable to gases, which can be a problem when working with cells. Moreover, cleanroom fabrication is also expensive. Hence, it is more practical to make flow paths, valves and pumps in a compliant polymer, such as dimethylsiloxane (PDMS). Whitesides and colleagues established PDMS as a material of choice for fabricating microfluidic devices[74]. Nowadays, microfluidic channels are commonly etched in silicon and sealed with a PDMS cover in conformal contact[75].

Due to the various optical properties, glass transition temperatures, chemical resistance and permeability to gases and liquids, a wide variety of polymers are available for microfluidic flow paths. And they usually using hot embossing for prototyping or injection molding for mass fabrication to produce disposable one-use devices[75]. Thermoplastic polymers, especially poly (methylmethacrylate) (PMMA) and cyclic olefin copolymer (COC), have emerged as the most used polymers for microfluidic chips. Their properties rely on the grade of the plastic[76] chosen for the application requirements, such as device operation temperature, transparency and autofluorescence[76].

2. Fabrication

Different from conventional fabrication techniques, microfluidic fabrication uses a variety of materials, the key factors are low manufacturing cost and the ability to rapidly prototype new designs. Fabrication approaches such as printing, embossing, molding and nanofabrication techniques[77] have been developed and used to fabricate disposable microfluidic devices[78].

Photolithography plays an important role in fabrication steps, which is used to produce polymer hot embossing masters and stamp thermoplastic polymers chips[78]. For example, multilayer PMMA microfluidic devices are made using a paraffin sacrificial layer[79]. Ultraviolet curable polymers are patterned directly to rapidly prototype microfluidic devices. A number of photocurable polymers have been developed and used to fabricate microfluidic chips such as perfluoropolyether[80], PDMS containing benzophenone[81], thiolene-basedmresin (NOA 81)[82, 83] and polyurethane-methacrylate. Rapid microfluidic chip prototypes are produced with maskless photolithography using a liquid crystal display projector to pattern photoresist[84] and direct lithographic patterning of photoresist using a collimated and focused ultraviolet light emitting diode[85].

Several approaches have been proposed to fabricate microfluidic devices using minimal infrastructure, as cleanroom environments for microfabrication are not always available for research prototyping. For example, the rapid prototyping of PDMS devices is fabricated by using a laser printer to produce masters on copper printed circuit board substrates, followed by etching, and replica moldingx[86]. And laser ablation is used to micromachine PMMA chips[87]. A sandwich of inkjet printer patterned transparent film has been used to fabricate rapid prototypes of paper based microfluidic devices[88]. By hand by placing a glass fiber between PDMS and a silicon wafer, microfluidic channels are quickly fabricated[89]. Thermally shrinkable polystyrene sheets are utilized to print Microfluidic flow paths onto[90]. Micropatterned light irradiation of a photoresponsive hydrogel are taken to form arbitrary microchannels[91]. The methods mentioned above are flexible and efficient to produce microstructures for microfluidics. However, the rough surfaces in microchannels may act as pinning sites for liquids, which may induce the creeping of liquids along corner/sidewalls and lead to the formation of air bubbles.

4.4.2 Flow control in chips

Different valves are used in microfluidic systems to control liquid. They can be classified as active valves or passive valves according to the actuation[92]. Active valves actuate using peripheral energy while passive valves exploit energy potential in the device to actuate. Different types of active and passive valves are shown in Figure 4.18 (see the Color Inset, p.18) and Figure 4.19.

1. Active valves

Active valves are often used to repeatedly dose liquid and control liquid at high pressure, as they can switch between an open and closed state with fast switching times and low leakage (Figure 4.18, see the Color Inset, p.18). PDMS valves developed by Quake and colleagues are widely used, due to their easy steps to fabricate and massive integration. An elastomeric membrane is formed in these valves. When pressure is applied in the control channel, the PDMS membrane is pressesed down and the flow channel is closed. This approach has been widely used in pneumatic latches[93]. And simillarily, normally closed valves are opened using vacuum filled control channels[94]. These PDMS valves hase ben integrated on microfluidic chips for immunoassays[95] and nucleic acid analysis[96]. But the matched peripheral equipment, including pneumatic tubes and external pumps, is usually large. In order to minimize peripheral equipment, valves based on the opening and closing of PDMS channels have been made using solenoids or manual mechanical actuation[97]. Screws embedded in a layer of polyurethane can be turned to collapse PDMS microchannels and remain in an opened or closed state[98]. Similarly, screws, pneumatic tubes and solenoids can be embedded directly into PDMS microfluidic devices[99].

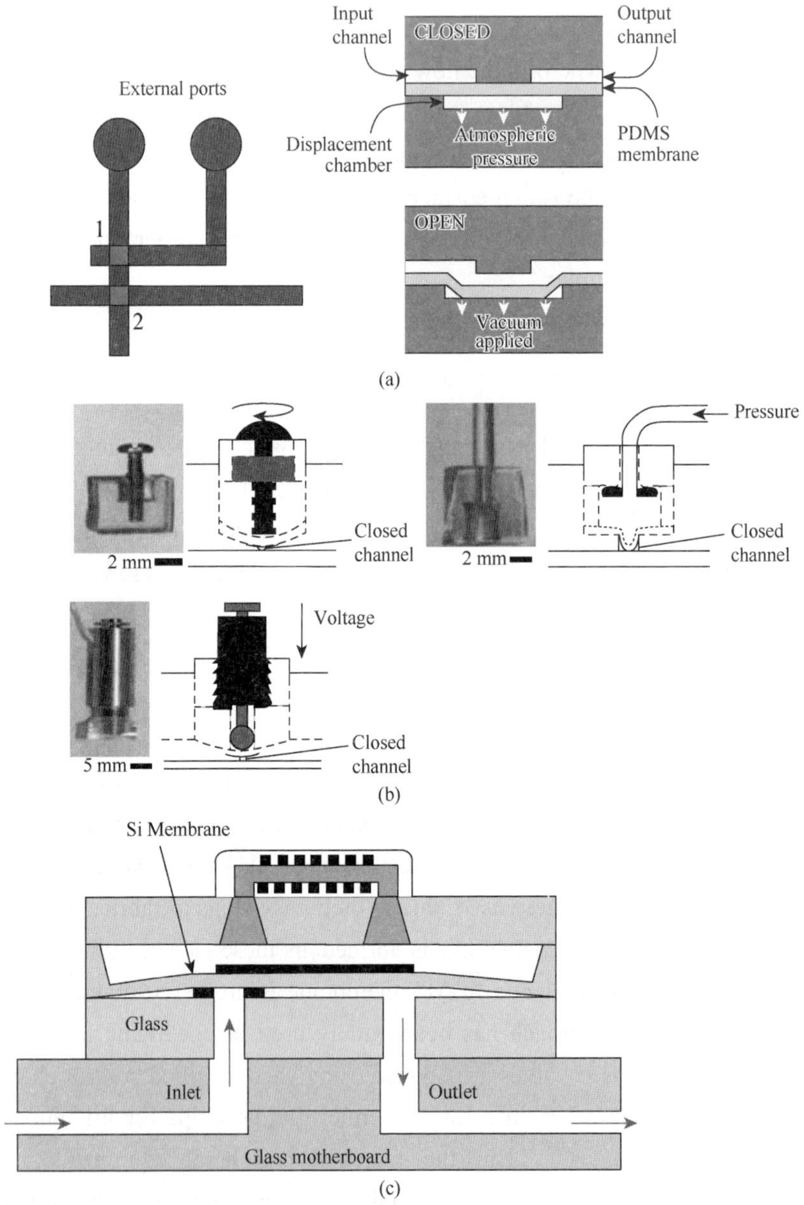

Figure 4.18 Active valves (a) pneumatic latches and valves in PDMS; (b) screws, pneumatic tubes and solenoids; and (c) magnetic inductors

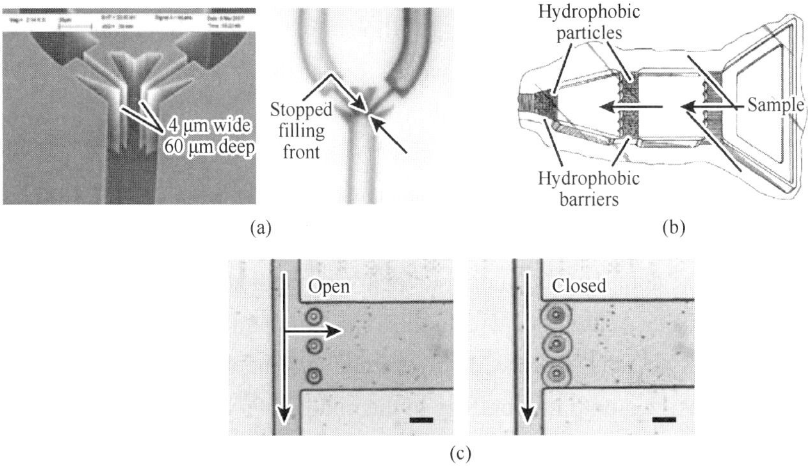

Figure 4.19 Passive valves using (a) an abrupt change in the curvature of a filling front in a capillary channel; (b) hydrophobic barriers with hydrophobic particles that are gradually made hydrophilic by the sample; and (c) posts of pH sensitive polymers that close a channel in the presence of a solution

The actuator coupled to the flexible membrane in active valves can be electrical, magnetic, piezoelectric, thermal or optical. For example, when applied a magnetic field, a magnetic inductor can produce a force to pull up a silicon membrane electroplated with NiFe permalloy[100] or magnetic field heating Fe_3O_4 nanoparticles[101], which allows fluid to flow. Phase changing materials have alse been utilzed to make valves. For instance, PEG heated by resistive heaters of Pt/Ti will change its phase from solid to liquid with a large change in volume to actuate a PDMS membrane microvalve[102]. And shape memory alloy wires of Ni/Ti looped around PDMS channels can be actuated to squeeze channels closed[103].

2. Passive valves

Passive valves usually work as one directional switches to pause the flow of liquid. Passive valves are cast into the design and cannot be opened or closed without changing the geometry (Figure 4.19). They could carry out incubation steps and sequential steps in POC applications. For example, a capillary valve can be made using an abrupt opening in a silicon microchannel[104]. Capillary valves are simple to fabricate and integrate but they depend on the chemical homogeneity of the surface. Trigger valves can be made when a second parallel channel brings liquid to the stopped meniscus and merges with flow continuing in a central channel.

The hydrophobic barrier developed by Biosite is a successful commercial example of a passive valve[105]. A hydrophobic valve, formed by a short section of a capillary channel, composes of a hydrophobic surface patterned with hydrophobic particles of latex or hydrophobic polymers of 10 nm and 10 μm in diameter. Proteins, polypeptides, polymers and detergents can bind to the particles and the hydrophobic barrier is changed to a hydrophilic zone, hence the reaction mixture can flow over it. The rate and concentration of the components binding to the hydrophobic barrier contribute to the time the hydrophobic barrier can hold, which allows optimal amount of time for reagents to incubate with analytes. pH sensitive hydrogel can also be made of autonomous valves when photopolymerized onto micropillars in the middle of a channel[106]. As a solution of pH 11 in the side channel, the hydrogel expands and blocks the entry of the channel.

3. Active pumping

Pumps perform accurate flow control in microfluidics by determining the overall flow rate and volume of liquids used. Flow control of liquids in microfl uidics can be achieved using active of passive pumps as is illustrated in Figure 4.20 (see the Color Inset, p.19) and Figure 4.21 (see the Color Inset, p.19)[107].In active pumping, the flow rate is controlled by using external power sources and can be driven mainly by mechanical displacement, electrical fields, magnetic fields or centripetal force (Figure 4.20, see the Color Inset, p.19). A large variety of active pumps are fabricated using membrane valves moving back and forth. Such pumps can control microliter volumes of liquids in a broad range of flow rates, pressures, viscosities, molecular weights and compositions.

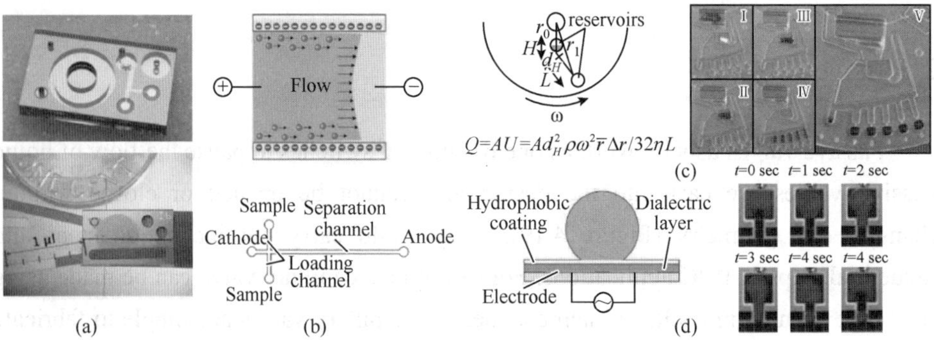

Figure 4.20 Active pumps function by (a) reciprocating movement of a membrane valve; (b) capillary electrophoresis using electro-osmotic flow; (c) centrifugal force in lab on a CD devices; (d) electrowetting

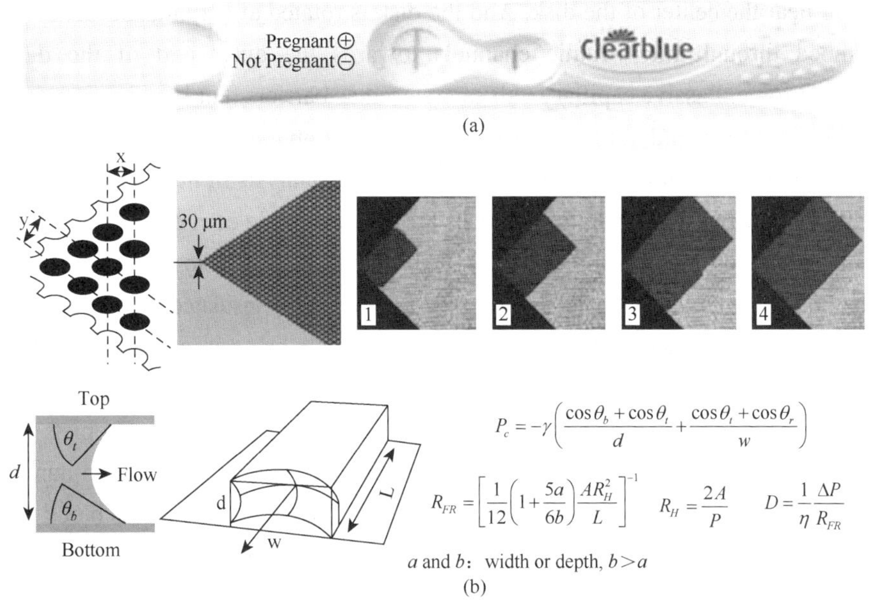

Figure 4.21 Passive pumps function using (a) porous capillary membranes such as in the Clearblue pregnancy test; (b) microfabricated capillary pumps

The syringe pump, as the simplest microfluidic pump, is used routinely in microfluidics laboratories. A stepper motor is used to move the plunger in a syringe at a setted speed defining the flow rate, with tubings connecting the syringe to a microfluidic chip. The advantages of syringe pumps is the wide range of volumes and flow rates, while the disadvantages are large dead volumes and significant flow hysteresis. Microfabricated mechanical displacement pumps actuate a membrane valve in a reciprocating movement. This pump usually uses a piezoelectric actuator compressing and decompressing fluid in the pumping chamber. The membrane valve is placed between a pair of inlet and outlet check valves for the flow of liquid.

Electro-osmotic pumping is based on a chemical equilibrium formed at the interface between a liquid and a solid that results in a charged surface and counter ions in the liquid. An applied electric field causes counter ions to move with the electrical current. The ion drag creates a pressure gradient and flow of liquid. One of the early demonstrations of a compact format electro-osmotic pump is made in 1997[108]. Since then, several highly compact electro-osmotic pumps have become commercially available.

As active pumping elements are often not integrated into the disposable device. One approach to keep the device minal is the Lab on a CD[109]. Liquid is loaded into a reservoir near the center of the disk. And the disk is rotated at varying speeds, so liquid is pumped through microfluidic channels towards the outer part of the disk by centrifugal force. Passive capillary valves are used as barriers that liquid can pass when the spinning speed and force is sufficiently high. Lab on a CD has been used to perform immunoassays from whole blood[110] and the analysis of nucleic acids[111]. Lab on a CD diagnostics and laboratory devices are now available commercially.

Electrowetting devices are often called digital microfluidics. They manipulate droplets of reagents and sample on an array of electrodes insulated by a hydrophobic dielectric layer. They apply pulses of electrical potentials between ground and actuation electrodes to create electrostatic and dielectrophoresis forces so that the wetting properties of the surface is temporarily changed from hydrophobic to hydrophilic. Droplets can be moved, split, merged and dispensed from reservoirs. A POC diagnostic immunoassay system has been developed using these principles[112]. The system dispenses blood from a reservoir, mixes the droplet with reagents and performs the optical detection.

4. Passive pumping

The flow rates in passive pumping are encoded in the design, which is much like in passive valves. The driving forces for propelling liquids in passive microfluidics are typically chemical gradients on surfaces, osmotic pressure, permeation in PDMS or capillary forces (Figure 4.21, see the Color Inset, p.19). These devices are clearly attractive because of their low cost advantage, low dead volume and zero power consumption. A disadvantage is that flow rates of fabricated devices are set in the design and thus can only be varied to a limited degree.

Porous capillary membranes such as a filter membrane, a nitrocellulose membrane for spotting antibodies and a wicking membrane for pumping are sandwiched together to form an immunoassay strip test such as the Clearblue pregnancy test. An alternative to this approach is to fabricate microchannels in plastic. This provides more reproducible flow rate between devices and volumes based on the dimensions of microchannels. Such devices are manufactured by Biosite[112] and perform a number of POC immunoassays such as for drug testing, cardiovascular disease and infection disease.

Capillary pumps have been fabricated using arrays of micropillars[113]. They can be precisely controlled for filling and flow rates can be set in the design by changing the dimensions and the wettability of the surface. Regular arrays of asymmetrical micropillars can be used to directionally control wetting[114]. The evaporation rate of liquid in the capillary pump is controlled to tun flow rates and the total volume[115]. However, when the dimensions of microchannels are large, capillary forces do not dominate and gravity starts to play a role. When dimensions are small, resistance dominates, evaporation in loading pads can be a challenge, and microchannels can block due to defects and dust particles. Nevertheless, if designed properly, capillary driven microfluidic systems have a strong chance of becoming low cost, reliable and state of the art POC devices.

4.4.3 Microparticles in chips

Solid microparticles can be filtered from a stream of liquid using constrictions [Figure 4.22 (a), see the Color Inset, p.20]. Cooper et al. used pillars inside a microchannel to create a microparticle bed for antibody-coated, 10 μm-diameter polystyrene microparticles and capture proteins from lysed cells [Figure 4.22 (a), left, see the Color Inset, p.20][116]. Capillary-driven microfluidic chips with microparticle traps were fabricated in silicon and were used to assemble monolayers of streptavidin-coated microspheres with diameters of 10 μm [Figure 4.22 (a), right, see the Color Inset, p.20][117]. McDevitt et al. introduced a bio-nano-chip in which agarose microparticles with a pore size of 140 nm were inserted into an array of pyramidal cavities for immunoassays [Figure 4.22 (b), see the Color Inset, p.20][118, 119]. Microfluidic traps, which rely on mechanical confinements to retain microparticles, will increase flow resistance of the microfluidic device once the traps are filled. While microfluidic traps based on hydrodynamic confinements also have been developed in which microparticles could be expelled from the traps using optically induced microbubbles[120].

A variety of recently developed microfluidic devices for complex, biochemical amplification and detection systems make use of magnetic microparticles[121, 122]. In these devices, microparticles are mainly used to keep receptors in place during purification and washing steps[123-125]. Viovy et al. manipulate magnetic microparticles inside microfluidic channels for cell sorting[126] and capturing of circulating tumor cells[127]. In their system, columns of biofunctionalized magnetic microparticles are created inside a microchannel using an external magnetic field and anchored to a magnetic pattern at the bottom of the microfluidic channel [Figure 4.22 (c), see the Color Inset, p.20].

The integration of hydrogel particles into a microfluidic device was recently demonstrated by Wheeler et al. A digital microfluidic device used hydrogel particles carrying several reagents, including fluorophores, enzymes and entire cells for reagent exchange, enzyme microreactors and 3D cell culture[128]. In particular, hydrogel particles with immobilized enzymes used for digestion of proteins in the context of proteomics, showed better performance than the conventional in-solution digestion[129].

Functionalized microparticles of sizes larger than the pore size of the hydrogel can be used to physically trap the microparticles inside the hydrogel so that immobilization of biomolecules is simplified. This approach was successfully implemented by Baba et al. in their immuno-pillar chip [Figure 4.22 (d), see the Color Inset, p.20][123]. Hydrogel precursors mixed with 1 μm-diameter functionalised polystyrene microparticles were filled into a microchannel. Using UV irradiation, pillars of 200 μm diameter were photocrosslinked inside the channel and residual reagents were flushed out. The designed chip performed a multiplexed immunoassay for C-reactive protein, α-fetoprotein and prostate-specific antigen.

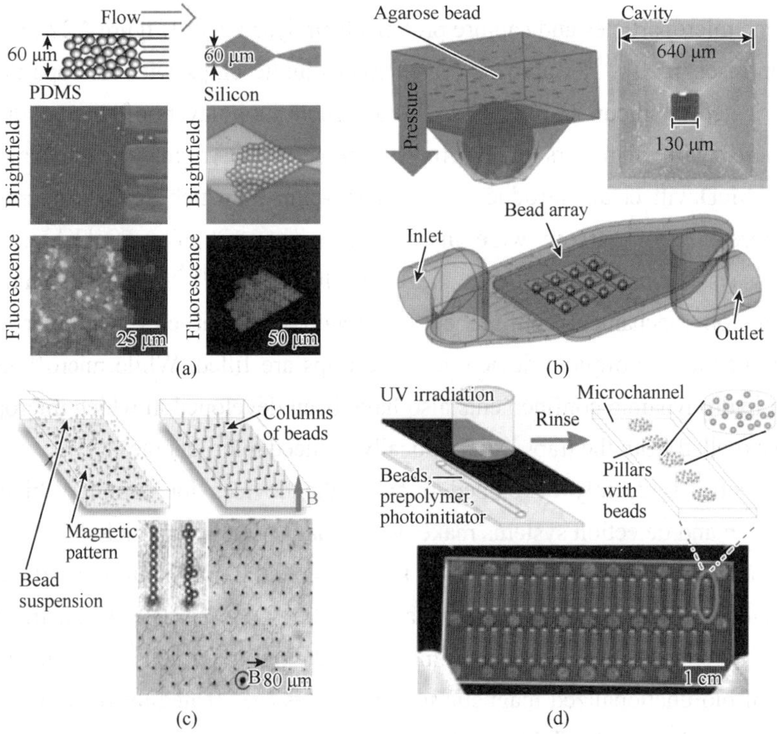

Figure 4.22 Microparticles in microfluidic chips. (a) Polystyrene beads are retained inside a microchannel by physical constrictions; (b) Agarose beads are flushed through pyramidal cavities, in which they are trapped; (c) Beads having a magnetic core are trapped along magnetic field lines; (d) Beads are immobilized inside hydrogel pillars that were crosslinked using UV irradiation

4.5 Instrumentation for Microparticles Analysis

4.5.1 Interation of microscope for analysis

Microscopic inspection of specimen, such as tissue, sputum, or blood film, is still regarded as the gold standard for diagnosis of many diseases, especially for infectious diseases[130, 131]. Among various optical microscopy platforms, fluorescent microscopy is particular important due to its high sensitivity and specificity[132]. However, conventional bright-field and fluorescent microscopes are lager in size and high in cost and costly. There has been a considerable effort to develop compact and cost-effective microscopy modalities to address these limitations. Richards-Kortum et al. developed a Global Focus microscope [Figure 4.23 (a), see the Color Inset, p.20], which is a portable, compact (7.5 cm × 13 cm × 18 cm) and light-weight (less than 1 kg) inverted bright-field and fluorescence microscope[133]. A white LED is used for bright-field illumination, while a blue LED is utilized for fluorescence excitation. The global focus microscope achieves a spatial resolution of ~ 0.8 μm at 1000 × magnification, which provides sufficient sensitivity to identify malaria parasites in bright-field mode and tuberculosis bacilli in fluorescent mode. More recently, Schnitzer et al. also introduced a miniature integrated fluorescent microscope that was made of mass-producible parts, including simple LEDs and a CMOS sensor [Figure 4.23 (b), see the Color Inset, p.20][134, 135]. It has a maximum field-of-view (FOV) of 600 μm × 800 μm, and an optical magnification of 5 × with a lateral resolution of 2.5 μm. This integrated microscope has been utilized to image mouse brain over an active area of ~0.5 mm^2, and it also holds potential for POCT applications.

4.5.2 Cellphone based devices

The use of cell-phone based bright-field and fluorescent microscopes has opened new opportunities for global health applications using different approaches. With the popularization of cellphone-based devices (CBDs), users will have access to cost-effective and compact bioanalytical technologies at any time and place. Such mobile healthcare (mHealthcare) technologies based on CBDs will further improve the self-management of chronic patients, and will enable the supervision of physically disabled, mentally ill or elderly individuals with minimum interference in their daily lives. For example, images taken by a cell-phone based microscope can be immediately processed and analyzed with the help of smart algorithms. And these images could be transmitted to remote locations through wireless for telemedicine applications. that will enable.

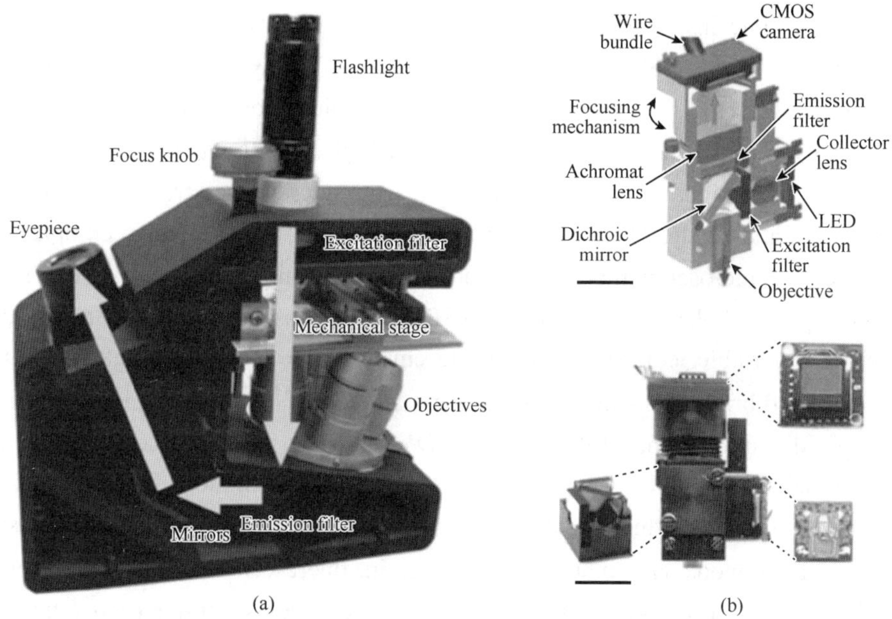

Figure 4.23 (a) A picture of the global focus microscope. The yellow arrows show the trans-illumination light path of the microscope. (b) Miniature integrated fluorescence microscope

In 2008, Ozcan et al. initially implemented the lensfree holographic imaging technique on a USB-powered stand-alone unit[136], which weighs about 46 grams [Figure 4.24 (a), see the Color Inset, p.21]. The sample is loaded using a sample tray located on the side of the microscope. And a single LED together with a large pinhole (an aperture of about 100 μm in front of the source) is utilized to illuminate the sample of interest. Spatially filtered LED light interacts with the sample and creates holographic signatures of individual particles/cells which are recorded by a complementary metal-oxide semiconductor (CMOS) image sensor-array. The scattered object fields are then rapidly processed using custom-developed reconstruction algorithms to provide amplitude and phase images of the samples of interest. The dimensions of the device is smaller than 4.2 cm×4.2 cm×5.8 cm and the spatial resolution is about 1.5 m over a FOV of ~24 mm^2. This is an order of magnitude larger than the imaging area of a typical 10× objective lens. Moreover, by using a simple hardware attachment (about 38 grams) on a commercially-available cellphone device, it is modified to implement the same imaging concept[137] [Figure 4.24 (b), see the Color Inset, p.21]. Another cellphone-based technique was developed for the quantification of colorimetric paper test strips i.e. commercially-available urine test strips and pH paper tests[138] [Figure 4.24 (c), see the Color Inset, p.21]. The technique utilizes chromaticity values in acquired images for the determination of the analyte concentration. It can be potentially useful for the development of various colorimetric modules for POCT applications.

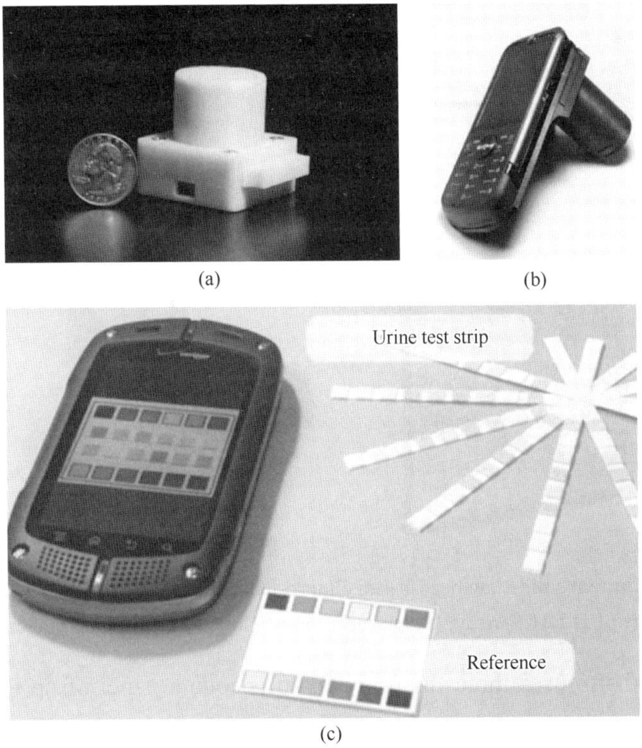

Figure 4.24 (a) A lensfree stand-alone microscope based on partially-coherent digital in-line holography; (b) A lensfree cellphone microscope; (c) Cellphone-based technique for colorimetric measurements in urine test strips

4.5.3 Central laboratory instrumentation

The instrumentations mentioned above are mostly custom designed to meet the demands of POCT applications, as they're portable, cost-effective and easy-to-use. While central laboratory testing is done mostly on clinical analyzers (Figure 4.25). The main selection criteria are throughput and cost per test. Samples are sent from the patient to the central laboratory and placed in a priority queue. It will take several minutes to hours for the results and it is usually not critical. Clinical analyzers have a large variety of analysis capabilities and can detect hundreds of analytes. However, the main drawback is that machines can be meters in size and weigh more than one ton.

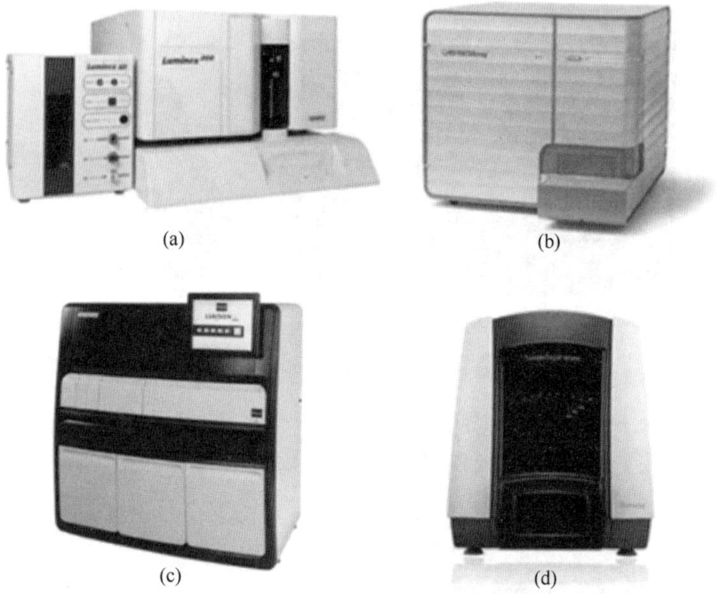

Figure 4.25 Current solid-state bead-based clinical laboratory analyzers: (a) Luminex 100/200; (b) BD FACSArray; (c) Diasorin Liaison; (d) Illumina Bead Xpress

Luminex Corp. uses three fluorophores to encode a panel of up to 500 different beads with size of 5.5 μm. Each bead type is matched to a specific capture probe. When the beads deliver through a suspension array, they are quickly decoded and their intensities are measured by using a 2-laser system[139]. Similarly, BD Biosciences offers fluorescently dyed 7.5 μm beads of different concentrations[140-143]. The BD FACSArray has four different spectral wavelengths for multiplexed detection. The use of these spectrally encoded beads in a 96-well plate on these platforms have been demonstrated for the detection of single nucleotide polymorphisms[144], cytokines[145-147], bacterial pathogens[148, 149], and infectious diseases[150].

Similarly, Illumina developed a optical fiber microwell array with a high density[151-154]. The tips of these glass optical fibers are etched with hydrofluoric acid to create a 5 μm well. And these fibers form an array containing 50000 fibers with a diameter of 1-2 mm when bundled together[152]. When immersed in a solution of spectrally encoded beads, tens of thousands of 3 μm beads randomly disperse and assemble onto the etched microwell array. After removing excess solution and microspheres, an imaging system decodes and quantifies the signal on each microbead. Comparing to the planar microarray, its test density is significantly higher. Hence, only a small volume of sample is required when running tens of thousands of tests in a single run.

4.6 Conclusions and Prospective

As the rapid developments of genomics and proteomics continue, there is an increasing demand in assays for multiplex and high throughput in order to save the sample, time, labor and cost. And in the last decades, many kinds of technologies emerged taking advantages of encoding microparticles to release the biomolecular binding events from the fixed plenary substrates. Some of them have been commercialized, while few of them are clinically used. The majority of them are still staying in the lab, facing a huge gap between the research results and the requirements of real applications.

As many researches put much emphasis on the encoding number while their applications and concomitant problems are neglected. For example, in biological applications, the high encoding capacities of pattern encoding and graphic encoding will be further limited by the non-specific reactions due to the cross reaction of antibodies. While in gene express or profiling assays, the advantages of pattern or graphic encoded microparticles need to be reconsidered. Moreover, the smaller size will result in higher surface to volume ratio and endow the assay with features like rapid homogeneous reaction, higher sensitivity and broader dynamic range. In this way, the color encoding is more flexible than pattern or graphic encoding because of its versatility in complex encoding and feasibility in combining with functional molecules like aptamers or DNA tiles.

On the other hand, the convenience of assays operation based on encoding microparticles is usually overlooked. The higher throughput will benefit from easier operation and higher atomization. And washing-free or label-free detection will have great prospects on the condition that the sensitivity or accuracy is not balanced.

Moreover, the decoding and detection signal acquisition should be included in the design of encoding strategy. As mentioned above, it will greatly reduce the complexity of assays to obtain the two kinds of data at the same time by only one readout. And microfluidic chips have been demonstrated good optional platforms to integrate the assay operations. Hence, automatization instrument or detection platform, which is also a very promising direction in the development of future encoding technology.

In summary, the multiplex and high throughput detection technology is in a relatively early stage of development, and more novel methods and materials are expected to be explored in the encoding of multiplex bioassays not only *in vitro*, but also in cell and *in vivo*. Also, the encoding microparticles are anticipated to be more widely applied in a variety of practical assays with high speed, throughput, accuracy, and automation becoming increasingly achievable in the next few years.

References

[1] Nadon R, Shoemaker J. Statistical issues with microarrays: Processing and analysis. *Trends in Genetics*, 2002, 18 (5): 265-271.

[2] Miklos G L G, Maleszka R. Microarray reality checks in the context of a complex disease. *Nature Biotechnology*, 2004, 22 (5): 615-621.

[3] Canales R D, Luo Y L, Willey J C, et al. Evaluation of DNA microarray results with quantitative gene expression platforms. *Nature Biotechnology*, 2006, 24 (9): 1115-1122.

[4] Jin S Q, Ye B C, Huo H, et al. Multiplexed bead-based mesofluidic system for gene diagnosis and genotyping. *Analytical Chemistry*, 2010, 82 (23): 9925-9931.

[5] Hu L, Zuo P, Ye B C. Multicomponent mesofluidic system for the detection of veterinary drug residues based on competitive immunoassay. *Analytical Biochemistry*, 2010, 405 (1): 89-95.

[6] Jin S Q, Yin B C, Ye B C. Multiplexed bead-based mesofluidic system for detection of food-borne pathogenic bacteria. *Applied and Environmental Microbiology*, 2009, 75 (21): 6647-6654.

[7] Zhang D, Zuo P, Ye B C. Bead-based mesofluidic system for residue analysis of chloramphenicol. *Journal of Agricultural and Food Chemistry*, 2008, 56 (21): 9862-9867.

[8] Mandy F F, Nakamura T, Bergeron M, et al. Overview and application of suspension array technology. *Clinics in Laboratory Medicine*, 2002, 21 (4): 713-729, vii.

[9] Pietz B C, Warden M B, du Chateau B K, et al. Multiplex genotyping of human minor histocompatibility antigens. *Human Immunology*, 2005, 66 (11): 1174-1182.

[10] Gijs M A M, Lacharme F, Lehmann U. Microfluidic applications of magnetic particles for biological analysis and catalysis. *Chemical Reviews*, 2009, 110 (3): 1518-1563.

[11] Herrmann M, Roy E, Veres T, et al. Microfluidic ELISA on non-passivated PDMS chip using magnetic bead

transfer inside dual networks of channels. *Lab on a Chip*, 2007, 7 (11): 1546-1552.

[12] Afshar R, Moser Y, Lehnert T, et al. Three-dimensional magnetic focusing of superparamagnetic beads for on-chip agglutination assays. *Analytical Chemistry*, 2011, 83 (3): 1022-1029.

[13] Shen F S, Hwang H, Hahn Y K, et al. Label-free cell separation using a tunable magnetophoretic repulsion force. *Analytical Chemistry*, 2012, 84 (7): 3075-3081.

[14] Yang S F, Gao B Z, Tsai H Y, et al. Detection of c-reactive protein based on a magnetic immunoassay by using functional magnetic and fluorescent nanoparticles in microplates. *Analyst*, 2014, 139 (21): 5576.

[15] Sivagnanam V, Bouhmad A, Lacharme F, et al. Sandwich immunoassay on a microfluidic chip using patterns of electrostatically self-assembled streptavidin-coated beads. *Microelectronic Engineering*, 2008, 86 (4): 1404-1406.

[16] Chung S H, Baek C Y, Cong V T, et al. The microfluidic chip module for the detection of murine norovirus in oysters using charge switchable micro-bead beating. *Biosensors and Bioelectronics*, 2015, 67: 625-633.

[17] Jokerst J V, Chou J, Camp J P, et al. Location of biomarkers and reagents within agarose beads of a programmable bio-nano-chip. *Small*, 2011, 7 (5): 613-624.

[18] Pernodet N, Maaloum M, Tinland B. Pore size of agarose gels by atomic force microscopy. *Electrophoresis*, 1997, 18 (1): 55-58.

[19] Plieva F M, Ekström P, Galaev I Y, et al. Monolithic cryogels with open porous structure and unique double-continuous macroporous networks. *Soft Matter*, 2008, 4: 2418-2428.

[20] Lee J, Hernandez P, Lee J, et al. Exciton-plasmon interactions in molecular spring assemblies of nanowires and wavelength-based protein detection. *Nature Materials*, 2007, 6 (4): 291-295.

[21] Dubertret B, Calame M, Libchaber A J. Single-mismatch detection using gold-quenched fluorescent oligonucleotides. *Nature Biotechnology*, 2001, 19 (4): 365-370.

[22] Seferos D S, Giljohann D A, Hill H D, et al. Nano-flares: Probes for transfection and mRNA detection in living cells. *Journal of the American Chemical Society*, 2007, 129 (50): 15477-15479.

[23] Maxwell D J, Taylor J R, Nie S M. Self-assembled nanoparticle probes for recognition and detection of biomolecules. *Journal of the American Chemical Society*, 2002, 124 (32): 9606-9612.

[24] Yang R H, Jin J Y, Chen Y, et al. Carbon nanotube-quenched fluorescent oligonucleotides: Probes that fluoresce upon hybridization. *Journal of the American Chemical Society*, 2008, 130 (26): 8351-8358.

[25] Zhang J, Wang L H, Zhang H, et al. Aptamer-based multicolor fluorescent gold nanoprobes for multiplex detection in homogeneous solution. *Small*, 2009, 6 (2): 201-204.

[26] Song S, Liang Z Q, Zhang J, et al. Gold-nanoparticle-based multicolor nanobeacons for sequence-specific DNA analysis. *Angewandte Chemie International Edition*, 2009, 48 (46): 8670-8674.

[27] He S J, Song B, Li D, et al. A graphene nanoprobe for rapid, sensitive, and multicolor fluorescent DNA analysis. *Advanced Functional Materials*, 2010, 20 (3): 453-459.

[28] Li Y G, Tseng Y D, Kwon S Y, et al. Controlled assembly of dendrimer-like DNA. *Nature Materials*, 2004, 3 (1): 38-42.

[29] Li Y G, Cu Y T H, Luo D. Multiplexed detection of pathogen DNA with DNA-based fluorescence nanobarcodes. *Nature Biotechnology*, 2005, 23 (7): 885-889.

[30] Geiss G K, Bumgarner R E, Birditt B, et al. Direct multiplexed measurement of gene expression with color-coded

probe pairs. *Nature Biotechnology*, 2008, 26 (3): 317-325.

[31] Su Y H, Li E, Geiss G K, et al. A perturbation model of the gene regulatory network for oral and aboral ectoderm specification in the sea urchin embryo. *Developmental Biology*, 2009, 329 (2): 410-421.

[32] Nicewarner-Pena S R, Freeman R G, Reiss B D, et al. Submicrometer metallic barcodes. *Science*, 2001, 294 (5540): 137-141.

[33] Qian X M, Peng X H, Ansari D O, et al. *In vivo* tumor targeting and spectroscopic detection with surface-enhanced Raman nanoparticle tags. *Nature Biotechnology*, 2007, 26 (1): 83-90.

[34] Doering W E, Piotti M E, Natan M J, et al. SERS as a foundation for nanoscale, optically detected biological labels. *Advanced Materials*, 2007, 19 (20): 3100-3108.

[35] Xu H X, Aizpurua J, Käll M, et al. Electromagnetic contributions to single-molecule sensitivity in surface-enhanced Raman scattering. *Physical Review E*, 2000, 62 (3): 4318.

[36] Jiang J, Bosnick K, Maillard M, et al. Single molecule raman spectroscopy at the junctions of large Ag nanocrystals. *The Journal of Physical Chemistry B*, 2003, 107 (37): 9964-9972.

[37] Cao Y W C, Jin R C, Mirkin C A. Nanoparticles with Raman spectroscopic fingerprints for DNA and RNA detection. *Science*, 2002, 297 (5586): 1536-1540.

[38] Nam J M, Thaxton C S, Mirkin C A. Nanoparticle-based bio-bar codes for the ultrasensitive detection of proteins. *Science*, 2003, 301 (5641): 1884-1886.

[39] Nam J M, Park S J, Mirkin C A. Bio-barcodes based on oligonucleotide-modified nanoparticles. *Journal of the American Chemical Society*, 2002, 124 (15): 3820-3821.

[40] Nam J M, Stoeva S I, Mirkin C A. Bio-bar-code-based DNA detection with PCR-like sensitivity. *Journal of the American Chemical Society*, 2004, 126 (19): 5932-5933.

[41] Goluch E D, Stoeva S I, Lee J S, et al. A microfluidic detection system based upon a surface immobilized biobarcode assay. *Biosensors and Bioelectronics*, 2009, 24 (8): 2397-2403.

[42] Goluch E D, Nam J M, Georganopoulou D G, et al. A bio-barcode assay for on-chip attomolar-sensitivity protein detection. *Lab on a Chip*, 2006, 6 (10): 1293-1299.

[43] Wang J, Liu G D, Merkoçi A. Electrochemical coding technology for simultaneous detection of multiple DNA targets. *Journal of the American Chemical Society*, 2003, 125 (11): 3214-3215.

[44] Wang J, Liu G D, Munge B, et al. DNA-based amplified bioelectronic detection and coding of proteins. *Angewandte Chemie International Edition*, 2004, 43 (16): 2158-2161.

[45] Braeckmans K, de Smedt S C, Leblans M, et al. Encoding microcarriers: present and future technologies. *Nature Reviews Drug Discovery*, 2002, 1 (6): 447-456.

[46] Wilson R, Cossins A R, Spiller D G. Encoded microcarriers for high-throughput multiplexed detection. *Angewandte Chemie International Edition*, 2006, 45 (37): 6104-6117.

[47] Han M, Gao X, Su J Z, et al. Quantum-dot-tagged microbeads for multiplexed optical coding of biomolecules. *Nature Biotechnology*, 2001, 19 (7): 631-635.

[48] Lin C X, Liu Y, Yan H. Self-assembled combinatorial encoding nanoarrays for multiplexed biosensing. *Nano Letters*, 2007, 7 (2): 507-512.

[49] Taubes G. Photonic crystal made to work at an optical wavelength. *Science*, 1997, 278 (5344): 1709-1710.

[50] Cunin F, Schmedake T A, Link J R, et al. Biomolecular screening with encoded porous-silicon photonic crystals. *Nature Materials*, 2002, 1 (1): 39-41.

[51] Meade S O, Yoon M S, Ahn K H, et al. Porous silicon photonic crystals as encoded microcarriers. *Advanced Materials*, 2004, 16 (20): 1811-1814.

[52] Meade S O, Chen M Y, Sailor M J, et al. Multiplexed DNA detection using spectrally encoded porous SiO_2 photonic crystal particles. *Analytical Chemistry*, 2009, 81 (7): 2618-2625.

[53] Zhao X W, Liu Z B, Yang H, et al. Uniformly colorized beads for multiplex immunoassay. *Chemistry of Materials*, 2006, 18 (9): 2443-2449.

[54] Zhao X, Cao Y, Ito F, et al. Colloidal crystal beads as supports for biomolecular screening. *Angewandte Chemie International Edition*, 2006, 45 (41): 6835-6838.

[55] Zhao Y W, Zhao X J, Sun C, et al. Encoded silica colloidal crystal beads as supports for potential multiplex immunoassay. *Analytical Chemistry*, 2008, 80 (5): 1598-1605.

[56] Sun C, Zhao X W, Zhao Y J, et al. Fabrication of colloidal crystal beads by a drop-breaking technique and their application as bioassays. *Small*, 2008, 4 (5): 592-596.

[57] Zhao X W, Zhao Y J, Hu J, et al. Sintering photonic beads for multiplex biosensing. *Journal of Nanoscience and Nanotechnology*, 2010, 10 (1): 588-594.

[58] Zhao Y J, Zhao X W, Hu J, et al. Encoded porous beads for label-free multiplex detection of tumor markers. *Advanced Materials*, 2009, 21 (5): 569-572.

[59] Li J, Zhao X W, Zhao Y J, et al. Quantum-dot-coated encoded silica colloidal crystals beads for multiplex coding. *Chemical Communications*, 2009 (17): 2329-2331.

[60] Li J, Zhao X W, Zhao Y J, et al. Colloidal crystal beads coated with multicolor CdTe quantum dots: Microcarriers for optical encoding and fluorescence enhancement. *Journal of Materials Chemistry*, 2009, 19 (36): 6492-6497.

[61] Zhao Y J, Zhao X W, Pei X P, et al. Multiplex detection of tumor markers with photonic suspension array. *Analytica Chimica Acta*, 2008, 633 (1): 103-108.

[62] Ganesh N, Zhang W, Mathias P C, et al. Enhanced fluorescence emission from quantum dots on a photonic crystal surface. *Nature Nanotechnology*, 2007, 2 (8): 515-520.

[63] Qian W P, Gu Z Z, Fujishima A, et al. Three-dimensionally ordered macroporous polymer materials: An approach for biosensor applications. *Langmuir*, 2002, 18 (11): 4526-4529.

[64] Shamah S M, Cunningham B T. Label-free cell-based assays using photonic crystal optical biosensors. *The Analyst*, 2011, 136 (6): 1090-1102.

[65] Ke Y, Lindsay S, Chang Y, et al. Self-assembled water-soluble nucleic acid probe tiles for label-free RNA hybridization assays. *Science*, 2008, 319 (5860): 180-183.

[66] Qin L D, Banholzer M J, Millstone J E, et al. Nanodisk codes. *Nano Letters*, 2007, 7 (12): 3849-3853.

[67] Banholzer M J, Osberg K D, Li S Z, et al. Silver-based nanodisk codes. *ACS Nano*, 2010, 4 (9): 5446-5452.

[68] Pregibon D C, Toner M, Doyle P S. Multifunctional encoded particles for high-throughput biomolecule analysis. *Science*, 2007, 315 (5817): 1393-1396.

[69] Appleyard D C, Chapin S C, Doyle P S. Multiplexed protein quantification with barcoded hydrogel microparticles. *Analytical Chemistry*, 2010, 83 (1): 193-199.

[70] Pregibon D C, Doyle P S. Optimization of encoded hydrogel particles for nucleic acid quantification. *Analytical Chemistry*, 2009, 81 (12): 4873-4881.

[71] Lee H, Kim J, Kim H, et al. Colour-barcoded magnetic microparticles for multiplexed bioassays. *Nature Materials*, 2010, 9 (9): 745-749.

[72] Kim H, Ge J P, Kim J, et al. Structural colour printing using a magnetically tunable and lithographically fixable photonic crystal. *Nature Photonics*, 2009, 3 (9): 534-540.

[73] Verpoorte E, de Rooij N F. Microfluidics meets MEMS. *Proceedings of the IEEE*, 2003, 91 (6): 930-953.

[74] Abgrall P, Gue A M. Lab-on-chip technologies: making a microfluidic network and coupling it into a complete microsystem—a review. *Journal of Micromechanics and Microengineering*, 2007, 17 (5): R15.

[75] Gervais L, Delamarche E. Toward one-step point-of-care immunodiagnostics using capillary-driven microfluidics and PDMS substrates. *Lab on a Chip*, 2009, 9 (23): 3330-3337.

[76] Diaz-Quijada G A, Peytavi R, Nantel A, et al. Surface modification of thermoplastics—towards the plastic biochip for high throughput screening devices. *Lab on a Chip*, 2007, 7 (7): 856-862.

[77] Gates B D, Xu Q B, Stewart M, et al. New approaches to nanofabrication: molding, printing, and other techniques. *Chemical Reviews*, 2005, 105 (4): 1171-1196.

[78] Lorenz R M, Edgar J S, Jeffries G D M, et al. Microfluidic and optical systems for the on-demand generation and manipulation of single femtoliter-volume aqueous droplets. *Analytical Chemistry*, 2006, 78 (18): 6433-6439.

[79] Fuentes H V, Woolley A T. Phase-changing sacrificial layer fabrication of multilayer polymer microfluidic devices. *Analytical Chemistry*, 2008, 80 (1): 333-339.

[80] Bhagat A A S, Jothimuthu P, Papautsky I. Photodefinable polydimethylsiloxane (PDMS) for rapid lab-on-a-chip prototyping. *Lab on a Chip*, 2007, 7 (9): 1192-1197.

[81] Cygan Z T, Cabral J T, Beers K L, et al. Microfluidic platform for the generation of organic-phase microreactors. *Langmuir*, 2005, 21 (8): 3629-3634.

[82] Wägli P, Homsy A, Guélat B Y, et al. Microfluidic devices made of UV-curable glue (NOA81) for fluorescence detection based applications//14th International Conference on Miniaturized Systems for Chemistry and Life Sciences, 2010: 1937-1939.

[83] Kuo J S, Ng L, Yen G S, et al. A new USP Class VI-compliant substrate for manufacturing disposable microfluidic devices. *Lab on a Chip*, 2009, 9 (7): 870-876.

[84] Itoga K, Kobayashi J, Tsuda Y, et al. Second-generation maskless photolithography device for surface micropatterning and microfluidic channel fabrication. *Analytical Chemistry*, 2008, 80 (4): 1323-1327.

[85] Guijt R M, Breadmore M C. Maskless photolithography using UV LEDs. *Lab on a Chip*, 2008, 8 (8): 1402-1404.

[86] Abdelgawad M, Watson M W L, Young E W K, et al. Soft lithography: Masters on demand. *Lab on a Chip*, 2008, 8 (8): 1379-1385.

[87] Tan H Y, Loke W K, Tan Y T, et al. A lab-on-a-chip for detection of nerve agent sarin in blood. *Lab on a Chip*, 2008, 8 (6): 885-891.

[88] Martinez A W, Phillips S T, Wiley B J, et al. FLASH: A rapid method for prototyping paper-based microfluidic devices. *Lab on a Chip*, 2008, 8 (12): 2146-2150.

[89] Huang T T, Taylor D G, Sedlak M, et al. Microfiber-directed boundary flow in press-fit microdevices fabricated

from self-adhesive hydrophobic surfaces. *Analytical Chemistry*, 2005, 77 (11): 3671-3675.

[90] Sollier K, Mandon C A, Heyries K A, et al. "Print-n-Shrink" technology for the rapid production of microfluidic chips and protein microarrays. *Lab on a Chip*, 2009, 9 (24): 3489-3494.

[91] Sugiura S, Szilágyi A, Sumaru K, et al. On-demand microfluidic control by micropatterned light irradiation of a photoresponsive hydrogel sheet. *Lab on a Chip*, 2009, 9 (2): 196-198.

[92] Oh K W, Ahn C H. A review of microvalves. *Journal of Micromechanics and Microengineering*, 2006, 16 (5): R13.

[93] Melin J, Quake S R. Microfluidic large-scale integration: the evolution of design rules for biological automation. *Annual Review of Biophysics and Biomolecular Structure*, 2007, 36: 213-231.

[94] Grover W H, Ivester R H C, Jensen E C, et al. Development and multiplexed control of latching pneumatic valves using microfluidic logical structures. *Lab on a Chip*, 2006, 6 (5): 623-631.

[95] Kartalov E P, Zhong J F, Scherer A, et al. High-throughput multi-antigen microfluidic fluorescence immunoassays. *BioTechniques*, 2006, 40 (1): 85-90.

[96] Ottesen E A, Hong J W, Quake S R, et al. Microfluidic digital PCR enables multigene analysis of individual environmental bacteria. *Science*, 2006, 314 (5804): 1464-1467.

[97] Vestad T, Marr D W M, Oakey J. Flow control for capillary-pumped microfluidic systems. *Journal of Micromechanics and Microengineering*, 2004, 14 (11): 1503.

[98] Weibel D B, Kruithof M, Potenta S, et al. Torque-actuated valves for microfluidics. *Analytical Chemistry*, 2005, 77 (15): 4726-4733.

[99] Hulme S E, Shevkoplyas S S, Whitesides G M. Incorporation of prefabricated screw, pneumatic, and solenoid valves into microfluidic devices. *Lab on a Chip*, 2009, 9 (1): 79-86.

[100] Oh K W, Han A, Bhansali S, et al. A low-temperature bonding technique using spin-on fluorocarbon polymers to assemble microsystems. *Journal of Micromechanics and Microengineering*, 2002, 12 (2): 187.

[101] Satarkar N S, Zhang W, Eitel R E, et al. Magnetic hydrogel nanocomposites as remote controlled microfluidic valves. *Lab on a Chip*, 2009, 9 (12): 1773-1779.

[102] Kaigala G V, Hoang V N, Backhouse C J. Electrically controlled microvalves to integrate microchip polymerase chain reaction and capillary electrophoresis. *Lab on a Chip*, 2008, 8 (7): 1071-1078.

[103] Vyawahare S, Sitaula S, Martin S, et al. Electronic control of elastomeric microfluidic circuits with shape memory actuators. *Lab on a Chip*, 2008, 8 (9): 1530-1535.

[104] Zimmermann M, Hunziker P, Delamarche E. Valves for autonomous capillary systems. *Microfluidics and Nanofluidics*, 2008, 5 (3): 395-402.

[105] Buechler K F. Diagnostic devices for the controlled movement of reagents without membranes: U. S. Patent 5, 458, 852. 1995-10-17.

[106] Beebe D J, Moore J S, Bauer J M, et al. Functional hydrogel structures for autonomous flow control inside microfluidic channels. *Nature*, 2000, 404 (6778): 588-590.

[107] Laser D J, Santiago J G. A review of micropumps. *Journal of Micromechanics and Microengineering*, 2004, 14 (6): R35-R64.

[108] Ramsey R S, Ramsey J M. Generating electrospray from microchip devices using electroosmotic pumping. *Analytical Chemistry*, 1997, 69 (13): 2617.

[109] Duffy D C, Gillis H L, Lin J, et al. Microfabricated centrifugal microfluidic systems: characterization and multiple enzymatic assays. *Analytical Chemistry*, 1999, 71 (20): 4669-4678.

[110] Lee B S, Lee J N, Park J M, et al. A fully automated immunoassay from whole blood on a disc. *Lab on a Chip*, 2009, 9 (11): 1548-1555.

[111] Lutz S, Weber P, Focke M, et al. Microfluidic lab-on-a-foil for nucleic acid analysis based on isothermal recombinase polymerase amplification (RPA). *Lab on a Chip*, 2010, 10 (7): 887-893.

[112] Sista R, Hua Z, Thwar P, et al. Development of a digital microfluidic platform for point of care testing. *Lab on a Chip*, 2008, 8 (12): 2091-2104.

[113] Zimmermann M, Schmid H, Hunziker P, et al. Capillary pumps for autonomous capillary systems. *Lab on a Chip*, 2006, 7 (1): 119-125.

[114] Jokinen V, Leinikka M, Franssila S. Microstructured surfaces for directional wetting. *Advanced Materials*, 2010, 21 (47): 4835-4838.

[115] Zimmermann M, Bentley S, Schmid H, et al. Continuous flow in open microfluidics using controlled evaporation. *Lab on a Chip*, 2005, 5 (12): 1355-1359.

[116] Sedgwick H, Caron F, Monaghan P B, et al. Lab-on-a-chip technologies for proteomic analysis from isolated cells. *Journal of the Royal Society Interface*, 2008, 5 (Suppl 2): S123-S130.

[117] Hitzbleck M, Delamarche E. Reagents in microfluidics: an "in" and "out" challenge. *Chemical Society Reviews*, 2013, 42 (21): 8494-8516.

[118] Oyama Y, Osaki T, Kamiya K, et al. A glass fiber sheet-based electroosmotic lateral flow immunoassay for point-of-care testing. *Lab on a Chip*, 2012, 12 (24): 5155-5159.

[119] Jokerst J V, Chou J, Camp J P, et al. Location of biomarkers and reagents within agarose beads of a programmable bio-nano-chip. *Small*, 2011, 7 (5): 613-624.

[120] Tan W H, Takeuchi S. A trap-and-release integrated microfluidic system for dynamic microarray applications. *Proceedings of the National Academy of Sciences of the United States of America*, 2007, 104 (4): 1146-1151.

[121] Lacharme F, Vandevyver C, Gijs M A M. Magnetic beads retention device for sandwich immunoassay: Comparison of off-chip and on-chip antibody incubation. *Microfluidics and Nanofluidics*, 2009, 7 (4): 479-487.

[122] Gijs M A M, Lacharme F, Lehmann U. Microfluidic applications of magnetic particles for biological analysis and catalysis. *Chemical Reviews*, 2009, 110 (3): 1518-1563.

[123] Ikami M, Kawakami A, Kakuta M, et al. Immuno-pillar chip: A new platform for rapid and easy-to-use immunoassay. *Lab on a Chip*, 2010, 10 (24): 3335-3340.

[124] Wang C H, Lien K Y, Wu J J, et al. A magnetic bead-based assay for the rapid detection of methicillin-resistant Staphylococcus aureus by using a microfluidic system with integrated loop-mediated isothermal amplification. *Lab on a Chip*, 2011, 11(8):1521-1531.

[125] Ramadan Q, Jafarpoorchekab H, Huang C B, et al. NutriChip: Nutrition analysis meets microfluidics. *Lab on a Chip*, 2012, 13 (2): 196-203.

[126] Saliba A E, Saias L, Psychari E, et al. Microfluidic sorting and multimodal typing of cancer cells in self-assembled magnetic arrays. *Proceedings of the National Academy of Sciences of the United States of America*, 2010, 107 (33): 14524-14529.

[127] Horák D, Svobodová Z, Autebert J, et al. Albumin-coated monodisperse magnetic poly (glycidyl methacrylate) microspheres with immobilized antibodies: Application to the capture of epithelial cancer cells. *Journal of Biomedical Materials Research Part A*, 2012, 101a(1):23-32.

[128] Fiddes L K, Luk V N, Au S H, et al. Hydrogel discs for digital microfluidics. *Biomicrofluidics*, 2012, 6 (1): 14112.

[129] Luk V N, Fiddes L K, Luk V M, et al. Digital microfluidic hydrogel microreactors for proteomics. *Proteomics*, 2012, 12 (9): 1310-1318.

[130] Moody A. Rapid diagnostic tests for malaria parasites. *Clinical Microbiology Reviews*, 2002, 15 (1): 66-78.

[131] Steingart K R, Henry M, Ng V, et al. Fluorescence versus conventional sputum smear microscopy for tuberculosis: A systematic review. *The Lancet Infectious Diseases*, 2006, 6 (9): 570-581.

[132] Lichtman J W, Conchello J A. Fluorescence microscopy. *Nature Methods*, 2005, 2 (12): 910-919.

[133] Miller A R, Davis G L, Oden Z M, et al. Portable, battery-operated, low-cost, bright field and fluorescence microscope. *PLoS One*, 2010, 5 (8): e11890.

[134] Flusberg B A, Nimmerjahn A, Cocker E D, et al. High-speed, miniaturized fluorescence microscopy in freely moving mice. *Nature Methods*, 2008, 5 (11): 935-938.

[135] Ghosh K K, Burns L D, Cocker E D, et al. Miniaturized integration of a fluorescence microscope. *Nature Methods*, 2011, 8 (10): 871-878.

[136] Mudanyali O, Tseng D, Oh C, et al. Compact, light-weight and cost-effective microscope based on lensless incoherent holography for telemedicine applications. *Lab on a Chip*, 2010, 10 (11): 1417-1428.

[137] Tseng D, Mudanyali O, Oztoprak C, et al. Lensfree microscopy on a cellphone. *Lab on a Chip*, 2010, 10 (14): 1787-1792.

[138] Shen L, Hagen J A, Papautsky I. Point-of-care colorimetric detection with a smartphone. *Lab on a Chip*, 2012, 12 (21): 4240-4243.

[139] Nolan J P, Yang L, Heyde H C. Reagents and instruments for multiplexed analysis using microparticles. *Current Protocols in Cytometry*, 2006: Chapter 13. 8. 1-13. 8. 10.

[140] Zander K A, Saavedra M T, West J, et al. Protein microarray analysis of nasal polyps from aspirin-sensitive and aspirin-tolerant patients with chronic rhinosinusitis. *American Journal of Rhinology and Allergy*, 2009, 23 (3): 268-272.

[141] Vignali D A A. Multiplexed particle-based flow cytometric assays. *Journal of Immunological Methods*, 2000, 243 (1): 243-255.

[142] Derveaux S, Stubbe B G, Braeckmans K, et al. Synergism between particle-based multiplexing and microfluidics technologies may bring diagnostics closer to the patient. *Analytical and Bioanalytical Chemistry*, 2008, 391 (7): 2453-2467.

[143] Fulton R J, McDade R L, Smith P L, et al. Advanced multiplexed analysis with the FlowMetrix[TM] system. *Clinical Chemistry*, 1997, 43 (9): 1749-1756.

[144] Taylor J D, Briley D, Nguyen Q, et al. Research report flow cytometric platform for high-throughput single nucleotide polymorphism analysis. *BioTechniques*, 2001, 30: 661-669.

[145] Oliver K G, Kettman J R, Fulton R J. Multiplexed analysis of human cytokines by use of the FlowMetrix system. *Clinical Chemistry*, 1998, 44 (9): 2057-2060.

[146] Prabhakar U, Eirikis E, Miller B E, et al. Multiplexed cytokine sandwich immunoassays clinical applications//Joos T O, Fortina P. *Microarrays in Clinical Diagnostics*. Totowa: Springer Science & Business Media: Humana Press Inc., 2005: 223-232.

[147] Prabhakar U, Eirikis E, Davis H M. Simultaneous quantification of proinflammatory cytokines in human plasma using the LabMAP™ assay. *Journal of Immunological Methods*, 2002, 260 (1): 207-218.

[148] Dunbar S A, Zee C A V, Oliver K G, et al. Quantitative, multiplexed detection of bacterial pathogens: DNA and protein applications of the Luminex LabMAP™ system. *Journal of Microbiological Methods*, 2003, 53 (2): 245-252.

[149] Dunbar S A, Jacobson J W. Rapid Screening for 31 Mutations and Polymorphisms in the Cystic Fibrosis Transmembrane Conductance Regulator Gene by Luminex® xMAP™ Suspension Array//Joos T O, Fortina P. Microarrays in Clinical Diagnostics. Totowa: Springer Science & Business Media: Humana Press Inc. 2005: 147-171.

[150] Yan X M, Zhong W W, Tang A J, et al. Multiplexed flow cytometric immunoassay for influenza virus detection and differentiation. *Analytical Chemistry*, 2005, 77 (23): 7673-7678.

[151] Walt D R. Fibre optic microarrays. *Chemical Society Reviews*, 2010, 39 (1): 38-50.

[152] Gorris H H, Blicharz T M, Walt D R. Optical-fiber bundles. *The FEBS Journal*, 2007, 274 (21): 5462-5470.

[153] Stitzel S E, Aernecke M J, Walt D R. Artificial noses. *Annual Review of Biomedical Engineering*, 2011, 13: 1-25.

[154] Walt D R. Fiber optic array biosensors. *BioTechniques*, 2006, 41 (5): 529, 531, 533.

Chapter 5 Biomedical Applications of Zwitterionic Antifouling Polymers Based Nanoparticles

Wenchen Li, Qingsheng Liu, Lingyun Liu

5.1 Introduction

Materials that can resist protein, cell or bacterial adhesion play an important role in biomedical and engineering applications, such as biomedical devices[1], drug delivery[2]and ship hulls[3]. Some adverse consequences may be caused because of proteins adsorbing onto surfaces. For example, the non-specific protein adsorption on biosensor chips may cause the sensitivity decrease[4]. The proteins adsorbing on drug carriers can result in fast recognition of the particles by the body, leading to decreased blood circulation time and poor drug efficiency. For implanted devices, nonspecific protein adsorption may cause the formation of fibrous capsules, and finally the device failure[1]. Bacterial adhesion on the implanted medical devices may cause significant infection, sometimes fatal, and shorten the life time of the devices.

Ethylene glycol (EG) based polymers such as poly (ethylene glycol) (PEG) and oligo (ethylene glycol) (OEG) are the mostly applied antifouling materials. However, the presence of oxygen and transition metal ions found in most biochemically relevant solutions will cause their degradation[5], which limits their sustainable applications. The zwitterionic polymers, such as those based on phosphorylcholine (PC) [6, 7], sulfobetaine (SB) [8], and carboxybetaine (CB) [9, 10](all with the positive and negative charges in series on the same side chain), amino acids[11-13] (with positive and negative charges in parallel on the same side chain), and equimolar mixed cationic-anionic pairs[14] (with positive and negative charges on two different side chains), have emerged as promising alternatives for antifouling materials, due to their ultralow fouling properties. The homogenously mixed positively and negatively charged groups can bind water molecules strongly via electrostatically induced hydration, which contributes to reducing protein adsorption[15].

Because of the unique structures and superior properties of zwitterionic materials, their potential biomedical applications have been explored by many groups. This chapter will review the biomedical applications of zwitterionic polymers in the physical form of nanoparticles. Here, we classify the structure of nanoparticles as inorganic/polymer composite, nanogel, and amphiphilic copolymer. The applications of each structure will mainly focus on circulation, bioimaging, biosensing, or drug/gene delivery. Schematic illustration of the chapter is shown in Figure 5.1.

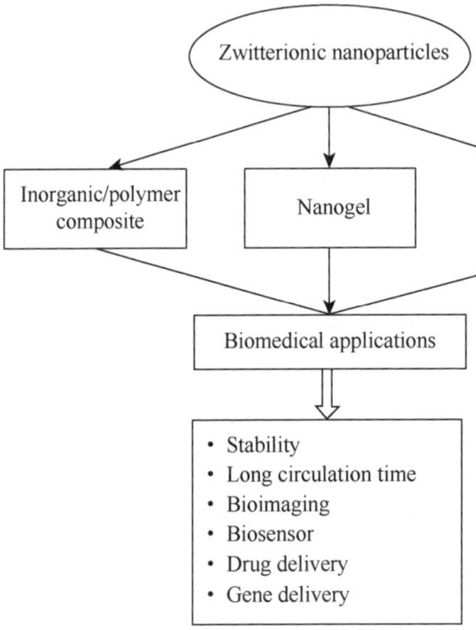

Figure 5.1　Biomedical applications of zwitterionic nanoparticles

5.2 Biomedical Applications of Zwitterionic Nanoparticles

5.2.1 Inorganic /polymer composite

Zwitterionic polymers can be coated on the surfaces of inorganic nanoparticles such as gold nanoparticles (GNPs), magnetic nanoparticles (MNPs), and quantum dots (QDs). Such hybrid structure is here referred to as inorganic/polymer composite, which possesses the properties of both the core particles and shell/surface polymer coatings.

1. Stability

The zwitterionic materials have been widely used to stabilize inorganic nanoparticles, like gold nanoparticles[16], gold nanorods[17], magnetic nanoparticles[18], silica nanoparticles[19]. The unique properties of inorganic nanoparticles lead to their wide applications in drug delivery, bioimaging, diagnosis, and biological labeling. The suitable surface modification renders them stability in biological solutions, like single protein solutions and complex media (e.g., blood serum or plasma).

The stability of gold nanoparticles coated with poly (carboxybetaine acrylamide) (pCBAA-GNPs) was evaluated in 10% and 100% blood serum, along with the bare GNPs, PEG5000 coated gold nanoparticles (PEG-GNPs) and OEGMA coated gold nanoparticles (OEGMA-GNPs)[16]. As shown in Figure 5.2, in 10% blood serum, the bare GNPs showed an increase of ~60 nm in size in 1 h and increased to ~80 nm after 72 h. In contrast, all pCBAA coated GNPs showed no agglomeration without obvious size increase during the test period of 72 h. In 100% blood serum, only pCBAA-GNPs showed no agglomeration, whereas other three samples showed obvious aggregation with diameter increment.

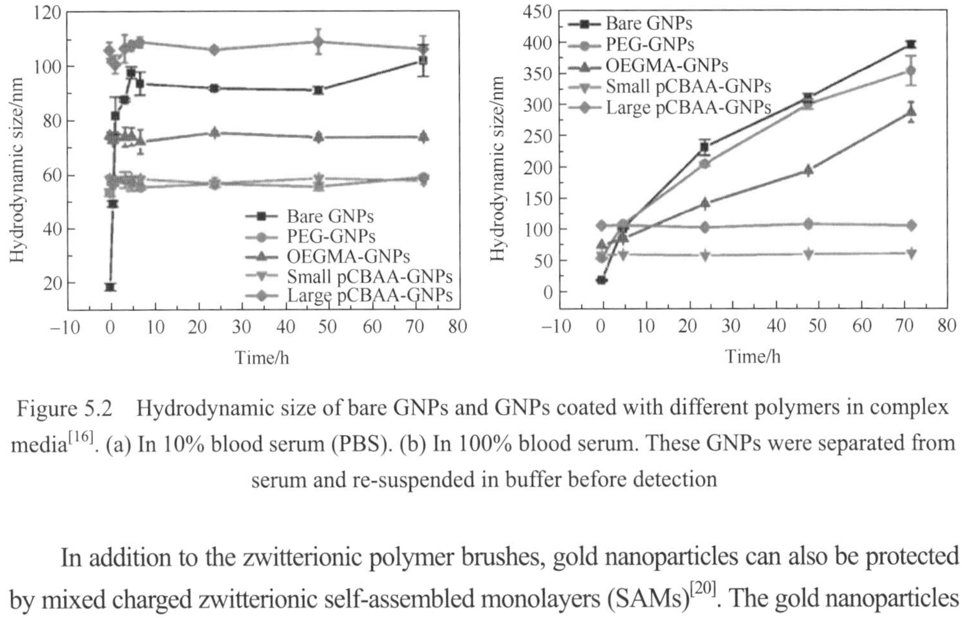

Figure 5.2 Hydrodynamic size of bare GNPs and GNPs coated with different polymers in complex media[16]. (a) In 10% blood serum (PBS). (b) In 100% blood serum. These GNPs were separated from serum and re-suspended in buffer before detection

In addition to the zwitterionic polymer brushes, gold nanoparticles can also be protected by mixed charged zwitterionic self-assembled monolayers (SAMs)[20]. The gold nanoparticles with diameters ranging from 16 nm to even 100 nm were stabilized via a simple place exchange reaction with 1∶1 mixed negatively charged sodium 10-mercaptodecanesulfonic acid (HS-C10-S) and positively charged (10-mercaptodecyl) trimethylammonium bromide (HS-C10-N4). The GNPs with 16 nm diameter showed superior stability under a wide variety of conditions, including a wide pH range, high salt concentration, and physiological phosphate buffered saline (PBS), and even cell culture medium with 10% fetal bovine serum (FBS). The larger GNPs with diameters of 50 nm and 100 nm also showed better stability in salt solution than the HS-C11-EG4 protected ones.

Mesoporous silica nanoparticles (MSNs) have been extensively used in biomedical applications, like drug delivery[21, 22], gene delivery[23] and diagnosis[24, 25]. To lower macrophage cell uptake and improve specific cell uptake efficiency, pCBAA was grafted by atom transfer radical polymerization (ATRP) from silica nanoparticles, which had been modified by 2-bromo-2-methyl-N-[3-(trimethoxysilyl) propyl] propanamide (BrTMOS) initiator. The prepared silica particle sizes are 135.5 mm (pCBAA-SiP1) and 221.3 mm (pCBAA-SiP2). The stability of pCBAA-SiP was tested by incubating in lysozyme and bovine serum albumin (BSA) solutions in PBS. All particles showed excellent stability without an obvious size increase during a 72 h incubation period. Such results demonstrated that pCBAA polymer layer can efficiently improve biointerfacial properties of silica nanoparticles[26].

In addition to poly (carboxybetaine) (pCB), poly [2-(methacryloyloxy) ethyl phosphorylcholine] (pMPC) was also employed to stabilize inorganic nanoparticles. Yuan et al.[27]prepared poly [2-(methacryloyloxy) ethyl phosphorylcholine]-block-(glycerol monomethacrylate) (pMPC-pGMA) as stabilizer of magnetite sols, with pGMA as the adsorbing block and pMPC as the stabilizing block. Compared to the PEG block copolymer, $pMPC_{30}$-$pGMA_{30}$ copolymer appeared to confer enhanced colloidal stability.

2. Long circulation time

The prolonged circulation of nanoparticles will increase the therapeutic efficiency by delivering the loadings to the targeted sites, and reduce the accumulation in normal tissues. The blood is a heterogeneous complex medium that contains salts, proteins, cells, etc. To prolong the circulation of nanoparticle, the stealth surface of nanoparticle that can resist nonspecific plasma protein adsorption (i.e., minimize opsonization) is required, which can escape the recognition by the reticuloendothelial system[28]. The size and mechanical properties of nanoparticle will also affect its circulation time *in vivo*. Bypassing *in vivo* clearance barriers, such as the splenic filtration and the renal clearance[29], also improves circulation time.

In Yang et al.'s work, pCB coated GNPs exhibited superior blood retention compared to PEG coated GNPs[30]. After the first dose, the blood clearance profile showed that the pCB-GNPs exhibited 50% overall retention, whereas the PEG-GNPs only showed 3% retention. After the second dose, the elimination half-life was 55.6 h and 5.2 h for pCB-GNPs and PEG-GNPs, respectively.

PCB nanogels were also prepared by an inverse microemulsion polymerization method, loaded with GNPs[31]. Two types of pCB nanogels (one stiff and one soft) were prepared by varying the monomer amount and the crosslinking density. The softest nanogel showed longest circulation half-life as long as 19.6 h and minimal accumulation in spleen, when the bulk gel modulus was 0.18 MPa. Those results show that the pCB polymer can effectively protect GNPs from clearance. Furthermore, softer nanogel has strong capability to deform to pass through *in vivo* barriers.

3. Bioimaging/labeling

Bioiamging/labeling is an important biomedical application of inorganic nanoparticles. Magnetic nanoparticles (MNPs) have attracted significant attention because of their excellent properties, such as low toxicity and unique magnetic properties[32, 33]. The zwitterionic polymer coatings on MNPs can provide them better stability and functional groups for conjugation of targeting ligands.

In a previous study, pMPC was prepared by ATRP and grafted to the MNP surface by Peacock et al[34]. They found pMPC coatings did not lower the MNP relaxivity with an R2 value of 127 mM/s, which is close to the commercial clinical contrast agent. To evaluate the cell labeling capability, particle uptake study was performed in a kidney- derived murine stem cell line, exposed to 1-50 μgFe/mL. At low loading of Fe, like 5 μgFe/mL, pMPC-MNPs exhibited a lower uptake compared to Molday Ion (the commercial MNP coated with dextran), which is due to nature of the polymer coating. At 50 μgFe/mL, high levels of uptake were achieved, which was shown under microscopy. The results demonstrate that the pMPC-coated MNPs are potential candidates for use as cell-labelling agents.

MNPs can also be used for non-invasive imaging to monitor the targeting and therapeutic efficiency. In Zhang's work[2], MNPs was encapsulated in pCB nanogels as a magnetic resonance imaging (MRI) contrast reagent using a reduction-sensitive crosslinker. The degradation of nanogels was monitored by MRI test. The R2 relaxivity of the degraded samples decreased from 113.12 mM/s to 41.56 mM/s, which is close to the R2 relaxivity of monodisperse MNPs, indicating the complete disassembly of the encapsulated monodisperse MNPs. Magnetic resonance images were also consistent with the T2 tests, presenting different contrast ability of different samples.

Quantum dots (QDs) are powerful fluorescent tools for biological applications because of their narrow emission range and excellent photo-stability[35, 36]. To explore their potential applications, Liu et al. modified CdSe/ZnS QDs with zwitterionic 11-mercaptoundecylphosphorylcholine (HS-PC)[37]. The PC modified QDs (PC-QDs) showed good stability in the saline solution, pH varying solution and 100% human plasma. Common mouse macrophage RAW 264.7, normal endothelial cell HUVEC and typical cancer cell HepG2 were used to evaluate the cellular interaction between QDs and cells. Compared to the traditional mercaptoundecanoic acid (MUA) modified QDs (MUA-QDs), PC-QDs could effectively suppress the nonspecific interaction of QDs with phagocytic and non-phagocytic cells. Tat peptide, which can interact with cells strongly, was also used to functionalize PC-QDs. After incubated with HepG2 and RAW 264.7 for 12 h, the cellular distribution of the Tat conjugated nanoparticles can be visualized under confocal microscope. They were observed to prefer localizing at perinuclear region, but not enter the nucleus, as shown in Figure 5.3.

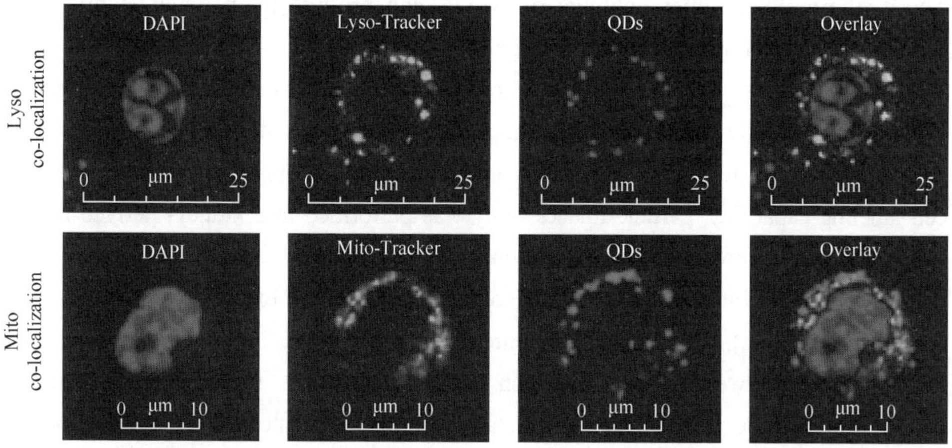

Figure 5.3 Confocal microscopy images of RAW 264.7 macrophages after incubation with Tat peptide functionalized PC-QDs for 12 h[37]. Nucleus was stained by DAPI (blue), lysosome was stained by lysotracker (green, upper), mitochondrion was stained by mitotracker (green, bottom), and QDs emitted red photoluminescence (red). The right two images are overlay of cell nuclei, QDs, with lysosomes or mitochondria

To develop biofunctionalized QDs, poly [2-(methacryloyloxy) ethyl phosphorylcholine (MPC)-*co*-*n*-butyl methacrylate (BMA)-*co*-ω-methacryloyloxy poly (ethylene oxide) oxycarbonyl 4-nitrophenol (MEONP)](PMBN) and poly (L-lactic acid) (PLA) were used to fabricate the QD-containing polymeric nanoparticles[38].

Hydrophobic PLA here was to increase interactions between QDs and polymer domains. The subsequently coated PMBN layer reduced nonspecific protein adsorption and cell uptake via its PC domains, as well as created a specific affinity by ligand molecules immobilized on the surface MEONP domains. To evaluate the ability of PMBN/PLA/QD to protect against nonselective cellular uptake, glycine was immobilized on PMBN/PLA/QD, which helped to avoid the chemical reactions between active ester groups in the MEONP units and proteins in the culture medium. It was hard to find any fluorescence emitted from cells. On the other hand, the octaarginine (R8)-immobilized PMBN/PLA/QD helped nanoparticles translocate through cell membranes. After incubated with HeLa cells, the internalized R8-modified PMBN/PLA/QD was able to be visualized under fluorescence microscope.

4. Biosensor

The highly sensitive detection of biomolecules is of great importance for medical diagnoses. Zwitterionic modification of nanoparticles helps to minimize non-specific adsorption. In this way, target protein binding events can be effectively transduced and amplified.

Park et al.[39] performed ligand exchange on CdSe/CdZnS QDs with lipoic acid and zwitterionic form lipoic acid. Then streptavidin (SA) and biotin were conjugated on QDs to form zwitterion-decorated SA-QD conjugate [(z)-SA-QD] and zwitterions-decorated biotin-QD conjugate [(z)-B-QD]. Because of the high specificity between SA and biotin, layer-by-layer self-assembly of (z)-SA-QD and (z)-B-QD was prepared using biotinylated agar beads. Through a simple alternating dipping process, fluorescence signal was amplified to reach conventional ELISA detection levels in half the time ELISA requires. The immunoassay and QD self-assembly-based signal amplification yielded a subattomolar limit of detection of myoglobin in 100% serum. Both experimental and computational results suggested that such highly sensitive signal amplification was led to by the zwitterionic surfaces, which provided equilibrium constants 5 orders of magnitude larger for specific binding than for nonspecific binding.

5.2.2 Zwitterionic nanogel

Zwitterionic nanogels composing of crosslinked hydrophilic polymers and water possess high water content, excellent biocompatibility, and desirable stability. Therefore, the potential applications of zwitterionic nanogels in the field of drug delivery have received extensive attentions in nanomedicine[40]. Zwitterionic nanogels are highly stable in ionic solutions, single protein solutions, or complex media (e.g., blood serum) for *in vitro* and *in vivo* biomedical applications. Two representative examples are discussed below. Zwitterionic poly (carboxybetaine methacrylate) (pCBMA) nanogels retained their original sizes in 100% FBS after 18 h incubation[41]. By tuning the mechanical property of pCBMA nanogels, Zhang et al. further investigated blood circulation time of pCBMA nanogels *in vivo*[31]. It was found that softer pCBMA nanogels achieved the circulation half-life of 19.6 h in rat. The softer nanogels passed through physiological barriers, especially the splenic filtration, more easily than their stiffer counterparts, consequently leading to longer circulation half-life and lower splenic accumulation. Zwitterionic poly (lysine methacrylamide) (pLysAA) and poly (ornithine methacrylamide) (pOrnAA) nanogels, both derived from natural amino acids—lysine and ornithine, maintained their initial hydrodynamic sizes even after 24 h incubation in undiluted blood serum[12]. In contrast, the control cationic nanogels aggregated quickly, with size increased from 140 nm to 800 nm after 1 h incubation with same media. Furthermore, the encapsulated model drug FITC—dextran exhibited controlled release from the pOrnAA nanogels. The nanogels made from amino acid-based zwitterionic polymers, including pLysAA and pOrnAA, showed minimal cell toxicity[42].

5.2.3 Zwitterionic copolymer nanoparticle

Random or block zwitterionic copolymers with amphiphilic character exhibit a large solubility difference between hydrophilic and hydrophobic segments, having a tendency to self-assemble into micelles in selective solvents[43-45]. In an aqueous solution, core-shell structure of micelles is formed: the insoluble hydrophobic segments aggregate to form core, which is surrounded by shell composed of hydrophilic segments. The self-assembling property of amphiphilic zwitterionic copolymers provides their high utility in many fields[46, 47]. Here we focus on biomedical applications of nanoparticles of zwitterionic based copolymers.

1. Stability

Antifouling materials were widely used to modify or compose nanoparticles to achieve their "stealth" properties. Jiang and coworkers prepared the poly (lactic-*co*-glycolic acid) (PLGA)-*b*-pCBMA (PLGA-PCB) block polymer[48]. The self-assembled nanoparticles of PLGA-PCB were shown pretty stable in protein solutions, maintaining their initial size over 5 days in both 100 mg/mL BSA and 100% FBS media, while unmodified PLGA particles aggregated immediately after incubation in these media. In addition, Chen et al. reported highly nonfouling micelles composed of CBMA as hydrophilic segment and 2-(methacryloyloxy) ethyl lipoate as hydrophobic segment[49]. No size increase of micelles in 1 mg/mL fibrinogen PBS solution and 50% FBS was observed after 3-day incubation.

A systematic study on how morphological changes of poly (N-isopropylacrylamide)-block-poly (sulfobetaine methacrylate) (pNIPAAm-b-pSBMA) copolymers with well-controlled molecular weights affected hemocompatibility in human blood showed that there was no obvious size increase of pNIPAAm-*b*-pSBMA micelles in fibrinogen solution[50]. In contrast, hydrophobic pNIPAAm nanoparticles aggregated dramatically, with size increased by 800% under the same condition. The nonfouling stability of pNIPAAm-b-pSBMA correlated well with its excellent hemocompatiblity (i.e., high anticoagulant activity and antihemolytic activity) over the explored temperatures from 4℃ to 40℃, suggesting potential in blood-contacting applications. The PLA nanoparticles modified with MPC polymer had excellent blood compatibility[51]. The amount of BSA adsorbing on the pMPC-modified nanoparticles was significantly smaller compared with the unmodified PLA nanoparticles.

2. Long circulation time and drug delivery

It is well known that one of the most undesired properties for nanoparticles is the rapid clearance from blood due to rapid protein adsorption *in vivo*. Polyzwitterionic materials are ideal candidates to achieve long circulation nanoparticles.

PCBMA based copolymer-DOX conjugates were prepared as nano drug vehicles to deliver anticancer drug[52]. The blood clearance of DOX and polymer-DOX was monitored in female mice. Results showed that DOX was rapidly cleared from blood due to its small molecular size and hydrophobicity. Nevertheless, the circulation time of pCBMA based copolymer-DOX micelles was greatly prolonged as compared to the free DOX. These micelles also exhibited low cytotoxicity and promoted DOX release under mild acidic conditions, indicating its great potential for cancer therapy. Chen et al. prepared four kinds of star CB copolymers of different molecular weights via ATRP[53]. These copolymers had excellent circulation time in mice, particularly for the largest polymer (123 kDa). The circulation half-life in mice was prolonged up to 40 h *in vivo*, with no appreciable damage or inflammation observed in the major organ tissues.

Phosphorylcholine (PC) is a zwitterionic polar group of phospholipids commonly found in the outer leaflet of cellular membranes. The pMPC-containing polymers were extensively studied to prolong circulation time of nanoparticles *in vivo*. PMPC-DOX nanocarriers with biodegradable polycarbonate backbones were prepared by conjugation of DOX to pMPC copolymers via a pH-sensitive hydrazine bond[54]. The circulation time of nanocarriers was further investigated in mice. Half life time of pMPC-DOX nanoparticles *in vivo* was 8 times longer than that of the free DOX because of excellent stealth property of pMPC. An amphiphilic random copolymer bearing pMPC was recently developed by Zhao et al. The copolymer was composed of 2-(methacryloyloxy) ethyl phosphorylcholine (MPC), stearyl methacrylate (SMA), and trimethoxysilylpropyl methacrylate (TSMA), as shown in Figure 5.4[55]. The blood circulation half-life of the formed polymeric micelles following intravenous administration in rabbits was investigated. The PC-bearing micelles showed significant prolongation of circulation time in blood with the half-life up to 90.5 h, while micelles bearing positive charges exhibited a rapid clearance from blood stream and the remaining dose at 1 h after administration was 46.6%. Compared to the short half-lives (0.1-1 h) of most nanoparticles, for example, PLA nanoparticles of less than 5 min[56], cationic and partially deacetylated chitin nanoparticles of 0.7 h[57], PLGA nanoparticles of 30 min[58] and so on, the long half-life of these micelles bearing PC groups is promising for the development of drug delivery systems.

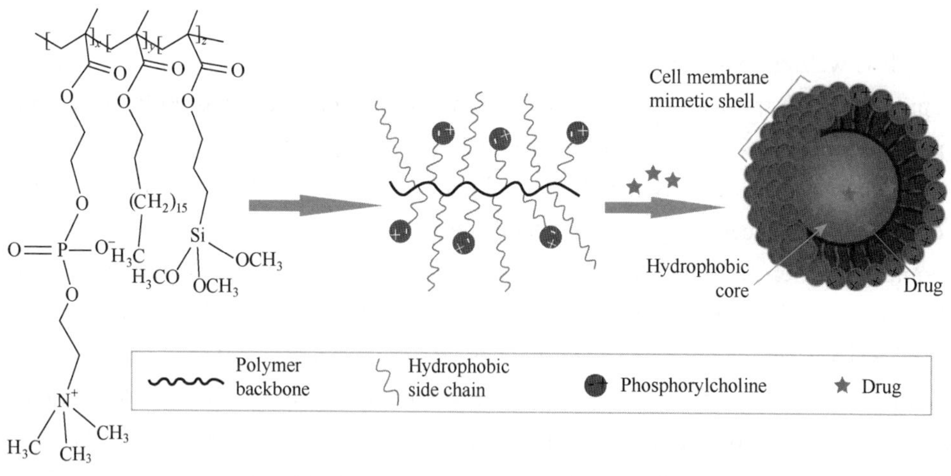

Figure 5.4　Schematic diagram of polymeric micelle formation of an amphiphilic random copolymer bearing MPC zwitterions by self-assembly in aqueous solution[55]

Sulfobetaine is another important zwitterionic molecule. Zhou et al. recently developed chitosan-*g*-pSBMA nanocarriers encapsulating radioprotective agent ferulic acid (FA)[59]. The pSBMA-modified nanoparticles exhibited prolonged retention behavior in blood, with a half-life of 10 h. In contrast, the half-life was only 10 min in blood for pure FA. The nanocarriers thus showed better radioprotective efficacy than the pure agent for irradiated mice. Zhai et al. synthesized an amphiphilic triblock copolymer containing zwitterionic SB, poly (3-caprolactone)-block-poly (diethylaminoethyl methacrylate)-block-poly (sulfobetaine methacrylate) (PCL-PDEA-PSBMA), as a novel pH-sensitive drug carrier[60]. A hydrophobic drug, curcumin, was chosen as the model drug to investigate the controlled drug release of this triblock copolymer at different pH values. The curcumin-loaded copolymer micelles were shown to significantly prolong the retention of curcumin in the circulation system *in vivo* with the drug circulation time extended to 24 h, much longer than that of the free curcumin solution, which was quickly removed from circulation.

3. Gene delivery

Gene delivery holds great promise as therapeutic treatment for many fatal diseases such as cancer and genetic disorder. The main challenge for gene delivery is to design suitable vectors possessing desirable properties including low cytotoxicity, high transfection efficiency, and complex stability. Due to their high resistance to protein adsorption, zwitterionic polymers, including CB-, SB-, and PC-based materials, are becoming attractive gene delivery vectors.

Jiang et al reported a new polymer for gene delivery, cationic hydrolytic pCB-based ester, as shown in Figure 5.5 [61]. The polymer can switch from cationic (DNA binding) to zwitterionic (DNA releasing) form while conferring nontoxicity. Interestingly, the hydrolysis-triggered DNA release induced gene expression 20 times more effectively than the branched polyethylenimine (PEI), yet without PEI's cytotoxic side effect. Therefore, the pCB-ester can be a good candidate as gene delivery vector to condense plasmid DNA into nanosized polyplexes for highly effective gene delivery with low toxicity. Similarly, a novel photolabile pCB-nitrobenzyl ester (pCB-NBE) was developed to allow rapid and externally controlled degradation[62]. PCB-NBE polyplexes released 72% of their entrapped DNA when polyplexes were exposed to 365 nm UV to rapidly degrade the side chain esters. A diblock copolymer poly (carboxybetaine methacrylate ethyl ester)-poly (carboxybetaine methacrylate) (pCBMAEE-pCBMA) was developed to form core-shell vector for gene delivery[63].

The hydrophobic convertible pCBMAEE segment helped to condense plasmid DNA (pDNA) into a hydrophobic core, which protected pDNA from nuclease attack and maintained the condensed structure against dilution. Furthermore, the hydrolysis of pCBMAEE released pDNA and reduced cytotoxicity of the cationic polymer. The pCBMA segment, a fixed polyzwitterionic antifouling material, was used to stabilize the gene vector in complex media. This polymeric gene vector has demonstrated very promising results for high and stable gene transfection even in complex growth media and difficult-to-transfect cell lines.

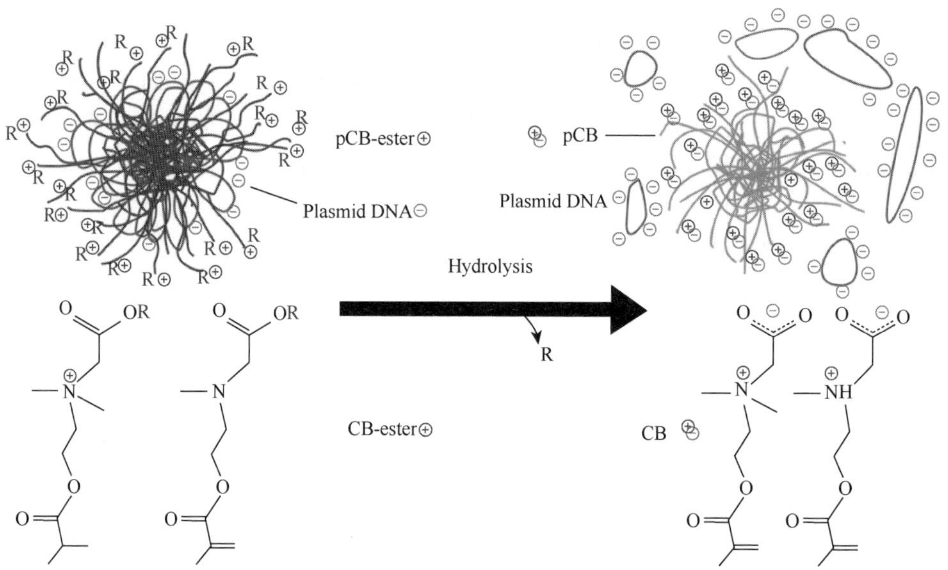

Figure 5.5 Polyplex formation and DNA release upon hydrolysis, using poly (carboxybetain) ester as gene delivery vehicle. Cationic CBMA ester polymer (pCB-ester) electrostatically condenses anionic DNA. Upon hydrolysis, DNA is released, leaving a zwitterionic polymer pCB[61]

The biomimetic PC-based polymers have shown effective stabilization to the polymer-DNA complexes and hence ensured good colloidal stability. Therefore, PC based polymers were widely used as non-virus gene delivery vectors. A novel supramolecular gene carrier system was self-assembled from two building blocks: a redox-sensitive β-cyclodextrin based cationic star polymer and an adamantyl end capped zwitterionic PC based polymer[64]. The assembly was based on the host-guest interaction between β-cyclodextrin and adamantyl. The supramolecular gene carrier can be degraded via disulfide-based reduction and shows excellent protein stability, serum tolerance, cellular uptake, and intracellular DNA release properties with low cytotoxicity and exceptionally high gene transfection efficiency, showing great potential for cancer gene therapy. In another work, a pH-responsive diblock copolymer of poly (methacryloyloxy ethyl phosphorylcholine)-block-poly (diisopropanolamine ethyl methacrylate) (PMPC-*b*-PDPA) was developed to deliver siRNA-mouse double minute 2 (MDM2) for the non-small cell lung cancer (NSCLC) treatment[65]. Sufficient MDM2 knockdown could inhibit tumor growth. With minimal surface charge and particle size of 50 nm, PMPC-*b*-PDPA/siRNA-MDM2 complex nanoparticles induced significant cell apoptosis and growth inhibition of NSCLC H2009 cells *in vitro*, and effectively inhibited the growth of H2009 tumor in nude mice *in vivo* via repeated injection of the complex nanoparticles.

The SB monomer was grafted onto PEI with different grafting ratios to form a new gene delivery vector[66]. The PEI-SB conjugates showed much lower protein adsorption and cytotoxicity compared with PEI with little hemolytic effect. Moreover, the PEI-SB conjugates could effectively condense DNA, and the PEI-SB/DNA complexes exhibited remarkably higher gene transfection efficiency than that of PEI. A comb-shaped vector, consisting of a dextran backbone and disulfide-linked cationic poly [(2-dimethyl amino) ethyl methacrylate] side chains, was functionalized with zwitterionic CB or SB to improve biophysical properties for efficient gene delivery[67]. The incorporation of zwitterionic betaines reduced cytotoxicity of the vector without destroying its DNA condensation capability. Cellular internalization results showed that the betaine functionalization improved the vector stability in serum and enhanced cellular uptake.

5.3 Conclusions and Outlooks

Zwitterionic nanoparticles are critically important for many biomedical applications due to excellent anti-biofouling properties of zwitterionic polymers. Significant advancements of zwitterionic antifouling polymer based nanoparticles have been achieved both *in vitro* and *in vivo* during the last two decades. In our opinion, future directions for zwitterionic nanoparticles may be aimed towards the following areas: (1) Multi-functional zwitterionic nanoparticles. The nanoparticles with targeting, bioimaging and drug delivery capabilities simultaneously are highly desirable. (2) Biodegradable zwitterionic nanoparticles. This helps the eventual cleanup of the nanoparticles from body. Currently, most of the zwitterionic polymers are nondegradable, which limits the scope of their use especially for applications *in vivo*. (3) Ultrathin yet effective zwitterionic antifouling coatings on nanoparticles. The coatings are expected to not significantly increase the nanoparticle size, which again helps to clean up the particles from body. (4) Zwitterionic nanoparticles with minimal cytotoxicity.

References

[1] Ratner B D, Bryant S J. Biomaterials: Where we have been and where we are going. *Annual Review of Biomedical Engineering*, 2004, 6: 41-75.

[2] Zhang L, Xue H, Cao Z, et al. Multifunctional and degradable zwitterionic nanogels for targeted delivery, enhanced MR imaging, reduction-sensitive drug release, and renal clearance. *Biomaterials*, 2011, 32 (20): 4604-4608.

[3] Rosenhahn A, Schilp S, Kreuzer H J, et al. The role of "inert" surface chemistry in marine biofouling prevention. *Physical Chemistry. Chemical Physics*, 2010, 12 (17): 4275-4286.

[4] Kirk J T, Brault N D, Baehr-Jones T, et al. Zwitterionic polymer-modified silicon microring resonators for label-free biosensing in undiluted human plasma. *Biosensors and Bioelectronics*, 2013, 42: 100-105.

[5] Luk Y Y, Kato M, Mrksich M. Self-assembled monolayers of alkanethiolates presenting mannitol groups are inert to protein adsorption and cell attachment. *Langmuir*, 2000, 16 (24): 9604-9608.

[6] Hirota K, Murakami K, Nemoto K, et al. Coating of a surface with 2-methacryloyloxyethyl phosphorylcholine (MPC) co-polymer significantly reduces retention of human pathogenic microorganisms. *FEMS Microbiology Letters*, 2005, 248 (1): 37-45.

[7] Wei F, Zhu S P, Kazuhiko I, et al. Adsorption of fibrinogen and lysozyme on silicon grafted with poly (2-methacryloyloxyethyl phosphorylcholine) via surface-initiated atom transfer radical polymerization. *Langmuir*, 2005, 21 (13): 5980-5987.

[8] Zhang Z, Chen S F, Chang Y, et al. Surface grafted sulfobetaine polymers via atom transfer radical polymerization as superlow fouling coatings. *The Journal of Physical Chemistry B*, 2006, 110 (22): 10799-10804.

[9] Zhang Z, Chao T, Chen S F, et al. Superlow fouling sulfobetaine and carboxybetaine polymers on glass slides. *Langmuir*, 2006, 22 (24): 10072-10077.

[10] Jon L, Zhang Z, Chen S F, et al. Zwitterionic polymers exhibiting high resistance to nonspecific protein adsorption from human serum and plasma. *Biomacromolecules*, 2008, 9 (5): 1357-1361.

[11] Liu Q S, Singh, A, Liu L Y. Amino acid-based zwitterionic poly (serine methacrylate) as an antifouling material. *Biomacromolecules*, 2013, 14 (1): 226-231.

[12] Liu Q S, Li W C, Singh A, et al. Two amino acid-based superlow fouling polymers: Poly (lysine methacrylamide) and poly (ornithine methacrylamide). *Acta Biomater*, 2014, 10 (7): 2956-2964.

[13] Li W C, Liu Q S, Liu L Y. Antifouling gold surfaces grafted with aspartic acid and glutamic acid based zwitterionic polymer brushes. *Langmuir*, 2014, 30 (42): 12619-26.

[14] Bernards M, Cheng G, Zhang Z, et al. Nonfouling polymer brushes via surface-initiated, two-component atom transfer radical polymerization. *Macromolecules*, 2008, 41 (12): 4216-4219.

[15] He Y, Hower J, Chen S F, et al. Molecular simulation studies of protein interactions with zwitterionic phosphorylcholine self-assembled monolayers in the presence of water. *Langmuir*, 2008, 24 (18): 10358-10364.

[16] Yang W, Zhang L, Wang S L, et al. Functionalizable and ultra stable nanoparticles coated with zwitterionic poly (carboxybetaine) in undiluted blood serum. *Biomaterials*, 2009, 30 (29): 5617-5621.

[17] Chen X J, Lawrence J, Parelkar S, et al. Novel zwitterionic copolymers with dihydrolipoic acid: Synthesis and preparation of nonfouling nanorods. *Macromolecules*, 2013, 46 (1): 119-127.

[18] Zhang L, Xue H, Gao C, et al. Imaging and cell targeting characteristics of magnetic nanoparticles modified by a functionalizable zwitterionic polymer with adhesive 3, 4-dihydroxyphenyl-l-alanine linkages. *Biomaterials*, 2010, 31 (25): 6582-6588.

[19] Zhu Y, Sundaram H S, Liu S, et al. A robust graft-to strategy to form multifunctional and stealth zwitterionic polymer-coated mesoporous silica nanoparticles. *Biomacromolecules*, 2014, 15 (5): 1845-1851.

[20] Liu X S, Huang H Y, Jin Q, et al. Mixed charged zwitterionic self-assembled monolayers as a facile way to stabilize large gold nanoparticles. *Langmuir*, 2011, 27 (9): 5242-5251.

[21] Trewyn B G, Giri S, Slowing I I, et al. Mesoporous silica nanoparticle based controlled release, drug delivery, and biosensor systems. *Chemical Communications*, 2007, (31): 3236-3245.

[22] Zhu Y F, Shi J L, Shen W H, et al. Stimuli-responsive controlled drug release from a hollow mesoporous silica sphere/polyelectrolyte multilayer core-shell structure. *Angewandte Chemie International Edition*, 2005, 44: 5083-5087.

[23] Liu J W, Stace-Naughton A, Brinker C J. Silica nanoparticle supported lipid bilayers for gene delivery. *Chemical Communications*, 2009, (34): 5100-5102.

[24] Lin Y S, Tsai C P, Huang H Y, et al. Well-ordered mesoporous silica nanoparticles as cell markers. *Chemistry of Materials*, 2005, 17 (18): 4570-4573.

[25] Rosenholm J, Sahlgren C, Lindén M. Cancer-cell targeting and cell-specific delivery by mesoporous silica nanoparticles. *Journal of Materials Chemistry*, 2010, 20 (14): 2707-2713.

[26] Jia G W, Cao Z Q, Xue H, et al. Novel zwitterionic-polymer-coated silica nanoparticles. *Langmuir*, 2009, 25 (5): 3196-3199.

[27] Yuan J J, Armes S P, Takabayashi Y, et al. synthesis of biocompatible poly[2-(methacryloyloxy) ethyl phosphorylcholine]-coated magnetite nanoparticles. *Langmuir*, 2006, 22 (26): 10989-10993.

[28] Perrault S D, Walkeyc, Jennings T, et al. Mediating tumor targeting efficiency of nanoparticles through design. *Nano Letters*, 2009, 9 (5): 1909-1915.

[29] Moghimi SM, H A C, Murray J C. Long-circulating and target-specific nanoparticles: theory to practice. *Pharmacological Reviews*, 2001, 53 (2): 283-318.

[30] Yang W, Liu S J, Bai T, et al. Poly (carboxybetaine) nanomaterials enable long circulation and prevent polymerspecific antibody production. *Nano Today*, 2014, 9: 10-16.

[31] Zhang L, Cao Z Q, Li Y T, et al. Softer zwitterionic nanogels for longer circulation and lower splenic accumulation. *ACS Nano*, 2012, 6 (8): 6681-6686.

[32] Chen F, Zhang M Q. Multifunctional magnetic nanoparticles for medical imaging applications. *Journal* of *Materials Chemistry*, 2009, 19: 6258-6266.

[33] Arruebo M, Fernández-Pacheco R, Ibarra M R, et al. Magnetic nanoparticles for drug delivery. *Nano Today*, 2007, 2 (3): 22-32.

[34] Peacock A K, Cauët S I, Taylor A, et al. Poly[2-(methacryloyloxy) ethylphosphorylcholine]-coated iron oxide

nanoparticles: Synthesis, colloidal stability and evaluation for stem cell labelling. *Chemical Communications*, 2012, 48 (75): 9373-9375.

[35] Dabbousi B O, Rodriguez-Viejo J, Mikulec F V, et al. (CdSe) ZnS core-shell quantum dots: Synthesis and characterization of a size series of highly luminescent nanocrystallites. *The Journal of Physical Chemistry B*, 1997, 101: 9463-9475.

[36] Bruchez M, Moronne M, Gin P, et al. Semiconductor nanocrystals as fluorescent biological labels. *Science*, 1998, 281 (5385): 2013-2016.

[37] Liu X S, Zhu H G, Jin Q, et al. Small and stable phosphorylcholine zwitterionic quantum dots for weak nonspecific phagocytosis and effective Tat peptide functionalization. *Advanced Healthcare Materials*, 2013, 2 (2): 352-360.

[38] Goto Y, Matsuno R, Konno T, et al. Artificial cell membrane-covered nanoparticles embedding quantum dots as stable and highly sensitive fluorescence bioimaging probes. *Biomacromolecules*, 2008, 9: 3252-3257.

[39] Joonhyuck P, Youngrong P, Sungjee K. Signal amplification via biological self-assembly of surface-engineered quantum dots for multiplexed subattomolar immunoassays and apoptosis imaging. *ACS Nano*, 2013, 7 (10): 9416-9427.

[40] Liu L Y, Li W C, Liu Q S. Recent development of antifouling polymers: Structure, evaluation, and biomedical applications in nano/micro-structures. *Nanomedicine and Nanobiotechnology*. 2014, 6 (6): 599-641.

[41] Cheng G, Mi L, Cao Z Q, et al. Functionalizable and ultrastable zwitterionic nanogels. *Langmuir*, 2010, 26 (10): 6883-6886.

[42] Li W C, Liu Q S, Liu L Y. Amino acid-based zwitterionic polymers: Antifouling properties and low cytotoxicity. *Journal of Biomaterials Science, Polymer Edition*, 2014, 25 (14-15): 1730-1742.

[43] Moffitt M, Khougaz K, Eisenberg A. Micellization of ionic block copolymers. *Accounts of Chemical Research*, 1996, 29 (2): 95-102.

[44] Kataoka K, Harada A, Nagasaki Y. Block copolymer micelles for drug delivery: Design, characterization and biological significance. *Advanced Drug Delivery Reviews*, 2001, 47 (1): 113-131.

[45] Tuzar Z, Kratochvil P. Micelles of block and graft copolymers in solutions. *Surface and Colloid Science*, 1993, 15 (1): 1-83.

[46] Kramarenko E Y, Potemkin I I, Khokhlov A R, et al. Surface micellar nanopattern formation of adsorbed diblock copolymer systems. *Macromolecules*, 1999, 32 (10): 3495-3501.

[47] Antonietti M, Göltner C. Superstructures of functional colloids: chemistry on the nanometer scale. *Angewandte Chemie International Edition*, 1997, 36 (9): 910-928.

[48] Cao Z Q, Yu Z Q, Xue H, et al. Nanoparticles for drug delivery prepared from amphiphilic PLGA zwitterionic block copolymers with sharp contrast in polarity between two blocks. *Angewandte Chemie International Edition*, 2010, 122 (22): 3859-3864.

[49] Lin W F, He Y Y, Zhang J, et al. Highly hemocompatible zwitterionic micelles stabilized by reversible cross-linkage for anti-cancer drug delivery. *Colloids and Surfaces B*: *Biointerfaces*, 2014, 115: 384-390.

[50] Shih Y J, Chang Y, Deratani A, et al. "Schizophrenic" hemocompatible copolymers via switchable thermoresponsive transition of nonionic/zwitterionic block self-assembly in human blood. *Biomacromolecules*, 2012, 13 (9): 2849-2858.

[51] Konno T, Kurita K, Iwasaki Y, et al. Preparation of nanoparticles composed with bioinspired 2-methacryloyloxyethyl phosphorylcholine polymer. *Biomaterials*, 2001, 22 (13): 1883-1889.

[52] Wang, Z, Ma G L, Zhang J, et al. Development of zwitterionic polymer-based doxorubicin conjugates: Tuning the surface charge to prolong the circulation and reduce toxicity. *Langmuir*, 2014, 30 (13): 3764-3774.

[53] Lin W F, Ma G L, Ji F Q, et al. Biocompatible long-circulating star carboxybetaine polymers. *Journal of Materials Chemistry B*, 2015, 3: 440-448.

[54] McRae Page S, Henchey E, Chen X J, et al. Efficacy of polyMPC-DOX Prodrugs in 4T1 tumor-bearing mice. *Molecular Pharmaceutics*, 2014, 11 (5): 1715-1720.

[55] Zhao J, Chai Y D, Zhang J, et al. Long circulating micelles of an amphiphilic random copolymer bearing cell outer membrane phosphorylcholine zwitterions. *Acta Biomaterialia* 2015, 16: 94-102.

[56] Shan X Q, Liu C S, Yuan Y et al. *In vitro* macrophage uptake and *in vivo* biodistribution of long-circulation nanoparticles with poly (ethylene-glycol)-modified PLA (BAB type) triblock copolymer. *Colloids and Surfaces B: Biointerfaces*, 2009, 72 (2): 303-311.

[57] Sheng Y, Liu C S, Yuan Y et al. Long-circulating polymeric nanoparticles bearing a combinatorial coating of PEG and water-soluble chitosan. *Biomaterials*, 2009, 30 (12): 2340.

[58] Parveen S, Sahoo S K. Long circulating chitosan/PEG blended PLGA nanoparticle for tumor drug delivery. *European Journal of Pharmacology*, 2011, 670 (23): 372-383.

[59] Zhou Y, Hua S, Yu J H, et al. A strategy for effective radioprotection by chitosan-based long-circulating nanocarriers. *Journal of Materials Chemistry B*, 2015, 3: 2931-2934.

[60] Zhai S Y, Ma Y H, Chen Y Y, et al. Synthesis of an amphiphilic block copolymer containing zwitterionic sulfobetaine as a novel pH-sensitive drug carrier. *Polymer Chemistry*, 2014, 5 (4): 1285-1297.

[61] Carr L R, Jiang S Y. Mediating high levels of gene transfer without cytotoxicity via hydrolytic cationic ester polymers. *Biomaterials*, 2010, 31 (14): 4186-4193.

[62] Sinclair A, Bai T, Carr L R, et al. Engineering buffering and hydrolytic or photolabile charge shifting in a polycarboxybetaine ester gene delivery platform. *Biomacromolecules*, 2013, 14 (5): 1587-1593.

[63] Zhang J, Wang Z, Lin W F, et al. Gene transfection in complex media using PCBMAEE-PCBMA copolymer with both hydrolytic and zwitterionic blocks. *Biomaterials*, 2014, 35 (27): 7909-7918.

[64] Wen Y T, Zhang Z X, Li J. Highly efficient multifunctional supramolecular gene carrier system self-assembled from redox-sensitive and zwitterionic polymer blocks. *Advanced Functional Materials*, 2014, 24 (25): 3874-3884.

[65] Yu H, Zou, Y, Jiang L, et al. Induction of apoptosis in non-small cell lung cancer by downregulation of MDM2 using pH-responsive PMPC-*b*-PDPA/siRNA complex nanoparticles. *Biomaterials*, 2013, 34 (11): 2738-2747.

[66] Sun J, Zeng F, Jian H, et al. Conjugation with betaine: A facile and effective approach to significant improvement of gene delivery properties of PEI. *Biomacromolecules*, 2013, 14 (3): 728-736.

[67] Xiu K M, Zhao N N, Yang W T, et al. Versatile functionalization of gene vectors via different types of zwitterionic betaine species for serum-tolerant transfection. *Acta Biomaterialia*, 2013, 9 (7): 7439-7448.

Chapter 6 Nucleosome Organization Around Transcriptional Sites and Its Role in Gene Regulation

Xiao Sun, Yue Hou, Huan Huang, Yumin Nie, Yiru Zhang, Honde Liu

6.1 Introduction

Nucleosomes are the basic unit of chromatin, into which a eukaryotic genome with huge genetic information is packaged. The linear chromatin comprises DNA periodically wrapped nearly twice in left-handed turns around histone octamers to form nucleosomes containing about 147 bp DNA. Nucleosomes are then further assembled into various higher order structures that can potentially be unwrapped as needed. Generally nucleosomes are arranged in spaced array, like a set of gapped beads threaded by a string, but the spacing between nucleosomes varies along a chromatin. For a given chromatin, nucleosomes are dense in most part while there are some regions free of nucleosomes. The arrangement of nucleosomes affects every process that occurs on DNA, e.g. DNA replication and DNA repair. Especially the positioning of nucleosomes with respect to DNA plays an important role in the regulation of gene transcription by managing DNA accessibility to regulatory proteins[1]. Moreover, a few of nucleosomes carry many types of histones modifi cations, which give more biological roles to nucleosomes in both transcription and replication. Understanding the rules and factors that determine how nucleosomes are positioned and how they influence gene transcription is one of the key questions in genome biology currently.

Nucleosome positioning describes the precise location of a nucleosome on DNA. It has been known for decades that nucleosomes are organized as nonrandom, regularly spaced arrays, with the spacing between nucleosomes varying between different organisms and cell types[2]. There is a key question related to the nonrandom nucleosomal organization: what determines nucleosome positions throughout a genome? The mechanism of nucleosome positioning along DNA remains controversial: some researchers focus the role of the DNA sequence, notably "sequence dependent positioning"[3, 4], where as others stress the effects of *in vivo* environment (e.g. protein factors, ATP-consuming remodeling enzymes), namely "statistical positioning" [5, 6].

Nucleosome positioning is partly determined by DNA sequence. Since structural properties of DNA, such as local bendability, depend on DNA sequence, one might expect that DNA sequence will at least partially contribute to nucleosome positioning[7]. The structure of poly-dA/dT sequences differs from the canonical double helix[8] and is presumed to be resistant to the distortions necessary for wrapping around nucleosomes. Conversely, sequences with AA/TT/TA dinucleotides spaced at ~10 bp intervals are intrinsically bendable and thus bind the histone octamer with higher affinity than random sequence[9, 10]. There is a concept of "chromatin code", i.e., DNA sequence modulates how and where the genomic DNA is packaged around nucleosomes. In yeast, it was reported that ~50% of nucleosomes can be predicted based on the principal of 10-bp periodicity of dinucleotides AA/TT/TA and GC[10]. Moreover, the 10-bp periodicity is pronounced across genomes of human, fly, worm and yeast[3, 10-12]. As nucleosome positioning is determined by DNA sequence in a certain extent, it could be predicted from genomic DNA sequence. We have proposed a curvature profile model for predicting nucleosome positions and applied it to the human genome. The results indicated that DNA sequences, through affecting the curvature of DNA double helix, partly determine nucleosome positions[13].

From the perspective of statistical positioning, nucleosome organization is mainly a result of "barrier". The barrier can be formed through the binding of transcription factors to its DNA sites, i.e., genomic DNA occupied by other proteins instead of histones. These fixed barriers affect nearby nucleosome positioning and result in sufficiently high nucleosome density such that one nucleosome sterically restricts the position of a neighboring nucleosome[14]. Transcription factor bound strongly to a specific DNA site might sterically prevent nucleosomes from occupying that location. The correlations in nucleosome positions near such a barrier can generate spatial variations of nucleosome occupancy, but without the necessity of any intrinsic DNA-sequence-dependence of histone-DNA interactions[15]. Previous studies indicated that nucleosome positions around transcription start sites *in vivo* are different from *in vitro* positions, which has led to the suggestion that transcription promotes nucleosome organization *in vivo*. Generally there is a nucleosome free region, head of transcription start site, which could be a barrier affecting the downstream nucleosome organization. Pre-initiation complexes are assembled in this kind of nucleosome free region.

We suggested that there is a default nucleosome position determined by DNA sequence, which could be repositioned in a specific cell environment. Nucleosome positioning is not determined by any single factor but rather by the combined effects. The intrinsic properties of DNA structure is the *cis* determinants of nucleosome positioning, while there are some other *trans* determinants of nucleosome positioning[7]. DNA-binding proteins, especially transcription factors, which compete with histones, are the major classes of *trans* determinants implicated in nucleosome positioning. Other *trans* determinants include chromatin remodelers and the RNA polymerase II (RNAP II) transcription machinery[10].

Over the past ten years, an increasing number of studies have made use of micrococcal nuclease (MNase) digestion with DNA microarray or DNA sequencing to map nucleosomes across the genomes of organisms ranging from budding yeast to humans[16-19]. Briefly, crosslinked chromatin is digested to mononucleosomes using micrococcal nuclease, which preferentially digests linker DNA and leaves nucleosomal DNA intact. Protected DNA is isolated, and nucleosomal fragments are assayed either by tiling microarray or by ultra high-throughput sequencing. These studies produced data of nucleosome occupancy, measuring the density of nucleosomes at a specific genomic region in a population. Currently there is new technique to depict the nucleosome scenery in a single cell and measure the real nucleosome positioning. Small et al. developed a technique to determine nucleosome positioning in single cells by virtue of the ability of the nucleosome to protect DNA from GpC methylation[20].

High-throughput whole genome nucleosome data make it possible for us to analyze the determinants of nucleosome positioning and its role in gene transcription regulation. Generally, in comparison to nucleosome depleted DNA, nucleosomal DNA is less accessible for DNA-binding proteins such as transcription factors, which means that the precise positioning and density of nucleosomes serves as a potent mechanism for controlling transcription and other DNA-templated processes.

There are different nucleosome arrangements in different regulatory regions. Usually in the upstream of an expressed gene, there is a nucleosome free region immediately head of transcription start site (TSS). Moreover, an array of highly positioned nucleosomes surrounds TSSs, with positioning generally decreasing with distance from the TSS. Nucleosome occupancy is relatively low at many enhancers, promoters, and transcription termination sites.

Investigating the precise locations of nucleosomes in a genome is the key to understanding the mechanism of gene regulation. In this chapter we will report the nucleosome organization around different kinds of transcriptional sites and its effect on gene transcriptional regulation. We got these results by analyzing high throughput DNA sequencing data. Our analysis of genome-wide nucleosome positioning data from $CD4^+$ T cells indicates that the first nucleosome upstream of TSSs is shifted to the 5′ direction, thus forming a broad nucleosome-free region near the TSSs in highly expressed genes and house-keeping genes. The nucleosome distributions around splice sites and polyadenylation sites influence alternative splicing and regulate 3′ end processing of protein-coding genes respectively. There are some special patterns of nucleosome occupancy around transcription factor binding sites, which are associated with the expression levels of target genes. While DNA sequence is the intrinsic determinant of nucleosome positioning, nucleosomes could be remodeled in the process of gene transcriptional regulation.

6.2 Nucleosome Depletion Upstream of Transcription Start Sites

It is well known that there exists a specific nucleosome organization in the vicinity of the transcription start site. There is a nucleosome-free (depleted) region (NFR or NDR), roughly 200 bp wide, flanked by two well-positioned nucleosomes that often contain the histone variant H2A.Z[21]. A −1 nucleosome is situated upstream of TSS, and a +1 nucleosome is situated downstream of TSS, in addition to an array of positioned nucleosomes throughout the gene body in the direction of transcription. The upstream nucleosome depletion is a conserved feature reported in eukaryotic genomes[11, 12] which is functionally important because it affects the accessibility for protein factors to bind to DNA target sites thus regulates initiation of gene transcription.

Saito et al.[22] examined the publicly available data of both nucleosome positions, which are estimated from MNase digestion of whole chromatin, and Pol II binding near representative TSSs in *C. elegans*. The finding is that the periodicity of the nucleosome positioning around representative TSSs was estimated as 160 bp, which is smaller than the 175-bp average spacing in the entire genome. These observations indicate that transcription correlates with more consistent packing of nucleosomes. In yeast, the size of the NFR varies, and expressed genes tend to have larger NFRs than unexpressed ones[23]. In human $CD4^+$ T cells, the first +1 nucleosome downstream of the TSS exhibits a differential positioning pattern in active and silent genes. H2A.Z-containing and histone-modified nucleosomes are preferentially lost from the first −1 nucleosome position upstream of TSS, suggesting the importance of the −1 nucleosome in transcription[19].

Previous studies indicate that the nucleosome organization nearby TSS is determined by interplay between DNA sequence-dependent positioning and statistical positioning. Zhang and colleagues suggested that the apparent statistical positioning observed near TSS barriers requires ATP[5]. However, some experiments suggested that DNA sequence does play a dominant role in positioning nucleosomes near TSS[3, 24]. Owing to ATP-driven remodeling, nucleosome organization is indeed highly dynamic. Jyotsana et al.[15] developed a theoretical description of nucleosome dynamics near barriers around TSSs and examine the effect of DNA sequence as well as ATP-dependent remodeling. They found that establishment of apparent statistical positioning on biologically relevant timescales requires active chromatin remodeling.

To investigate the interplay between sequence-dependent positioning and statistical positioning, we compared the nucleosome profiles derived both from computer prediction based on DNA curvature and from experiment data[19]. We have developed this computer model, which predicts the nucleosome signal from DNA double helix curvature[25]. We found that curvature of nucleosomal DNA shows a particular pattern, namely the curvature pattern. Given a DNA sequence, the curvature value is calculated at each position. The whole curvature of the sequence is called the curvature curve. The nucleosome positions can be predicted from the convolution of the curvature curve and the curvature pattern of a nucleosomal DNA helix. If a segment of the curve resembles the pattern signal, the convolution will peak at the corresponding position, indicating a nucleosome. Our analysis show that the curvature computation results in a ~150-bp NFR around the TSSs of 3571 protein-coding genes[13], which is consistent with experiment data. However, not all TSSs are nucleosome free *in vivo*. Using Zhao et al.'s experimental data[19] in the range of −150 bp to 50 bp from the TSS, 3571 TSSs were divided into two classes by a *k*-means clustering method. This resulted in 1080 occupied TSSs (class I) and 2491 nucleosome-free TSSs (class II). Both predicted nucleosome signal and the experimental nucleosome signal are consistent with nucleosome depletion at class II TSSs. Around the class I TSSs, no distinct positioning was observed in the predicted nucleosome profile, while a positioned nucleosome was suggested by the experiment data. This difference between the prediction and the experiment indicates that positioning is not completely determined by DNA sequences *in vivo*. We believe that the nucleosome-free state is the default configuration, partly determined by DNA sequences at a TSS; however, *in vivo*, due to the function of the remodeling complexes, some of TSSs are occupied. This is why there are differences between the prediction and the experiment. DNA sequences partly encode a default NFR around a TSS. Due to the requirement of gene transcription *in vivo*, some TSSs are occupied through chromatin remodeling, while others are still in NFRs. The positioned nucleosome at a TSS can block the binding sites of the pre-initiation complex, implying that a TSS-occupied gene should exhibit a lower expression level, which was confirmed by our analyses[13].

Nucleosome organization surrounding TSSs can affect gene transcription and gene expression. The question is whether different expressed genes have different nucleosome organization round TSSs, that is whether there is an obvious association between the nucleosome organization and the expression. We examined the nucleosome data from both resting and activated $CD4^+$ T cells, focusing on the NFRs upstream of TSSs, of different genes. The most important finding is that highly expressed genes have a broader NFR. Generally, there are two kinds of nucleosome patterns near TSSs, corresponding to gene expression levels. The nucleosome profiles of highly expressed genes show a similar shape regardless of the resting or activated cell states. For lowly expressed gene, the profiles also match in resting and activated cells, especially in a 0.4 kbp region surrounding the TSS[26]. These results suggest that most nucleosomes have stable positions even when the cell states change. Moreover, the +1 nucleosome is well positioned at ~+135 bp downstream of TSSs in both types of genes. In the "barrier" model for nucleosome organization[18, 11], the +1 nucleosome serves as a "barrier" and is involved in forming uniform positioning downstream of the TSS. The +1 nucleosome indicates the site of gene transcription initiation, because the +1 nucleosome is consistently aligned relative to the TSS. Significantly, the first −1 nucleosome upstream of TSS shows differential positioning. In low expressed genes, it positions at −100 bp, while in highly expressed genes, it shifts to −175 bp upstream of the TSS (Figure 6.1, see the Color Inset, p. 21). Considering the +1 nucleosome well positions at ~+135 bp downstream of TSSs in both highly expressed genes and lowly expressed genes, we can infer that the NFR in highly expressed genes is 75 bp broader than that in lowly expressed genes. The −1 nucleosome is shifted to the 5′ direction, thus forming a broad nucleosome-free region (NFR) near the TSS in highly expressed genes in $CD4^+$ T cells. This is consistent with the finding that a broad NFR at promoter regions correlates with highly expression of genes in yeast[23].

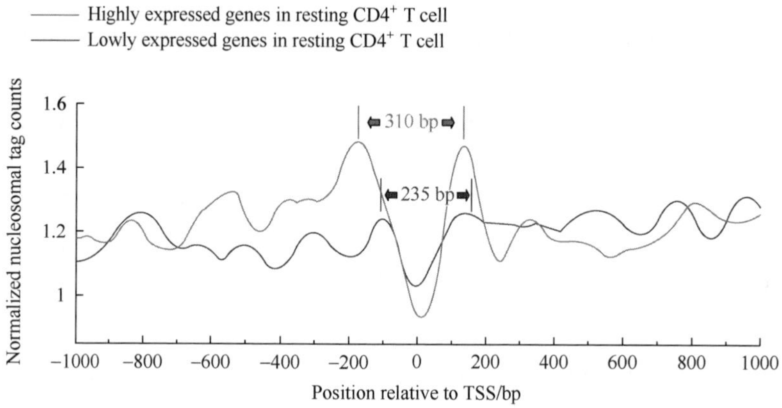

Figure 6.1 Profiles of nucleosomes near the TSSs of highly expressed genes and lowly expressed genes. It is obvious that highly expressed genes (the green curve) have a broader NFR than that of lowly expressed genes (the blue one)

Because of the important role of nucleosome occupancy around the TSSs, we further explored the distinct patterns of nucleosomes around the TSSs of the intronless and intron-containing genes. Our analysis revealed that nucleosome occupancy in the promoters of intronless genes is significantly lower than that of intron-containing genes. Furthermore, intron-containing genes always have deeper and wider NFR than intronless ones. We inferred that Poly (dA : dT) tracts are the important determinants of fuzzy nucleosomes in the promoters of intronless genes. However, nucleosome occupancy is significantly higher in intronless genes when it comes to the gene body. In intronless genes, nucleosome occupancy on gene body is significantly higher than that on flank area of gene body. And intronless genes always maintain a stable level for the nucleosomes along gene body, while nucleosome occupancy pattern on the gene body of intron-containing genes is not consistent with that showed in intronless genes. In intron-containing genes, the +1 nucleosome is well positioned and downstream nucleosome occupancy displays oscillations, with amplitude that decays with the increasing distance from the TSS. We concluded that the sequence preference, the DNA methylation and barrier nucleosome model all have influences on the nucleosomes along the gene body of intronless genes[27].

6.3 Nucleosome Occupancy Around Transcriptional Splice Sites

Nucleosome organization has functional importance in gene transcription. Previous studies indicated that nucleosome positioning influences transcription start, but its role in subsequent RNA processing is unclear. Alternative splicing is the characteristics of eukaryotic RNA processing, which is believed to be coupled with gene transcription. In this process, particular exons of a gene may be included or excluded from the final mRNA, resulting in multiple protein products from a single coding gene. In the human genome, ~95% of multiexonic genes undergo alternative splicing[28]. Regulation of alternative splicing is as important as regulation of transcription to determine cell and tissue specific features, normal cell functioning and responses of eukaryotic cells to external cues[29]. In the process of splicing, donor sites (5′ end of the intron) and acceptor sites (3′ end of the intron) are recognized and selected by the spliceosome, resulting in introns (also some exons) removed. Different pairings of donor and acceptor sites could be formed and different parts of a gene will be removed, named alternative splicing, by which a single gene producing multiple proteins. The selection of splice sites is the key to alternative splicing. Lines of evidences show that the regulation and selection of splice sites are influenced by nucleosome organization in the vicinity splice sites.

Results of recent studies demonstrate the general link between nucleosome organization and exon-intron architecture. Trifonov and colleagues analyzed DNA sequence pattern and nucleosome position preference around transcriptional splice sites and stated that splice sites correlate with nucleosome positions and splice sites are frequently located near the nucleosome dyad axis[30, 31]. Tilgner et al.[32] found nucleosome occupancy around splice sites is weak while stable within exons, by investigating the nucleosome organizations of humans and *Caenorhabditis elegans*. Their results also indicate that nucleosomes are positioned central to exons rather than proximal to splice sites and the ratio between nucleosome occupancy within and upstream from the exons correlates with exon-inclusion levels. By analyzing genome-wide nucleosome-positioning data sets from humans, flies and worms, Schwartz1 et al. found that exons show increased nucleosome occupancy levels with respect to introns[33]. It was also confirmed by Robin and colleagues[34]. that nucleosomes are well positioned in exons. Iannone et al. systematically explored potential links between nucleosome positioning and alternative splicing regulation upon progesterone stimulation of breast cancer cells. They confirmed preferential nucleosome positioning in exons and reported four distinct profiles of nucleosome density around alternatively spliced exons[35].

Our previous work shows that different types of splicing events has different chromatin organization[36]. But, the relationship of nucleosome occupancy and alternative events is still unclear. Some studies assume that nucleosomes might serve as "speedbumps" for RNAP II (Pol II), slowing down transcription rates in the vicinity of splice sites to improve the recognition capability of the splicing machinery[33, 37]. If this assumption is true, nucleosomes could also be important in regulating alternative events.

We have investigated the potential influence of nucleosome organization around genome-wide alternative events by analyzing high-throughput experimental data in the human genome[38]. The analysis of nucleosome occupancy around splice sites of different types of exons indicates that the constitutive sequences of alternative events have stronger nucleosome enrichment than that of skipped parts, suggesting the constitutive portion might have been in a more stable chromatin structure to protect the key region and facilitate exon recognition.

To further explore the relationship between nucleosome occupancy and alternative splicing, we calculated the nucleosome occupancy profiles across a 1000-nt window surrounding splice sites in human $CD4^+$ T cells. There is an obvious peak in alternative acceptor exons (AAEs), which grows with the alignment site from the proximal to distal alternative 3′ splice site ($p<0.01$). There is also an obvious peak of nucleosome occupancy in alternative donor exons (ADEs), and the peak surrounding distal site is significantly higher than that of proximal site ($p<0.01$). These results indicate that nucleosomes are preferentially positioned surrounding distal splice sites in ADEs and AAEs, which are relevant to intrinsic nucleosome sequence preferences. It suggests that nucleosomes could help in improving the distal alternative splice site recognition. Moreover, nucleosomes in ADEs tend to be positioned within alternative portions while, in AAEs, they are within constitutive portions (Figure 6.2), suggesting that nucleosome may have different roles in the recognition of alternative 3′ splice site and alternative 5′ splice site. This supports the assumption that a nucleosome well positioned in an exon could serve as a “speed-bump” for RNAP II, slowing down the transcription rate and increasing the chance of kinetic coupling of splicing factors carried by the RNAP II, which has been recruited to the pre-mRNA, and thus improving the downstream splicing site recognition.

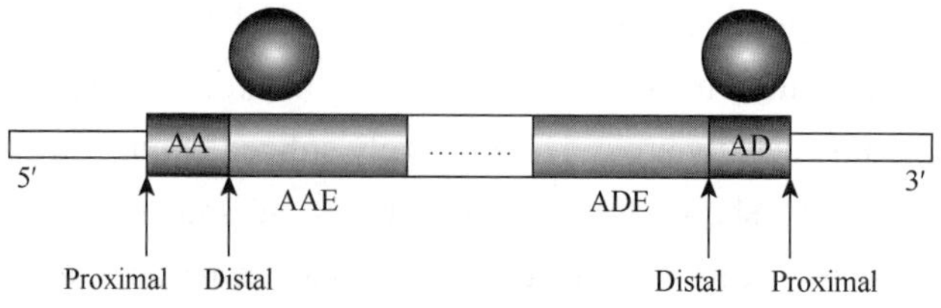

Figure 6.2 Nucleosomes in ADEs tend to be positioned within alternative portions while, in AAEs, they are within constitutive portions

6.4 Nucleosome Positioning in the Vicinity of Transcription Termination Sites

RNA polymerase II (RNAP II) is responsible for the transcription protein-coding genes and many non-coding RNAs. Some studies have showed that changes in chromatin are important for termination and suggested that RNAP II requires a specific chromatin structure to terminate efficiently[39].

Formation of 3′ end of precursor messenger RNA (pre-mRNA) is an essential step in the procedure of eukaryotic gene expression, and it directly impacts many other steps in the gene expression pathway, such as transcription termination, mRNA stability and export [40, 41]. 3′ end processing of mammalian cells involves two tightly coupled steps, cleavage and polyadenylation, and requires a polyadenylation signal [PAS, AAUAAA or something similar which is generally found ~10-30 bases upstream of the site of cleavage/polyadenylation (polyA site)] and a downstream sequence element (DSE, U or G/U rich sequence which is generally found ~14-70 bases downstream of polyA site)[42]. Transcription termination is triggered following recognition of the polyadenylation signal by RNAP II and subsequent pre-mRNA cleavage, which occurs at the polyadenylation site (polyA site). Over half of the genes in the human genome have alternative polyA sites[43]. Some researchers suggested that nucleosomes have a potential effect on pre-mRNA 3′ end formation or termination[44, 45]. It has been reported that nucleosome occupancy dropped precipitously near the polyadenylation site in many species. But the influences on nucleosome occupancy and the relationship between nucleosome organization and polyadenylation are still unclear.

We investigated the nucleosome organization near transcription termination sites (TTSs) in the human genome with integrated data of genome sequence, protein-coding genes and non-coding genes, nucleosome positions in $CD4^+$ T cells, Granulocytes and *in vitro*, gene expression derived from mRNA-seq and RNA polymerase II occupancy[46]. We found that different nucleosome distribution patterns near transcription termination sites of different types of genes, and protein-coding genes have a specific chromatin structure near transcription termination sites relative to non-coding genes, which is related to polyadenylation. Strikingly, protein-coding genes display an obvious nucleosome depleted region (NDR) in the upstream of TTSs, and stronger nucleosome occupancy immediately downstream of TTSs than that of non-coding genes in different types of cells and *in vitro*, suggesting that nucleosome occupancy around TTSs is not associated with the type of cells and is determined by DNA sequence to some extent. It is believed that DNA sequence plays an important role in nucleosome positioning and DNA sequences rich in AT disfavor core histones[3, 47]. We observed that nucleosome depletion near TTSs is very strongly positively correlated with the conservation of PAS signal. Therefore, we conjectured that nucleosome disfavoring sequences—PAS plays a mainly important role in producing nucleosome depletion in the upstream of TTSs. Our analysis also showed that the lowest GC content appears near the position of PAS and there is also a lower GC content near DSE, which is consistent with nucleosome depletion. The results suggest that GC content play a crucial role in nucleosome occupancy. While conservative sequence elements are the determinant of nucleosome depletion near TTSs, proteins binding to them also have effect on nucleosome arrangement. Furthermore, we showed that nucleosome occupancy is regulated by gene transcription and RNAP II occupancy. Indeed, highly expressed genes have a lower nucleosome level and higher RNAP II enrichment near TTSs than that of lowly expressed genes.

To address the relationship between nucleosome occupancy and polyadenylation, the polyA sites were divided into constitutive polyA sites and alternative sites. If a gene has only one polyA site, the site is called a constitutive site. If a gene has more than one polyA site, those sites are called alternative sites. We analyzed nucleosome distribution around constitutive and alternative polyA sites, and found that nucleosomes immediately downstream of constitutive polyA sites are intrinsically better positioned. Constitutive polyA sites displayed a significantly stronger nucleosome depletion near polyA sites than alternative sites did ($p<1\times10^{-19}$) *in vivo* but not *in vitro*. These results suggest that nucleosome profiles around polyA sites are partly determined by the DNA sequence and cellular trans-factors, *in vivo*.

The height of the nucleosome occupancy peak downstream of constitutive polyA sites is higher than that of downstream of alternative sites. Nucleosomes downstream of constitutive polyA sites are better positioned. Moreover, the nucleosome level around a constitutive site negatively correlates with the gene expression, and there is a proportional relationship between RNAP II occupancy and gene expression. This indicates that transcription activation can regulate nucleosome levels and RNAP II occupancy is a determinant of nucleosome dismissing.

We analyzed of nucleosome organization round alternative polyA site in detail. For high usage alternative polyA sites, the downstream nucleosome level decreases dramatically and RNAP II occupancy increases quickly, which is consistent with the constitutive sites. There is greater accumulation of RNAP II upstream of low-usage polyA sites, suggesting a different regulation of upstream nucleosomes[46].

In our analysis, alternative polyA sites were divided into proximal, distal and inbetween polyA sites, according to the position of the polyA sites relative to the TSSs. We observed different nucleosome occupancy patterns between different types of alternative polyA sites, especially, the proximal and distal sites. The proximal sites have higher upstream nucleosome occupancy than downstream nucleosome occupancy. The distal sites have better positioned downstream nucleosomes than upstream ones, which is similar to the constitutive polyA sites *in vivo*. We concluded that nucleosomes downstream of distal polyA sites are similar to those downstream of constitutive sites and have a similar influence on the regulation the usage of polyA sites and transcription termination, while proximal polyA sites have better positioned upstream nucleosomes to slow down RNAP II speed and improve the polyA sites utilization.

Based on these results, we proposed the dual pausing model of RNAP II and nucleosome occupancy patterns during regulation of the usage of different polyA sites and transcription termination (Figure 6.3, see the Color Inset, p. 22). For a given gene, the nucleosome upstream of the proximal site, which is adjustable and related to the site usage, is the first barrier pausing RNAP II. Meanwhile, it signals to RNAP II to prepare for a 3′ end processing event, by slowing down transcription speed. The nucleosome downstream of the distal or constitutive polyA site, which is intrinsically well positioned and regulated by transcription, is the second barrier pausing RNAP II to ensure transcription termination and recognition of the last polyadenylation sites if previous sites were missed.

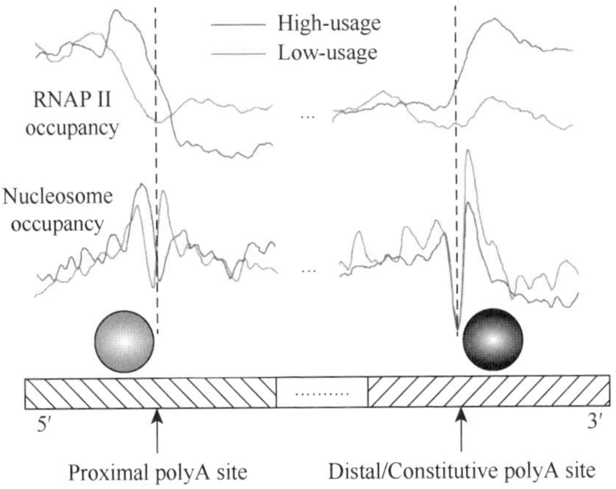

Figure 6.3 The dual pausing model of RNAP II and nucleosome occupancy patterns during regulation of the usage of different polyA sites and transcription termination. The nucleosome upstream of the proximal site is the first barrier pausing RNAP II, and the nucleosome downstream of the distal or constitutive polyA site is the second barrier pausing RNAP II to ensure transcription termination and recognition of the last polyadenylation sites if previous sites were missed. The blue curve represents high-usage sites and the red one represents low-usage sites

6.5 Nucleosome Occupancy Patterns Around Transcription Factor Binding Sites

Transcription factors (TFs) bind to specific DNA sequences and interact with components of the RNA polymerase complex, or with other complexes, to regulate transcription in a cell type-specific manner, and this process is highly dependent on the chromatin structure in eukaryotes[48, 49]. Nucleosomes can directly regulate the accessibility of TFs and transcriptional machinery to the DNA sequences[50]. Sequences in nucleosome-depleted regions are easier for proteins to access, while the accessibility of DNA within nucleosomes depends on nucleosome dynamics[51]. Although histone DNA complexes are very stable, histones are constantly evicted and reassembled onto DNA templates in a locus-specific manner. The occupancy patterns and dynamic positioning of nucleosomes around transcription factor binding sites (TFBSs) thus play crucial roles in regulating eukaryotic transcription.

The interaction between TFs and nucleosomes is very complex. Recent studies in vertebrate genomes show that in a specific cell many TFs preferentially bind to genomic sequences occupied by nucleosomes *in vitro*, suggesting that nucleosomes are gatekeepers of TF binding sites[52]. Moreover, these TFBSs are flanked by well-positioned and periodic nucleosomes. Generally, in the procedure of gene transcription, TF binding is associated with nucleosome removing or shifting, and functional binding is almost always associated reorganization of nucleosomes. But the cause and effect relationships remain unclear.

The binding of several TFs, such as the insulator binding protein CTCF[53], the RE1-silencing transcription factor (REST/NRSF) [11]and the multifunctional TF YY1[54], has been suggested to initiate nucleosome depletion at TF binding sites (TFBSs) and the phased nucleosome arrays in the flanking regions in human cells. In a recent study, Barozzi and colleagues try to address the hypothesis that in mammalian genomes the information controlling the ability of TFs to recognize cognate sites in cis-regulatory elements may at least partially coincide with the information that controls incorporation of the same sequence into nucleosomes. Their results show that TF recruitment and nucleosome deposition are controlled by overlapping DNA sequence features[52].

TF binding is accompanied by chromatin remodeling. Yadav et al. analyzed the relationships between TF binding and nucleosome positioning and found that TF binding and transcriptional activity are linked through local nucleosome repositioning, and promoter-proximal TF binding influences the expression of target gene if nucleosome repositioning occurs at or close to its binding site in most cases, while only in few cases change in target gene expression is found when TF binding occurred without local nucleosome reorganization[55].

In vertebrate genomes, the promoters, upstream gene transcription start sites, are often GC rich and are well bound by nucleosomes *in vitro*. These GC rich promoters will be bound by TFs and RNA polymerase II *in vivo*[3]. TFs and nucleosomes compete for binding to specific genomic sequences. An extension of this competition model is a collaborative competition model where two TFs can bind to DNA independently but together can cooperate and displace a nucleosome, which was analyzed by He et al.[56]. These analysis revealed an inherent competition between some TFs and nucleosomes for binding canonical TFBS. Our previous analysis of histone modifications around TFBSs also indicate that the existence of common mark features enabling TFs to bind with DNA[57].

To better understand the relationship between TF binding and nucleosome positioning, we did a detailed analysis, integrating data on TF binding, nucleosome occupancy and genomic sequence. We analyzed the CENTIPEDE-inferred binding sites for 519 TF binding motifs, representing up to a third of the human TF repertoire, and examined the nucleosome occupancy around these binding sites in human GM12878 and K562 cells [58]. CENTIPEDE is a computational method to infer binding sites in a particular cell type, incorporating cell-specific experimental data[59]. We also used a set of *in vitro* nucleosome occupancy data. *In vitro* nucleosome occupancy is affected mainly by the intrinsic specificity between histones and the DNA sequences, whereas *in vivo* occupancy is influenced by sequence preferences, TFs and chromatin remodelers[11].

Analyses of *in vivo* data show the nucleosome-depleted regions at TFBSs and an array of well-positioned nucleosomes in the flanking regions [Figure 6.4 (a)], which is consistent with the barrier model suggested in previous studies[60], while *in vitro* nucleosome occupancy profiles suggest that the binding sequences of TFs tend to form nucleosomes. TFBSs in the human genome have high GC content and intrinsic nucleosome occupancy, but low *in vivo* nucleosome occupancy. Binding of TFs can form barriers and other nucleosomes are stacked against them to generate the phased nucleosome arrays by ATP-dependent chromatin remodelers.

We classified the binding sites by the distances of sites relative to the nearest gene and the functions of the bound TFs, to test whether the nucleosome occupancy exhibited distinct patterns. TFBSs may be located proximal or distal to TSSs. Binding sites in the core or proximal promoter are typically located within one kilobase (kb), while distal sites may be situated up to several hundred kb from the core promoter. We observed lower *in vivo* nucleosome occupancy and fuzzier nucleosome positioning around proximal sites, which were more consistent with those around TSSs. On the other hand, distal sites are far from promoters, where the nucleosome occupancy is higher, the binding of TFs in the distal regions is therefore more likely to recruit ATP dependent chromatin remodelers to generate the phased nucleosome arrays.

Activators and repressors positively and negatively regulate transcription, respectively. Although nucleosome-depleted regions were observed at activator and repressor binding sites *in vivo*, nucleosomes flanking the repressors sites are better positioned compared with the activator sites. It was also found that the TFBSs of repressors have higher *in vitro* nucleosome occupancy, suggesting more complex nucleosome dynamics and chromatin remodeling, and this might contribute to the better-positioned nucleosomes flanking the repressor sites *in vivo*.

Some TFs can bind to the DNase I-resistant regions with high *in vivo* nucleosome occupancy, although most of TFs bind to nucleosome free sequences. This may provide a strategy to regulate gene expression in a cell-type-specific way. A traditional profile that averages over all TF binding sites will neglect these specific nucleosome occupancy patterns. To investigate nucleosome patterns around TFBSs in detail, we clustered the nucleosome occupancy profiles. The cluster analysis of all TFBSs indicates that only approximately a quarter of TFBSs are nucleosome depleted and flanked by well positioned nucleosomes on both sides. The majority of TFBSs show strong nucleosome positioning on one side, suggesting that asymmetric patterns of nucleosome occupancy are more pervasive around TFBSs. Moreover, some TFBSs are occupied by nucleosomes, indicating direct binding of TFs to genomic DNA occluded by nucleosomes [Figure 6.4 (b)].

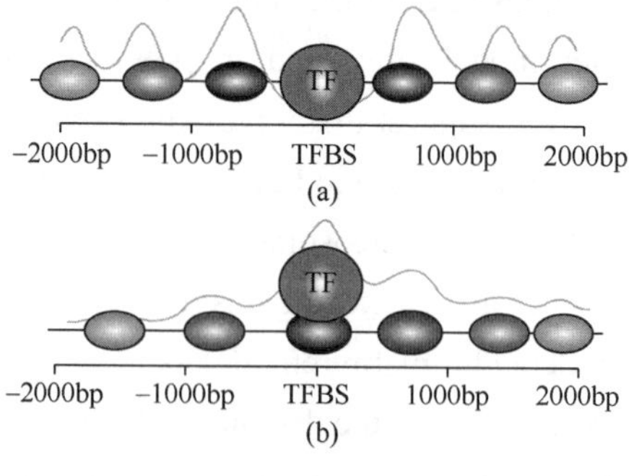

Figure 6.4 Nucleosome occupancy patterns around transcription factor binding sites.
(a) Nucleosome depleted region at TFBSs and an array of well-positioned nucleosomes in the flanking regions; (b) Asymmetric nucleosome occupancy around TFBSs and direct binding of TFs to genomic DNA occluded by nucleosomes

6.6 Conclusions

Nucleosome organization is important for gene transcriptional regulation. Nucleosome positioning around different ki of transcription sites play different role in gene transcription. By analyzing high throughput DNA sequencing data on nucleosome occupancy, we found that the first nucleosome upstream of transcription start sites is shifted to the 5′ end, forming a broad nucleosome-free region in highly expressed genes. The nucleosome distributions around splice sites and polyadenylation sites influence alternative splicing and regulate 3′ end processing of protein-coding genes respectively. Out results also indicate that there are some special patterns of nucleosome occupancy around transcription factor binding sites, which are associated with the expression levels of target genes. While DNA sequence is the intrinsic determinant of nucleosome positioning, nucleosome organization could be remodeled in the process of gene transcriptional regulation. Nucleosome organization is the result of the complicated interaction between protein factors and histones. Genomic DNA occupied by protein factors could be the barrier to nucleosome formation, resulting in a NFR surround by well-positioned nucleosomes, e.g. NFRs upstream of TSSs and NFRs around TFBSs. On the other hand, nucleosomes might serve as 'speed-bumps' for RNAP II, slowing down transcription rates to improve the recognition of special sites, e.g. alternative splice sites and polyadenylation sites. Currently only the nucleosome occupancy data are available, which is the measure of nucleosome positions by examining bulk populations of cells. It could be expected that nucleosome positioning data will be produced soon, due to the development of technique of single-cell nucleosome mapping.

References

[1] Henikoff S. Nucleosome destabilization in the epigenetic regulation of gene expression. *Nature Reviews Genetics*, 2008, 9 (1): 15-26.

[2] van Holde KE. *Chromatin: Springer Series in Molecular Biology. New York, Berlin*, Heidelberg, London, Paris , Tokyo: Springer-Verlag, 1988.

[3] Kaplan N, Moore I K, Fondufe-Mittendorf Y, et al. The DNA-encoded nucleosome organization of a eukaryotic genome. *Nature*, 2009, 458 (7236): 362-366.

[4] Brogaard K, Xi L Q, Wang J P, et al. A map of nucleosome positions in yeast at base-pair resolution. *Nature*, 2012, 486 (7404): 496-501.

[5] Zhang Z H, Wippo C J, Wal M, et al. A packing mechanism for nucleosome organization reconstituted across a

eukaryotic genome. *Science*, 2011, 332 (6032): 977-980.

[6] Sadeh R, Allis C D. Genome-wide "re"-modeling of nucleosome positions. *Cell*, 2011, 147 (2): 263-266.

[7] Radman-Livaja M, Rando O J. Nucleosome positioning: How is it established, and why does it matter? *Developmental Biology*, 2010, 339 (2): 258-266.

[8] Nelson H C, Finch J T, Luisi B F, et al. The structure of an oligo (dA)·oligo (dT) tract and its biological implications. *Nature*, 1987, 330 (6145): 221-226.

[9] Anselmi C, Bocchinfuso G, de Santis P, et al. Dual role of DNA intrinsic curvature and flexibility in determining nucleosome stability. *Journal of Molecular Biology*, 1999, 286 (5): 1293-1301.

[10] Struhl K. Segal E. Determinants of nucleosome positioning. *Nature Structural and Molecular Biology*, 2013, 20 (3): 267-273.

[11] Valouev A, Johnson S M, Boyd S D, et al. Determinants of nucleosome organization in primary human cells. *Nature*, 2011, 474 (7352): 516-520.

[12] Mavrich T N, Jiang C Z, Ioshikhes I P, et al. Nucleosome organization in the Drosophila genome. *Nature*, 2008, 453 (7193): 358-362.

[13] Liu H D, Duan X Y, Yu S X, et al. Analysis of nucleosome positioning determined by DNA helix curvature in the human genome. *BMC Genomics*, 2011, 12: 72.

[14] Kornberg R D, Stryer L. Statistical distributions of nucleosomes: Nonrandom locations by a stochastic mechanism. *Nucleic Acids Research*, 1988, 16 (14A): 6677-6690.

[15] Parmar J J, Marko J F, Padinhateeri R. Nucleosome positioning and kinetics near transcription-start-site barriers are controlled by interplay between active remodeling and DNA sequence. *Nucleic Acids Research*, 2014, 42 (1): 128-136.

[16] Yuan G C, Liu Y J, Dion M F, et al. Genome-scale identification of nucleosome positions in *S. cerevisiae*. *Science*, 2005, 309 (5734): 626-630.

[17] Johnson S M, Tan F J, McCullough H L, et al. Flexibility and constraint in the nucleosome core landscape of Caenorhabditis elegans chromatin. *Genome Research*, 2006, 16 (12): 1505-1516.

[18] Mavrich T N, Ioshikhes I P, Venters B J, et al. A barrier nucleosome model for statistical positioning of nucleosomes throughout the yeast genome. *Genome Research*, 2008, 18 (7): 1073-1083.

[19] Schones D E, Cui K R, Cuddapah S, et al. Dynamic regulation of nucleosome positioning in the human genome. *Cell*, 2008, 132 (5): 887-898.

[20] Small E C, Xi L, Wang J P, Widom J, et al. Single-cell nucleosome mapping reveals the molecular basis of gene expression heterogeneity. *Proceedings of the National Academy of Sciences of the United States of America*, 2014, 111 (24): E2462-E2471.

[21] Hartley P D, Madhani H D. Mechanisms that specify promoter nucleosome location and identity. *Cell*, 2009, 137 (3): 445-458.

[22] Saito T L, Hashimoto S, Gu S G, et al. The transcription start site landscape of *C. elegans*. *Genome Research*, 2013, 23 (8): 1348-1361.

[23] Zaugg J B, Luscombe N M. A genomic model of condition-specific nucleosome behavior explains transcriptional activity in yeast. *Genome Research*, 2012, 22 (1): 84-94.

[24] Raveh-Sadka T, Levo M, Shabi U, et al. Manipulating nucleosome disfavoring sequences allows fine-tune

regulation of gene expression in yeast. *Nature Genetics*, 2012, 44 (7): 743-750.

[25] Liu H D, Wu J S, Xie J M, et al. Characteristics of nucleosome core DNA and their applications in predicting nucleosome positions. *Biophysical Journal*, 2008, 94 (12): 4597-4604.

[26] Liu H D, Luo K, Wen H, et al. Quantitative analysis reveals increased histone modifications and a broad nucleosome-free region bound by histone acetylases in highly expressed genes in human $CD4^+$ T cells. *Genomics*, 2013, 101 (2): 113-119.

[27] Cheng X, Hou Y, Nie Y, et al. Nucleosome positioning of intronless genes in the human genome. *IEEE Transactions on Computational Biology and Bioinformatics* (In revision), 2015.

[28] Pan Q, Shai O, Lee L J, et al. Deep surveying of alternative splicing complexity in the human transcriptome by high-throughput sequencing. *Nature Genetics*, 2008, 40 (12): 1413-1415.

[29] Naftelberg S, Schor I E, Ast G, et al. Regulation of alternative splicing through coupling with transcription and chromatin structure. *Annual Review of Biochemistry*, 2015, 84: 165-198.

[30] Kogan S, Trifonov E N. Gene splice sites correlate with nucleosome positions. *Gene*, 2005, 352: 57-62.

[31] Denisov D A, Shpigelman E S, Trifonov E N. Protective nucleosome centering at splice sites as suggested by sequence-directed mapping of the nucleosomes. *Gene*, 1997, 205 (1-2): 145-149.

[32] Tilgner H, Nikolaou C, Althammer S, et al. Nucleosome positioning as a determinant of exon recognition. *Nature Structural and Molecular Biology*, 2009, 16 (9): 996-1001.

[33] Schwartz S, Meshorer E, Ast G. Chromatin organization marks exon-intron structure. *Nature Structural and Molecular Biology*, 2009, 16 (9): 990-995.

[34] Andersson R, Enroth S, Rada-Iglesias A, et al. Nucleosomes are well positioned in exons and carry characteristic histone modifications. *Genome Research*, 2009, 19 (10): 1732-1741.

[35] Iannone C, Pohl A, Papasaikas P, et al. Corrigendum Relationship between nucleosome positioning and progesterone-induced alternative splicing in breast cancer cells. *RNA*, 2015, 21 (3): 360-374.

[36] Huang H, Yu S X, Liu H D, et al. Notice of chromatin organization in different types of splicing events//*5th International Conference on Bioinformatics and Biomedical Engineering*, (*iCBBE*) 2011; 2011: 1-4.

[37] Schwartz S, Ast G. Chromatin density and splicing destiny: On the cross-talk between chromatin structure and splicing. *The EMBO Journal*, 2010, 29 (10): 1629-1636.

[38] Huang H, Yu S X, Liu H D, et al. Nucleosome organization in sequences of alternative events in human genome. *Bio Systems*, 2012, 109 (2): 214-219.

[39] Richard P, Manley J L. Transcription termination by nuclear RNA polymerases. *Genes and Development*, 2009, 23 (11): 1247-1269.

[40] Moore M J, Proudfoot N J. Pre-mRNA processing reaches back to transcription and ahead to translation. *Cell*, 2009, 136 (4): 688-700.

[41] Colgan D F, Manley J L. Mechanism and regulation of mRNA polyadenylation. *Genes and Development*, 1997, 11 (21): 2755-2766.

[42] MacDonald C C, McMahon K W. Tissue-specific mechanisms of alternative polyadenylation: Testis, brain, and beyond. *Wiley Interdisciplinary Reviews, RNA*, 2010, 1 (3): 494-501.

[43] Lee J Y, Yeh I, Park J Y, Tian B. PolyA_DB 2: mRNA polyadenylation sites in vertebrate genes. *Nucleic Acids*

Research, 2007, 35 (Database issue): D165-168.

[44] Spies N, Nielsen C B, Padgett R A, et al. Biased chromatin signatures around polyadenylation sites and exons. *Molecular Cell*, 2009, 36 (2): 245-254.

[45] Khaladkar M, Smyda M, Hannenhalli S. Epigenomic and RNA structural correlates of polyadenylation. *RNA Biology*, 2011, 8 (3): 529-537.

[46] Huang H, Liu H D, Sun X. Nucleosome distribution near the 3' ends of genes in the human genome. *Bioscience, Biotechnology, and Biochemistry*, 2013, 77 (10): 2051-2055.

[47] Chang G S, Noegel A A, Mavrich T N, et al. Unusual combinatorial involvement of poly-A/T tracts in organizing genes and chromatin in Dictyostelium. *Genome Research*, 2012, 22 (6): 1098-1106.

[48] Lenhard B, Sandelin A, Carninci P. Metazoan promoters: emerging characteristics and insights into transcriptional regulation. *Nature Reviews Genetics*, 2012, 13 (4): 233-245.

[49] Pan Y P, Tsai C J, Ma B Y, et al. Mechanisms of transcription factor selectivity. *Trends in Genetics*, 2010, 26 (2): 75-83.

[50] Wang X, Bai L, Bryant G O, et al. Nucleosomes and the accessibility problem. *Trends in Genetics*, 2011, 27 (12): 487-492.

[51] He H H, Meyer C A, Shin H, et al. Nucleosome dynamics define transcriptional enhancers. *Nature Genetics*, 2010, 42 (4): 343-347.

[52] Barozzi I, Simonatto M, Bonifacio S, et al. Coregulation of transcription factor binding and nucleosome occupancy through DNA features of mammalian enhancers. *Molecular Cell*, 2014, 54 (5): 844-857.

[53] Cuddapah S, Jothi R, Schones D E, et al. Global analysis of the insulator binding protein CTCF in chromatin barrier regions reveals demarcation of active and repressive domains. *Genome Research*, 2009, 19 (1): 24-32.

[54] Wang J, Zhuang J L, Iyer S, et al. Sequence features and chromatin structure around the genomic regions bound by 119 human transcription factors. *Genome Research*, 2012, 22 (9): 1798-1812.

[55] Yadav V K, Thakur R K, Eckloff B, et al. Promoter-proximal transcription factor binding is transcriptionally active when coupled with nucleosome repositioning in immediate vicinity. *Nucleic Acids Research*, 2014, 42 (15): 9602-9611.

[56] He X, Chatterjee R, John S, et al. Contribution of nucleosome binding preferences and co-occurring DNA sequences to transcription factor binding. *BMC Genomics*, 2013, 14: 428.

[57] Nie Y M, Liu H D, Sun X. The patterns of histone modifications in the vicinity of transcription factor binding sites in human lymphoblastoid cell lines. *PloS One*, 2013, 8 (3): e60002.

[58] Nie Y M, Cheng X F, Chen J, et al. Nucleosome organization in the vicinity of transcription factor binding sites in the human genome. *BMC Genomics*, 2014, 15 (1): 493.

[59] Pique-Regi R, Degner J F, Pai A A, et al. Accurate inference of transcription factor binding from DNA sequence and chromatin accessibility data. *Genome Research*, 2011, 21 (3): 447-455.

[60] Iyer V R. Nucleosome positioning: bringing order to the eukaryotic genome. *Trends in Cell Biology*, 2012, 22 (5): 250-256.

Chapter 7 Virtual and Augmented Reality in Medical and Biomedical Education

Zhuming Ai

7.1 Introduction

Patient safety has been a major issue in medical practice. It's been reviewed in recent years as an important area that needs improvement. The 1999 publication of *To Err is Human*[1] and the 2001 report of *Crossing the Quality Chasm*[2] brought to light the human toll and financial cost of medical error. In the years following, many patient safety initiatives have been introduced, and they have prevented tens of thousands of deaths each year.

How to improve patient safety has been discussed and debated for a long time. One of the areas that needs to improve is medical education. Medical education plays an important role in improving patient safety by incorporating safety-building methods into the curricula, assessment, and learning of health care professionals[3]. In particular, simulation-based medical training can be an effective tool to improve the health care system.

Many difficulties that medical education is facing are related to the complex nature of human anatomy and the lack of cadavers to help medical students to understand the human body. For example, a clear understanding of the intricate spatial relationships among the structures of the pelvic floor, rectum, and anal canal is essential for the treatment of numerous pathological conditions. The complexity of medical procedures is another reason why medical education is so challenging. Communication among health care providers is also an area that needs to improve. All these difficulties provide threedimensional (3D) computer graphics simulation a good opportunity to help in improving patient safety.

The advance of technology in virtual reality (VR) and augmented reality (AR) has inspired much research on improving medical training. Advanced simulation technologies, such as virtual reality, offer the potential of teaching skills that were rather difficult to learn in the past. Virtual reality technology allows improved visualization of 3D structures over conventional media because it supports stereoscopic-vision, viewer-centered perspective, large angles of view, and interactivity[4].

In the following sections, we will review some of the virtual reality and augmented reality techniques and the application of these techniques in medical and biomedical education.

7.2 Reality-Virtuality Continuum

Paul Milgram proposed the reality-virtuality continuum[5]. It is a continuous scale ranging between the completely virtual, a virtuality, and the completely real, a reality. This concept has been further extended into a two-dimensional plane of virtuality and mediality[6]. Computer technology has been used in different levels of Virtuality in medical education.

In particular, VR has been used in many applications to help medical students to understand the complex nature of human anatomy. AR is mainly used in medical procedure simulation and training, and it's also been used in actual medical procedures. We will also show in this chapter that AR can be used in medical modeling. Different haptics devices are also used in various situations. Among the many application areas of VR and AR, their applications in medical and biomedical related fields can be considered the most successful ones.

7.3 Virtual and Augmented Reality Applications in Medical and Biomedical Education

Since the beginning of the research of VR and AR, they have been recognized to be particularly useful in two areas, military and medicine. In biomedical and medical applications, the complex 3D nature of the human body makes it the perfect target to use such technology. In this section, we will first review how the human body parts can be modeled so that computers can handle it. Then we will discuss how the dynamics of bio-processes can be simulated in VR and AR environments. With the advance of high speed networks, VR and AR experiences have been extended to remote locations, and efforts have been made to make people at different locations work together efficiently. We will discuss the application of tele-immersion in remote medical education. One important area in medical and biomedical modeling is the visualization of volumetric data. We will discuss in the following section how volume rendering is working together with VR and AR, and even in a tele-immersive environment.

7.3.1 Biomedical modeling

The modeling of human anatomy is by itself a complicated process. One common procedure is to build the model based on medical imaging data, such as CT, MR, etc. However, very often it's difficult to identify different tissues and organs from such data. Our research group at the University of Illinois at Chicago (UIC), VRMedLab, has developed a process that combines automatic image processing and interactive modeling to create realistic 3D biomedical models[7].

The process starts with medical imaging data, such as CT, MR, or cadaver dissection photos, depending on which source of the data makes it easier to segment the body parts in the later steps. The parts of the body are isolated through a semiautomatic segmentation process. Then the marching cubes algorithm is applied to create a surface model. This surface model is refined using various software to make it smaller so that a computer can handle it in real time. Meanwhile, any mistakes in the model are corrected in the process.

This is a time consuming process, but delivers high quality models that can be used in VR and AR applications. Using this process, a virtual eye model, a virtual pelvic floor model (Figure 7.1, see the Color Inset, p. 22), a virtual nose model (Figure 7.2, see the Color Inset, p. 22), a virtual temporal bone model (Figure 7.3, see the Color Inset, p. 22), and other models are created and used in VR applications. These computerized models and VR applications have been used to facilitate teaching in a number of complex anatomical regions, such as the human temporal bone and pelvic floor.

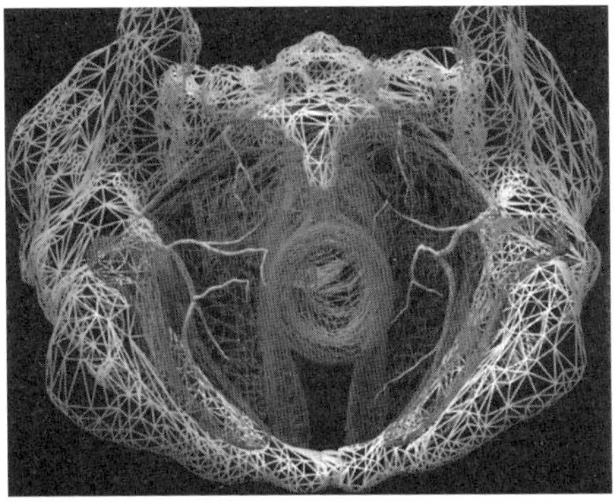

Figure 7.1 Virtual pelvic floor model

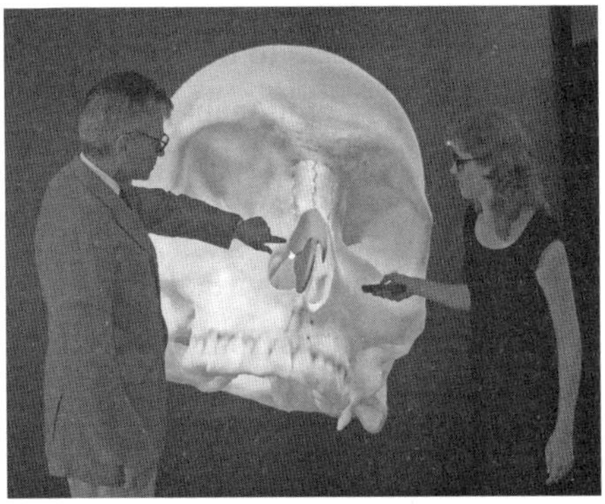

Figure 7.2 Virtual nose

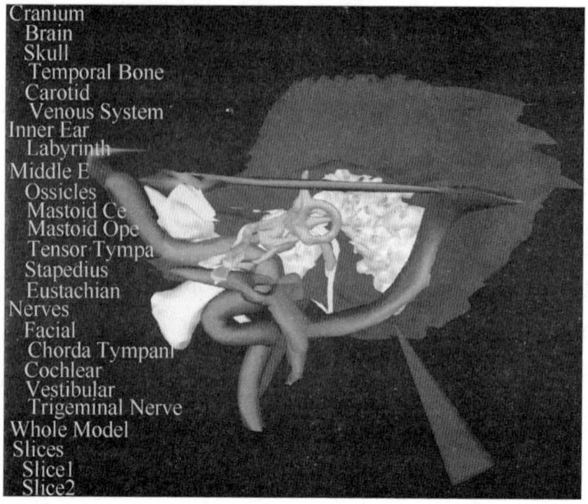

Figure 7.3 Virtual ear

Sometimes, when physical models are available, they can be used as a starting point for modeling. For example, when we started to build the pelvic floor model, a static physical model depicting the pelvic floor and anorectum was created and digitized at 1 mm intervals in a CT scanner. Multiple software programs were used along with endoscopic images to generate a realistic interactive computer model[4].

The 3D complex interplay of soft tissue, cartilaginous, and bony elements makes the mastery of nasal anatomy difficult. Conventional methods of learning nasal anatomy exist, but they often involve a steep learning curve. A 3D virtual reality model of the human nose is created and shown in Figure 7.2 (see the Color Inset, p. 22)[8].

Similar technology was used to build a virtual temporal bone model (Figure 7.3, see the Color Inset, p. 22).

Augmented reality can also be used in building medical models. We have developed an AR assisted process to build patient specific implants (Figure 7.4, see the Color Inset, p. 23). The process of implant design begins with CT data of the patient and the Personal Augmented Reality Immersive System (PARIS^{TM})[9]. The implant is designed by medical professionals in tele-immersive collaboration. In the PARIS^{TM} augmented reality system the users hands and the virtual images appear superimposed in the same volume so the user can see what he is doing. A haptic device supplies the sense of touch by applying forces to a stylus that the medical modeler uses to form the implant. After the virtual model of the implant is designed, the data is sent via network to a stereolithography rapid prototyping system that creates the physical implant model. After implant surgery, the patient undergoes a postoperative CT scan and results are evaluated and reviewed over the tele-immersive consultation system[10-12]. Besides the physical model, the virtual implant model can also be used for education in a tele-immersive environment.

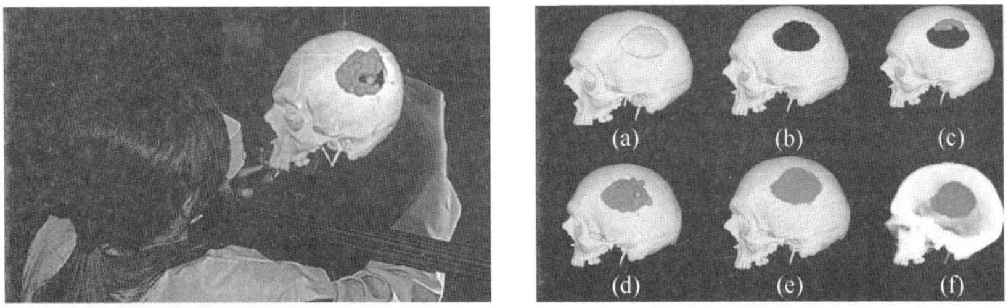

Figure 7.4 AR is used in building medical models

7.3.2 Biomedical simulation

Static models are very often not enough to show the dynamics of the human body. Dynamic simulation is a difficult task since each biomedical process is different and complex.

Molecular dynamics simulation has drawn much attention since it not only has educational benefit, it's also a powerful tool for drug discovery.

We have developed a dynamic simulation system in a cave automatic virtual environment (CAVE[13]). The system is connected to a molecular dynamics simulation engine at the back end. The front end is the CAVE in which user can not only watch the dynamic simulation in real time, but also interact with the molecular system (Figure 7.5, see the Color Inset, p. 23)[14].

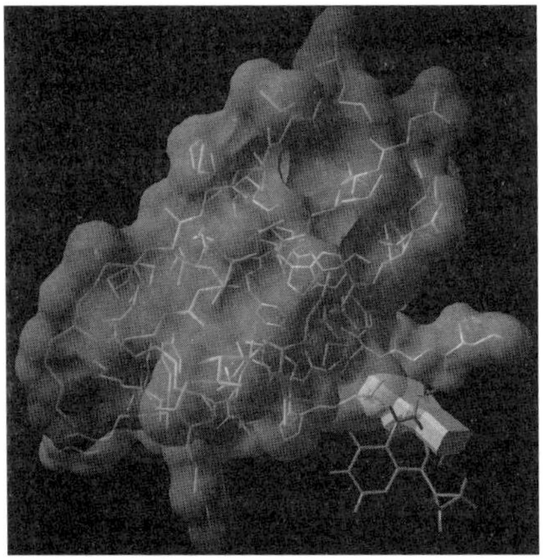

Figure 7.5 Molecular dynamic simulation in a CAVE virtual environment

Besides the molecular level simulations, human anatomy is also simulated dynamically. An agent-based model has been used to model biological systems[15]. A functional 3D bladder model has been developed using an agent-based modeling approach. The agent-based bladder model was able to void through the urethra with a reasonable number of bladder agents. The urethra in the model came from an MRI scan during continence. This modeling technique has potential for creating functional models of a variety of organs and tissues and could improve our understanding of human physiology.

A unique eye disease simulation VR system has been developed[16]. The purpose is mainly for education not only for health care providers, but also for patients and families to get better understanding of the diseases. The first-person interactive nature of VR provides mechanisms for these visualizations which are not present in other media such as video, film or print. This system has two components. One part is an eye anatomy model that shows the changes in eyes due to different conditions. The other part is a VR environment, a virtual apartment, in which these visual problems can be simulated in a familiar context (Figure 7.6, see the Color Inset, p. 23). The viewers can experience the difficulties the patient may experience in his/her daily life. Computer generated image masks are used to model visual field loss caused by glaucoma. Diplopia is created by supplying incorrect viewing directions in the virtual environment. Macular degeneration is simulated by warping the central area of the simulated environment. This first-person experience of the visual impairments can be used to teach medical professionals to recognize the kinds of problems their patients may be experiencing and to increase the sensitivity of friends and family to the patients' difficulties. It may even help to convince noncompliant patients of the serious nature of their disease and increase their adherence with medical protocols.

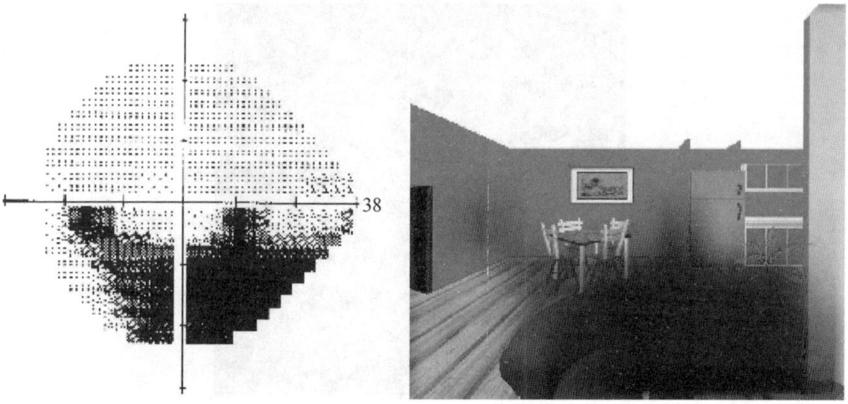

Figure 7.6 Eye disease simulation

In a similar virtual environment, color deficiency was also simulated[17].

Surgical simulation is very important part of medical simulation. It has been proven that it's a very efficient tool to improve patient safety. In one of our projects, a preliminary study has been done to use haptics together with volume rendering to simulate brain surgery.

7.3.3 Volume visualization

With the advancements in computerized tomography scanning technology, many of today's biomedical data are volumetric data. Volume rendering used to be a difficult task due to the size of the data. The advance of computer graphics hardware has rapidly changed the situation.

VolumePro is a hardware implementation of ray tracing algorithm developed in late 1990's. It supports real time rendering of a volume of the size of 256 voxel × 256 voxel × 256 voxel at 30 frames per second[18]. Since then, new computer graphics hardware and algorithms, especially graphics processing unit (GPU) based algorithms, have made it possible to render volumetric data without the need of dedicated hardware. GPU assisted ray casting can render volumes in real time on gaming graphics cards.

Using direct volume rendering and haptics, an implant design system in virtual environment was developed[10].

Large size volume rendering still poses a challenge since it is difficult to fit the whole volumetric data in a computer either in the main computer memory or in the graphics card memory. Different strategies have been used to deal with this problem. Data swapping is commonly used, which only loads part of the data that is needed for the current stage of the rendering. Computer cluster based algorithms are also powerful methods, which can potentially render volumetric data of very large size.

Jin et al. has developed a VR system with a remote computer cluster for interactive three-dimensional reconstruction and alignment of large confocal microscopy data (Figure 7.7, see the Color Inset, p. 24) [19-21]. The method can render a volume that is not possible to fit in one computer. It provides the flexibility and the accumulated power of a computer cluster for this specific application.

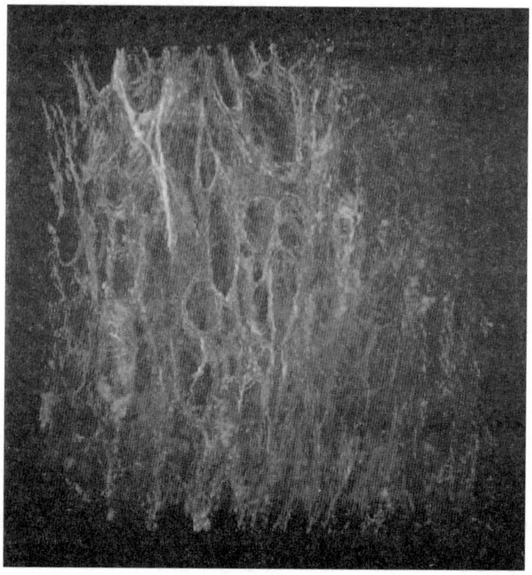

Figure 7.7 Volume rendering of large data set in virtual environment

7.3.4 VR educational environment and tele-immersion

With the wide adaptation of high speed internet, the concept of tele-immersion has emerged[22]. Tele-immersion is a technology that is implemented with high speed internet and VR and enables users in different geographic locations to come together in a simulated environment to interact. Users will feel like they are actually looking, talking, and meeting with each other face to face in the same room.

Tele-immersion could be very useful in medical and biomedical education. It allows experienced educators to reach a larger audience.

We have developed an environment that allows such experience for medical education purposes. It has a desktop setting as well as a conference room or classroom setting[18]. The major goal of this research is to develop a networked collaborative surgical system for tele-immersive consultation, surgical preplanning, implant design, post operative evaluation and education[23].

Volume rendering is also incorporated with VR and tele-immersion to allow remote education and consultation[10, 24, 25].

7.4 Curriculum and User Studies

VR has been quite successful in medical education. Some colleges have incorporated

VR courses into the medical school curriculum.

The effectiveness of VR in medical education has been tested through user studies[26]. The results have high internal validity, for the improved outcomes of VR compared to other methods of anatomy teaching. VR can be a solution to the problem of inadequate anatomy pedagogy.

Here is one example of a novel VR based model used to teach anorectal and pelvic floor anatomy, pathology, and surgery[4]. The pelvic floor model was designed to be viewed on a networked, interactive, VR display (CAVE or ImmersaDesk). A standard examination of ten basic anorectal and pelvic floor anatomy questions was administered to third-year and fourth-year surgical residents. A workshop using the virtual pelvic floor model was then given, and the standard examination was readministered so that it was possible to evaluate the effectiveness of the VR pelvic floor model as an educational instrument. The result shows that training on the VR model produced substantial improvements in the overall average test scores for the two groups, with an overall increase of 41 percent and 21 percent for third-year and fourth-year residents, respectively. Resident evaluations after the workshop also confirmed the effectiveness of understanding pelvic anatomy using the VR model.

7.5 Challenges

Although VR and AR have many advantages and have been proven effective in medical education, there are still many challenges facing the use of the technology as an educational tool.

The VR in medicine technology itself still needs to be improved. In particular, areas such as deformable tissues, blood, and force feedback in large space are some of the most challenging topics in VR technology that are needed to realistically model and simulate biomedical processes. Although there have been many improvements in recent years, because of the cost of hardware and software, to make such advanced techniques available to a larger audience, there is still a lot of work to do.

The other part of the challenge is how to make the VR technology available to the educational institutions. The VR and AR settings are very often complicated and difficult to maintain, and sometimes are costly. This restricts the use of the technology. Desktop low end VR systems seem to be more acceptable nowadays. Video game like settings are very welcome in educational environments. The education value of immersive VR has drawn more and more attention. With the advance of new VR/AR hardware, and with the unintentional help from gaming industry, VR/AR in medical and biomedical education is very promising.

7.6 Conclusions

Virtual and augmented reality have been proven to be useful tools for medical and biomedical education. VR/AR applications for medical education, courses using VR/AR technology, and curricula have been developed. New technology such as handheld devices provides new opportunities for VR in medical education.

VR and AR provide innovative interactive educational frameworks that allow educators to overcome some of the barriers to teaching surgical principles based on understanding highly complex three-dimensional anatomy. Using tele-immersive collaborative, shared virtual-reality environment, teachers and students can interact from locations world-wide to manipulate the components of different medical models to achieve educational goals.

7.7 Acknowledgement

Many of the research reviewed in this chapter were carried out when the author was working at the Department of Biomedical and Health Information Sciences, University of Illinois at Chicago.

References

[1] Kohn L T, Corrigan J M, Donaldson M S. *To Err is Human: Building a Safer Health System*. Washington, D. C. : National Academy Press, 1999.

[2] Committee on Quality of Health Care in America, Institute of Medicine U.S. *Crossing the Quality Chasm a New Health System for the 21st Century*. Washington, D. C. : National Academy Press, 2001.

[3] Ziv A, Small S D, Wolpe P R. Patient safety and simulation-based medical education. *Medical Teacher*, 2000, 22 (5): 489-495.

[4] Dobson H, Pearl R K, Orsay C P, et al. New method of teaching anorectal and pelvic floor anatomy. *Diseases of the Colon and Rectum*, 2003, 46 (3): 349-352.

[5] Milgram P, Takemura H, Utsumi A, et al. Augmented reality: A class of displays on the reality-virtuality continuum. *Telemanipulator and Telepresence Technologies*, 1994, 2351: 282-292.

[6] Mann S. Mediated reality with implementations for everyday life. *Presence: Teleoperators and Virtual Environments*, 2002, 6.

[7] Parikh M, Rasmussen M, Brubaker L, et al. Three dimensional virtual reality model of the normal female pelvic floor. *Annals of Biomedical Engineering*, 2004, 32 (2): 292-296.

[8] Vartanian A J, Holcomb J, Ai Z M, et al. The virtual nose: A 3-dimensional virtual reality model of the human nose. *Arhives of Facial Plastic Surgery*, 2004, 6 (5): 328-333.

[9] Johnson A D, Sandin D, Dawe G, et al. Developing the PARIS: Using the CAVE to prototype a new VR display// *Proceedings of IPT* 2000: *Immersive Projection Technology Workshop*, 2000.

[10] Ai Z M, Evenhouse R, Leigh J, et al. Biomedical modeling in tele-immersion//Cai Y. *Digital Human Modeling*. Springer, 2008: 47-70.

[11] Ai Z M, Evenhouse R, Rasmussen M. Haptic rendering of volumetric data for cranial implant modeling. *Annual International Conference of the IEEE Engineering in Medicine and Biology Society*, 2007, 5: 5124-5127.

[12] Ai Z M, Evenhouse R, Leigh J, et al. New tools for sculpting cranial implants in a shared haptic augmented reality environment. *Studies in Health Technology and Informati*, 2006, 119: 7-12.

[13] Cruz-Neira C, Sandin D J, Defanti T A. Surround-screen projection-based virtual reality: The design and impleme-ntation of the CAVE//*Proceedings of the 20th Annual Conference on Computer Graphics and Interactive techniques*. ACM, 1993: 135-142.

[14] Ai Z M, Fröhlich T. Molecular dynamics simulation in virtual environments. *Computer Graphics Forum*, 1998, 17 (3): C267-C273.

[15] Brubaker L, Parikh M, North M, et al. Agent-based functional 3-dimensional bladder model: Initial approach// *Proceedings International Continence Society*, 2002.

[16] Ai Z M, Gupta B K, Rasmussen M, et al. Simulation of eye diseases in a virtual environment. *Hawaii International Conference on System Sciences*, 2000, 5: 5024.

[17] Jin B, Ai Z M, Rasmusse M. Simulation of color deficiency in virtual reality. *Studies in Health Technology and Informatics*, 2005, 111: 223-226.

[18] Leigh J, Defanti T A, Johnson A, et al. Global tele-immersion: Better than being there. *Proceedings of ICAT '97, 7th Annual International Conference on Artificial Reality and Tele-Existence*, 1997: 10-17.

[19] Jin B, Ai Z M, Rasmussen M. Visualization of large-scale confocal data using computer cluster. *Studies in Health Technology and Informatics*, 2007, 125: 206-208.

[20] Chen X, Ai Z M, Rasmussen M, et al. Three-dimensional reconstruction of extravascular matrix patterns and blood vessels in human uveal melanoma tissue: Techniques and preliminary findings. *Investigative Ophthalmology and Visual Science*, 2003, 44 (7): 2834-2840.

[21] Ai Z M, Chen X, Rasmussen M, et al. Reconstruction and exploration of three-dimensional confocal microscopy data in an immersive virtual environment. *Computerized Medical Imaging and Graphics*, 2005, 29 (5): 313-318.

[22] Ai Z M, Rasmussen M. Desktop and conference room VR for physicians. *Studies in Health Technology and Informatics*, 2005, 111: 12-14.

[23] Ai Z M, Dech F, Silverstein J, et al. TeleImmersive medical educational environment. *Studies in Health Technology and Informatics*, 2002, 85: 24-30.

[24] Ai Z M, Jin B, Rasmussen M. Radiological tele-immersion//Kumar S, Krupinski E. *Teleradiology*. Berlin: Springer Science & Business Media, 2008: 49-64.

[25] Ai Z M, Dech F, Rasmussen M, et al. Radiological tele-immersion for next generation networks. *Studies in Health Technology and Informatics*, 2000, 70: 4-9.

[26] Yammine K, Violato C. A meta-analysis of the educational effectiveness of three-dimensional visualization technologies in teaching anatomy. *Anatomical Sciences Education*, 2014, 8 (6): 525-538.

Chapter 8 Relating Information in EEGs to Neurocognitive Processes

Li Zhang, Yuan Zhang, Philip R. Blue, Xiaolin Zhou

8.1 The Importance of EEG in Cognitive Neuroscience

8.1.1 The what and the why

Electroencephalogram (EEG), one of the most widely used techniques in cognitive neuroscience, is used to record the activity of nerve cells in the brain via multiple sensors located across the scalp. The recorded signals (also called EEGs) convey an abundance of information which may be related to the transferring and processing of multiple sensory modalities. Findings from clinical studies suggest that EEG can help doctors diagnose epilepsy, brain tumors, dementia, encephalitis, and other diseases related to encephalopathy. The extracted EEG components both provide reliable evidence and improve the accuracy of diagnosis. By making the exploration of neural mechanism underlying psychological processes possible, EEG research also sheds light on psychological studies related to attention, language, and social cognition.

1. A brief history of EEG development

In 1875, the British physicist Richard Caton first detected electric currents from the brains of dogs and apes by using bare unipolar electrodes placed on the cerebral cortex or on the surface of the skull[1]. However, development in this area plateaued before German psychiatrist Hans Berger recognized the importance of electrophysiological signals and reported the first EEG, most notably the discovery of Alpha waves[1]. Since then, more and more research has been focused on recording and analyzing EEG signals related to diseases and cognitive processes. It almost goes without saying that the discovery of EEG was a historical breakthrough that advanced the knowledge of neuroscience and neurologic diagnosis.

2. Cortical origins of EEG

The EEG recordings from the human scalp are known a mixture of activity at the surface of the cortex; this surface activity is affected by activity deep in the brain. The EEG rhythm activity reflects the discharges from large populations of neurons, which are electrically excitable cells. The neurons in the brain produce action potentials that move from one cell to another via the connections of synapses, where information is transmitted from the axon of one cell to the dendrite of another cell[2, 3]. Two types of neurotransmitters, one promoting information delivery to the next cell and the other preventing information delivery to the next cell, are responsible for signal transmission across the synapse. The brain maintains a balanced amount of the two types of neurotransmitters to ensure the successful transfer of information[4].

Electrical activity originating from neurons are classified into two types: action potentials and post-synaptic potentials[6]. Action potentials are short-lasting events ($<$10 ms) which appear as discrete voltage spikes and cause the electrical membrane potential in the axon to rise and fall rapidly. Post-synaptic potentials are longer changes (50-200 ms) in the electrical membrane potential of the post-synaptic receptor with a greater potential field and occur when the neurotransmitters bind to receptors on the post-synaptic membrane. Post-synaptic potentials are considered to be the primary generator of EEG[5, 6]. The cortex is primarily made up of pyramidal neurons which are close to the scalp of the human brain and are highly polarized; their primary orientation axis is perpendicular to the cortical surface[7]. As such, the pyramidal neurons play an important role in the formation of EEGs[6].

3. Spontaneous EEG and event-related potentials

1) Spontaneous EEG

Spontaneous EEG corresponds to the spontaneous electrical activity of a living being. The expansion of knowledge regarding EEG has allowed researchers to explore relatively large frequencies and bandwidths of brain activity. EEG recordings from the human scalp show oscillations at various frequencies. In clinical use and in scientific research, spontaneous EEG activity is conventionally classified into different wave types according to the frequency bands[8] (Figure 8.1).

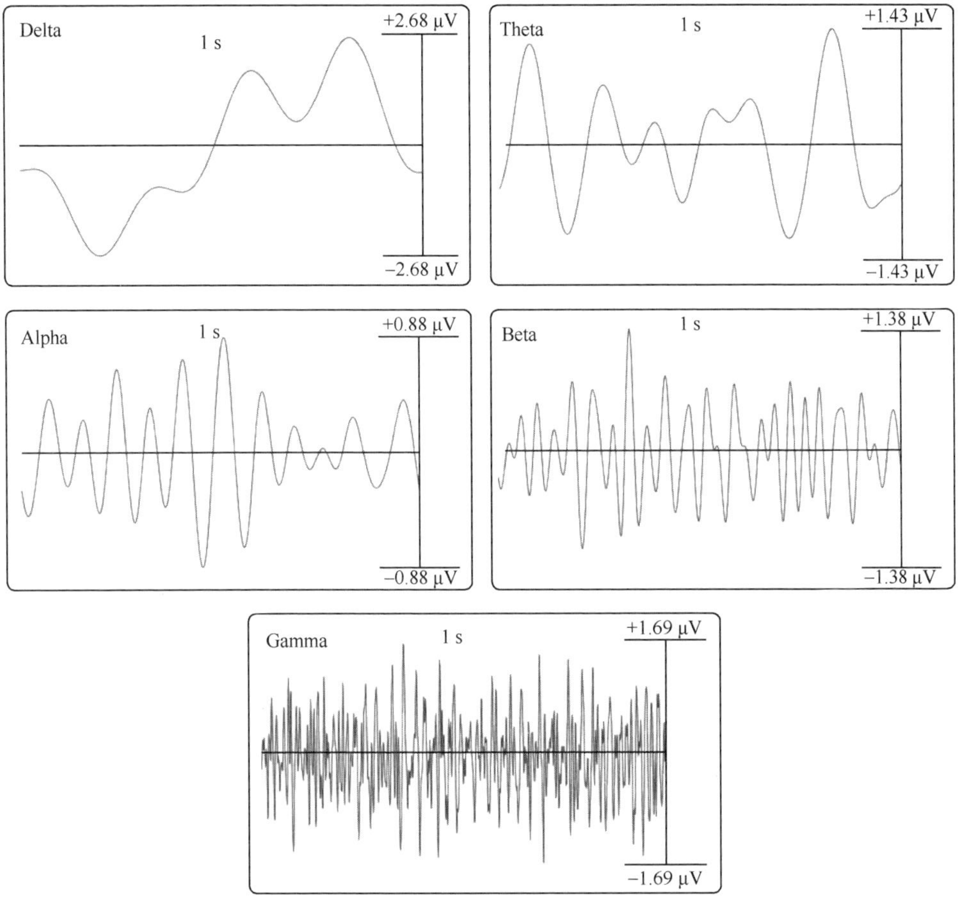

Figure 8.1 Wave patterns for delta, theta, alpha, beta and gamma band (data from Han et al. resting state data)

Sub-delta (0-0.5 Hz): In healthy individuals, this frequency band is regarded as an artifact in brain activity. In individuals with certain brain abnormalities, this frequency band represents interictal and ictal rhythm with focal seizures.

Delta (0.5-3.5 Hz): Delta waves are identified as slow waves with strong amplitude and low frequency. These waves are the primary waves found in infants (0-12 months old) and in the adult frontal cortex. In addition, delta waves are the most prominent waves in slow-wave sleep (stage 3 and stage 4 of sleep) and are believed to reflect the sustained attention during continuous performance task[9]. This frequency band can also inform us about the morphology and activity distribution related to different brain disorders[8].

Theta (3.5-8 Hz): Theta waves are considered slow wave activity; these waves are strengthened during stage 2 of sleep. They are the dominant brain rhythm in children and indicate abnormality in awake adults who have subcortical lesions. In addition, theta band reflects the response inhibition effect[9].

Alpha (8-13 Hz): In healthy individuals, alpha waves are the posterior dominant rhythm on both sides of the brain and are lateralized in amplitude on the dominant side, while centralized at rest. Alpha waves are stronger while relaxed or while closing one's eyes than when alert or with one's eyes open. These waves are also related to the inhibition control in top-down processing[10].

Beta (13-30 Hz): Beta waves are generally recorded at the frontal and central regions of the head, and distributed symmetrically on both sides. They are stronger when people are nervous and anxious. Beta waves are typically known as the normal brain rhythm. However, for those with pathological disorders, beta waves are associated with seizures.

Gamma (30-80 Hz): Gamma waves are fast brain activity and are associated with voluntary motor movement, learning, and memory. These waves are also related to seizures when pathological changes occur in the brain.

In addition, there are also ripples, which are fast brain activity with low amplitudes that important in brain functioning. There are three types of ripples: (1) "ordinary" ripples (80-250 Hz), which are related to cognitive processing, episodic memory consolidation, and interical and ictal seizure; (2) fast ripples (250-500 Hz), which are related to seizures in pathological disorders; and (3) very fast ripples (500-1000 Hz), which are associated with the acquisition of sensory information and with seizures resulting from pathological changes in the brain[8].

Brain rhythms are described and defined by their scalp distributions and provide a rich amount of biological meaning. In general, as the activity in the brain increases, EEG rhythm frequency increases while amplitude decreases[11]. However, properly extracting the biological meaning from EEG data is not easy, as EEGs are the combinations of multiple frequency bands. Moreover, EEG rhythm analysis can often lead to misdiagnosis of certain brain pathologies, as EEG rhythm patters are not always uniform for each brain disorder.

2) Event-related responses in EEG (ERPs, ERS, and ERD)

Event-related potentials (ERPs) are the neural signals extracted from EEG. ERPs are the responses to the presentation of transient sensory, motor, or cognitive events, which disturb the spontaneous EEG and are both time-locked and phase-locked to the onset of the sensory stimuli[12-14]. ERPs are a powerful non-invasive neurophysiologic technique for studying and interpreting brain activity. ERPs are usually characterized by polarity, latency, amplitude, and scalp distribution[12, 14, 15].

Spontaneous EEG amplitudes are extremely small, and the ERP signals extracted from these signals are even smaller (tens of microvolts)[16, 17]. Therefore, to identify ERPs, we turn to data analysis methods to increase the signal-to-noise ratio (SNR). We typically average the data across all trials in time domain to enhance SNR[18, 19].

Several models are used to interpret the generation of deflection in ERPs. For example, the evoked model states that the ERPs are additional activities and independent of the ongoing EEG[6, 20, 21]. The phase reset model states that ERPs are produced from the reorganization of phases of the ongoing EEG activity[21-23]. Nowadays, the neural fundamental of ERPs still remains a matter of debate; however, the combination of these two models may be more reasonable to explain the generation of ERPs[23].

ERPs illustrate the neural processing of sensory and motor events. Also, previous studies have stated that certain kinds of ERPs (e.g. long-latency ERPs) reflect the cortical processing of psychological activities that are time-locked to the onset of a stimulus[6, 17, 24]. With its high temporal resolution, ERP allows for a wide range of researchers, including physiologists, psychologists, and physicians, to make extraordinary discoveries in physiological, cognitive, and clinical fields[25-28].

In the time domain, data across trials are averaged; as a result, we can only identify ERPs which are phase-locked to the presentation of stimulus, thereby excluding all non-phase-locked event-related brain responses (i.e., event-related oscillation). Such event-related oscillations can be described in two ways: event-related synchronization (ERS) and event-related desynchronization (ERD). ERS and ERD show the frequency related information of the ongoing EEGs and represents the power increase and decrease in synchrony of the underlying neuronal populations[12, 13]. We can obtain an abundance of information from ERS and ERD, as they reveal novel, reliable, and detailed neural processing mechanisms.

4. EEG applications

EEG is typically employed to investigate the function of the human brain, both in clinical and non-clinical settings.

One of the main roles of EEG is to assist doctors in diagnosing epilepsy. In comparison with the plethora of sophisticated diagnostic techniques developed over the last few decades, EEG remains an important tool, as it is a convenient and relatively inexpensive way to inspect the physiological manifestations of epileptic seizures[29]. EEG recording provides the information characterized by spikes, sharp waves, and/or spike and wave complexes, which are abnormal in patients. EEG can also be used to monitor brain function, such as monitoring the depth of anesthesia. EEG is also an important technique in the evaluation of patient brain functioning in intensive care settings, such as monitoring convulsive status epilepticus[30-32].

Additionally, by using EEG techniques, research on the neural mechanisms of human brain processing has made considerable contributions to neuroscience, cognitive science, and psychophysiological science. Another interesting type of research is brain-computer interface (BCI), which directly connects the brain with an external device. Using BCI, EEG can disclose information from certain brain areas and then "tell" an external device to execute the corresponding action. For instance, Schwartz and his colleagues reported that monkeys could feed themselves with robotic arms controlled directly by information from their brains[33]. BCI provides a wonderful possibility for the rehabilitation of cognitive and sensory-motor functions.

5. Advantages and disadvantages

While EEG technique has been used widely across fields, the relationship between the scalp macroscopic EEG recordings and the microcosmic cortical processing in the brain is often very difficult to interpret. Also, it is critical to ensure that neurons are close to the scalp. In order to improve EEG research, it is crucial to understand both the advantages and disadvantages that arise in when recording and analyzing EEG data.

1) Temporal and spatial resolution

We have mentioned that EEG has a high temporal resolution (tenth of a millisecond). This is because the sampling rate of EEG recordings can reach as high as tens of kHz[34, 35]. Unfortunately, EEG has significant limitations in spatial resolution. The limitations are as follows: (1) volume conduction (which is caused by the skin, skull and meningeal layers interposed between the brain and electrodes) distorts and exerts a spatial low-pass filter on brain signals[36]; (2) the scalp signals are recorded in low spatial resolution (in the order of centimeters); (3) the inverse problem exists when determining source locations from the scalp topography[37, 38]. To address the shortcomings of EEG, researchers often combine EEG and fMRI to explore the neural processing in both high temporal and spatial resolution.

2) Cost, invasiveness, and sensitivity

In contrast to other techniques (e.g., MEG, PET, and fMRI), EEG is a very inexpensive tool for conducting neuroscience research and does not require participants to ingest foreign substances, making EEG a widely used and extremely popular technique for conducting non-invasive brain research. In addition, researchers using EEG can collect data and replicate experiments much easier than other research techniques.

EEG is not sensitive to the tangential dipoles which are caused by the close field. Even in the open field, if the dipoles are oriented perpendicular to the scalp electrodes, EEG cannot record the signals because of the insufficiently synchronous activity of neurons[36]. For all these reasons, some important neural information cannot be recorded in EEG. However, this also simplifies the analysis and evaluation processes of EEG signals.

8.1.2 Information in EEG/ERPs

As Niko Busch noted: (1) time domain analysis on ERPs shows when things (amplitudes) happen; (2) frequency domain analysis on ERPs shows the magnitudes and frequencies of waves; and (3) time-frequency analysis shows when which frequencies occur and with how much power. Generally, researchers apply different kinds of methods to investigate different types of information hidden in EEG/ERPs recordings.

1. The time domain and the frequency domain

The brain responses to a variety of sensory, motor, or cognitive stimuli contain a rich amount of information ranging from milliseconds to seconds[39]. ERPs can be classified into two categories: (1) early components with a peaking latency less than 100 ms are called "sensory" or "exogenous" because their attributes largely depend on the physical parameters of the stimuli; and (2) the later components are termed as "cognitive" or "endogenous" because they correspond to the information processes underlying the subjects' evaluations of stimuli[40]. We usually use these ERP components to describe and interpret neural processing in the human brain. Conventionally, ERP components are named by the polarity and the occurrence order [e.g., P1 (P: positive), N1 (N: negative), P2, N2, P3, etc., Figure 8.2] or by the polarity and the mean latency after the onset of a stimulus (e.g. N170, P360, P550, etc.)[32]. Examples are provided as follows[40]:

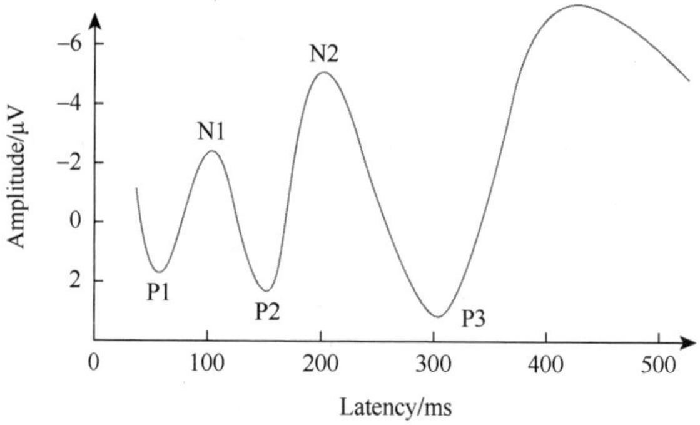

Figure 8.2　Simulation diagram of different ERP components

P1 occurs around 100 ms after the onset of a stimulus, and is typically associated with attention levels. P1 is maximal at the lateral occipital scalp and is usually related to the processing of visual stimuli, which are linked to the primary visual cortex and other visual areas in the brain.

N1 is a negative deflection peaking between 90 ms and 200 ms after the onset of an unexpected stimulus and is associated with general arousal. N1 is maximal at the vertex region, so it is also called the "vertex potential".

P2 refers to a positive deflection peaking around 100 ms to 250 ms after the onset of a stimulus. Studies show that N1/P2 components may be related to sensation-seeking behavior.

N2 is a negative deflection peaking around 200 ms after the onset of a stimulus and is typically used to investigate mismatch responses[41], executive control functions, and language processing[42, 43].

P3 is a positive deflection peaking around 250 to 500 ms after the onset of auditory stimuli. It is typically used to explore the endogenous processes involved in the evaluation or categorization of certain stimuli. P3 is consistently elicited in the Oddball paradigm, in which subjects are asked to respond to the infrequent (target) stimulus[44, 45].

N400 is a negative deflection peaking around 300 ms to 600 ms after the onset of a stimulus. It was first described in the context of semantic incongruity, as N400 is inversely related to the expectancy of a given word at the end of a sentence[40].

Signals can be presented in a variety of ways for different purposes. For instance, to investigate the event-related information, the engineering application signals are presented as a function of time (i.e. time-domain information). Exploring signals in the frequency domain is also important because many important features of the signal are better characterized in the frequency domain than in the time domain[46]. We typically apply spectral analyses to EEG signals to get the frequency-domain information. Spectral information at different frequency bands typically reveals different cognitive processes[32, 47-49]. Exemplar frequency bands are described above in the Spontaneous EEG section (Figure 8.1).

2. The time-frequency domain

EEG signals are complex, time-varying, and non-stationary, and contain different frequency components at different points in time. The most important and fundamental variables used to describe EEG signal are “time” and “frequency”. Information in the time domain indicates how the amplitude of a signal changes over time, and information in the frequency domain indicates how often these changes take place[46]. However, because of the uncertainty principle of the time domain and frequency domain, it is not possible to acquire high resolution in either the time or frequency domain. Time-frequency distribution is best used to describe the distribution of signal energy in time and frequency domains. The aim of time-frequency analysis is to find a density in time and frequency that indicates which frequencies are present in the signal and how they evolve over time. Furthermore, executing the time-frequency analysis on the signal can disclose two dimensions of information instead of one[50]. Fourier’s transformation is the bridge between time and frequency, and an increasing number of approaches for optimizing the algorithm are emerging[46]. Among these methods, some researchers apply discrete wavelet transformation[51-53] while others adopt continuous wavelet transform[54] to denoise the ERP waveforms. In time-frequency analysis, there is a tradeoff between time resolution and frequency resolution in a wide range of EEG rhythm frequencies[14, 55]. Time-frequency analysis can isolate the non-phase-locked event-related responses from the ongoing background EEG recordings and noise-related artifacts, which can enhance the SNR of single-trial ERP.

It is relatively complex to model signals such as EEG recordings using time-frequency methods. A detailed description of time-frequency analysis, including a variety of signal formulations and characteristics, can be found elsewhere[50]. Here we show the simplified time-frequency analysis method to describe the general idea of the data analysis.

Time-frequency distributions are usually calculated using a windowed Fourier or wavelet transformation. These calculations yield a complex time-frequency spectral estimation $F(t,f)$ at each point (t,f) of the time-frequency plane. The obtained time-frequency distributions are baseline corrected using the baseline interval to calculate the relative change of power which is expressed as ER% in the formula below:

$$\text{ER\%}(t,f)=\left[P(t,f)-R(f)\right]/R(f)\times 100 \tag{8.1}$$

$P(t,f)=\left|F(t,f)\right|^2$ is the power spectral density at one time-frequency point (t,f), and $R(f)$ is the averaged power spectral density of the signal enclosed within the baseline reference interval, for each estimated frequency f [14].

We can obtain evoked time-frequency distributions and induced time-frequency distributions by applying this method. Evoked time-frequency distributions only contain phase-locked responses (ERPs), which are estimated by executing time-frequency analysis on the averaged EEG responses. However, induced time-frequency distributions contain both phase-locked (ERPs) and non-phase-locked (ERS and ERD) neural activities, which can be obtained by performing time-frequency analysis on the single-trial waveform and averaging across trials[14]. Additionally, phase-locking value (PLV), a measure of signal phase synchrony across trials, can be estimated according to the formula below:

$$\text{PLV}(t,f)=\left|\frac{1}{N}\sum_{n=1}^{N}\frac{F_n(t,f)}{\left|F_n(t,f)\right|}\right|-\Psi(f) \tag{8.2}$$

where N is the number of trials, $F(t,f)$ is the time-frequency spectral estimation at each point (t,f), and $\Psi(f)$ is the average PLV of the baseline interval for each estimated frequency f [14, 56].

To explore the neural processing, we compare evoked time-frequency distributions, induced time-frequency distributions, and PLV. The information from these distributions reflect power changes in different patterns or synchronism for the EEG recordings.

3. Source separation and localization

To explore the relationship between brain structures and brain functions, one needs to isolate the mixture of EEG recordings into signals which are generated from functionally and neuro-anatomically specific brain regions[57]. EEG recordings have high temporal resolution and low spatial resolution. However, through source separation and localization methods, we can roughly explore the neural origins (i.e. brain regions) of the neuronal components. EEG signals have a mixture of neuronal sources and noise sources. Any field potential vector from the EEG recordings might consist of an infinite number of possible dipoles. Researchers have attempted to identify the independent components in the random vectors[57, 58]; one typical example is the so-called "cocktail party" problem, which is solved using blind source separation (BSS) and independent component analysis (ICA).

BSS problems are often referred to as blind signal decomposition or blind source extraction and work as a spatial filter. In addition, BSS is capable of exploring the spatial information contained in the multi-channel EEG recordings. Each electrode of EEG is one signal sensor; we observe the mixed sensor signal from the unknown multiple inputs of EEG recordings as

$$x(k)=\left[x_1(k),\cdots,x_m(k)\right]^T \tag{8.3}$$

where k is the discrete time and T represents the transpose vector. If the inverse system exists, which is stable, then we can estimate certain characteristics and primary source signals as

$$s(k)=\left[s_1(k),\cdots,s_n(k)\right]^T \tag{8.4}$$

The estimation is according to the sensor signals and output signals as

$$y(k)=\left[y_1(k),\cdots,y_n(k)\right]^T \tag{8.5}$$

This approach (i.e., estimating based on sensor signals) is used to find the original source from the sensor array without the transmission channel characteristics information. The second-order blind identification (SOBI) is a BSS algorithm used to decompose mixed EEG signals recorded from different tasks. SOBI is dependent on the detailed temporal structures and the choice of temporal delay parameters[59].

ICA is the most popular BSS method and is used by a growing number of investigators[60, 61]. ICA can recover independent sources based on mixed sensor signals and is an effective way of isolating stimulus-related and ongoing components in singletrial EEG recordings[59, 62, 63]. Compared with correlation-based transformations, such as principal component analysis (PCA), ICA can de-correlate the signals (2nd-order statistics) and reduce the higher-order statistical dependency, which increases the independence of the signals. Applied to multi-channel EEG recordings, ICA can decompose signals into a single linear combination of independent components with maximally-independent time courses and dipolar maps, without any consideration of head geometry or electrode locations[60].

It is important to note that performing source localizations leads to the typical ill-inverse problem, in which a given scalp topography has an infinite number of possible source configurations. We can only choose which method is better; therefore, the obtained source locations are potentially biased by our additional assumptions, thus limiting the potential applications of the findings[12].

8.2 Significant Findings in Attention, Language, and Social Cognition

Here we will introduce several studies in which EEG signals were collected during attention, language, and social cognition tasks, to highlight the importance of EEG in disclosing brain functions.

8.2.1 Attention

Human are living in a world full of information arising from variable and dynamic sources, but only certain pieces of information are processed by the brain. Attention is a top-down control, which is characterized by concentration on a discrete aspect of information in the environment. Attention provides for efficient processing of stimuli given finite cognitive resources. Different types of attention lead to different results: a positive attentional set promotes attention bias towards stimuli with specific properties, while a negative attentional set turns attention bias away from stimuli with particular properties[64, 65].

Zhang and colleagues explored whether the negative attentional set can impact online target processing in a rapid serial visual presentation (RSVP) task[66]. In this study, N2pc (N2 posterior contralateral), which is a component of ERPs evoked by lateralized targets, was used as an index of attentional selection. Behaviorally, they found that online processing was impaired by the negative attentional set elicited by a special distractor (D1). Compared with the D1 absent condition, D1 (both the digit and the Chinese number character) inhibited T2 (target) performance and delayed the latency of N2pc response to T2. Neurally, D1 elicited a fronto-central N2 peaking at around 300 ms post-onset of D1, suggesting that D1 is indeed an inhibition-evoking stimulus.

N2pc is the ERP component most commonly found in this kind of attention selection task. Other types of attention may be related to other components (e.g., P1 is related to spatial attention). From the above examples, we can see the importance of EEG in illuminating research on attention.

8.2.2 Language

Language contains multiple elements, which can reflect the variations of acoustic parameters. Prosody is a salient and crucial feature for language processing, as prosodic information has an immediate impact upon spoken sentence comprehension. However, little is known about the extent to which prosodic information constrains neurocognitive processes of written language processing.

Luo and colleagues investigated whether a particular prosodic constraint in Chinese (i.e., the rhythmic pattern of the verb-noun combination) affects sentence reading, and whether neural markers of rhythmic pattern processing are similar to those of prosodic processing in the spoken domain[67]. They manipulated the well-formedness of rhythmic pattern as well as the semantic congruency between the verb and the noun. They asked readers to make judgments about the acceptability of each visually presented sentence and found that the abnormal rhythmic pattern evoked both an N400-like effect and a late positivity effect in semantically congruent sentences, but elicited a posterior positivity effect (in the 300 to 600 ms time window) in semantically incongruent sentences. These findings suggest that information concerning rhythmic pattern is used rapidly and interactively to constrain semantic access/integration during Chinese sentence reading.

In addition to the EEG components, EEG coherence is also a useful method to quantitatively measure the linear dependency between two distant brain regions[68]. EEG is the primary technique used to explore the related brain activities and neural mechanisms underlying language comprehension and production.

8.2.3 Social cognition

Humans live in an interactive world; we perceive stimuli from the environment which result in different attitudes and behaviors. The mysteries of social cognition attract many researchers to explore the mechanisms of social processing in humans. Researchers use EEG to better understand how humans process, store, and apply information about other individuals and social situations. Many studies have applied the P300 as an index for cognitive processing and neurological/psychiatric disorders. For example, Zhang and colleagues combined source separation and source localization methods to investigate the cortical origins of the P300 elicited in a facial attractiveness judgment task[69]. For each participant, they applied SOBI to continuous EEG data to decompose the mixture of brain signals and noise. The equivalent current dipole (ECD) models were used to estimate the centrality of the SOBI-recovered P300. They found that the ECD models (which consist of dipoles in the frontal and posterior association cortices) account for $96.5\% \pm 0.5\%$ of variance in the scalp projection of the component. Given that the recovered dipole activities in different brain regions share the same time course with different weights, the conclusion is that the P300 originates from synchronized activity between anterior and posterior parts of the brain.

As seen above, EEG is a crucial tool for understanding how we think, feel, and interact with the world around us.

8.3 Summary

We have given an introduction to the EEG technique and presented several studies in the field of neuroscience that have used EEG to explore the functions of the brain. As one of the most important tools in diagnosing brain disorders and in understanding the neural mechanisms of the brain, the contributions of EEG to science cannot be ignored. While other techniques with better spatial localization properties have and are emerging, such as PET and fMRI, their time resolution is inferior to EEG. In addition, EEG rhythms have a unique and specific role in neural processing that cannot be distinguished by imaging techniques. Moreover, the greatest information content is usually connected with high-frequency rhythms (especially gamma) generated by small neuron pools, which makes this information even less accessible to imaging techniques. Therefore, these techniques are more likely to complement, not replace, EEG as a tool for neuroscience research.

References

[1] Berger H. 1929. Electroencephalogram in humans. *Arch Psychiat Nervenkr*, 1929, 87: 527-570.

[2] Hu L. *Chasing Evoked Potentials: Novel Approaches to Identify Brain EEG Responses at Single-trial Level*. Hong Kong: The University of Hong Kong, 2010.

[3] Squuire L R. *Fundamental Neuroscience*. 3rd ed. Amsterdam, Boston: Elsevier/Academic, 2008.

[4] Hildebrandt J, Smith D, Great Pacific Media. *The Nervous System: Neurons, Networks, and the Human Brain*. Colorado Spring, Colo: Great Pacific Media, 2008.

[5] Rowan A J, Tolunsky E. *Primer of EEG: With a Mini-atlas*. Philadelphia, Pa: Butterworth-Heinemann, 2003.

[6] Niedermeyer E, Lopes da Silva FH. *Electroencephalography: Basic principles, Clinical Applications, and Related Fields*. 5th ed. Philadelphia: Lippincott Williams & Wilkins, 2005.

[7] Kandel E R, Schwartz J H, Jessell T M. *Principles of Neural Science*. 4th ed. New York: McGraw-Hill Health Professions Division, 2000.

[8] Tatum W O, Ellen R. Grass lecture: Extraordinary EEG. The *Neurodiagnostic Journal*, 54 (1): 3-21.

[9] Kirmizi-Alsan E, Bayraktaroglu Z, Gurvit H, et al. Comparative analysis of event-related potentials during Go/NoGo and CPT: Decomposition of electrophysiological markers of response inhibition and sustained attention. *Brain Research*, 2006, 1104: 114-128.

[10] Klimesch W, Sauseng P, Hanslmayr S. EEG alpha oscillations: The inhibition-timing hypothesis. *Brain Research*

Reviews, 2007, 53 (1): 63-88.

[11] Hughes J R. *EEG in Clinical Practice*. 2nd ed. Boston: Butterworth-Heinemann, 1994.

[12] Mouraux A, Iannetti G D. Across-trial averaging of event-related EEG responses and beyond. *Magnetic Resonance Imaging*, 2008, 26 (7): 1041-1054.

[13] Pfurtscheller G, Lopes da Silva F H. Event-related EEG/MEG synchronization and desynchronization: Basic principles. *Clinical Neurophysiology*, 1999, 110 (11): 1842-1857.

[14] Zhang L, Peng W W, Zhang Z G, et al. Distinct features of auditory steady-state responses as compared to transient event-related potentials. *PLoS One*, 2013, 8 (7):e 69164.

[15] Callaway E, Tueting P, Koslow S H, et al. *Event-related Brain Potentials in Man*. New York: Academic Press, 1978.

[16] Hu L, Mouraux A, Hu Y, et al. A novel approach for enhancing the signal-to-noise ratio and detecting automatically event-related potentials (ERPs) in single trials. *Neuroimage*, 2010, 50 (1): 99-111.

[17] Rugg M D, Coles M G H. *Electrophysiology of Mind: Event-related Brain Potentials and Cognition*. Oxford, England: Oxford University Press, 1995.

[18] Dawson G D. A summation technique for detecting small signals in a large irregular background. *Journal of Physiology*, 1951, 115 (1): 2-3.

[19] Dawson G D. A summation technique for the detection of small evoked potentials. *Electroencephalogr and Clinical Neurophysiology*, 1954, 6 (1): 65-84.

[20] Jervis B W, Nichols M J, Johnson T E, et al. A fundamental investigation of the composition of auditory evoked potentials. *IEEE Transactions on Biomedical Engineering*, 1983, 30 (1): 43-50.

[21] Sauseng P, Klimesch W, Gruber W R, et al. Are event-related potential components generated by phase resetting of brain oscillations? A critical discussion. *Neuroscience*, 2007, 146 (4): 1435-1444.

[22] Makeig S, Westerfield M, Jung T P, et al. Dynamic brain sources of visual evoked responses. *Science*, 2002, 295 (5555): 690-694.

[23] Min B K, Busch N A, Debener S, et al. The best of both worlds: Phase-reset of human EEG alpha activity and additive power contribute to ERP generation. *International Journal of Psychophysiology*, 2007, 65 (1): 58-68.

[24] Garnsey S M. *Event-related Brain Potentials in the Study of Language*. Hove: Lawrence Erlbaum, 1993.

[25] Begleite H, Porjesz B, Gross M M. Cortical evoked potentials and psychopathology—a critical review. *Archives of General Psychiatry*, 1967, 17 (6): 755-758.

[26] Deletis V, Sala F. Intraoperative neurophysiological monitoring of the spinal cord during spinal cord and spine surgery: A review focus on the corticospinal tracts. *Clinical Neurophysiology*, 2008, 119 (2): 248-264.

[27] Duncan C C, Barry R J, Connolly J F, et al. Event-related potentials in clinical research: Guidelines for eliciting, recording, and quantifying mismatch negativity, P300, and N400. *Clinical Neurophysiology*, 2009, 120 (11): 1883-1908.

[28] Gonzalez A A, Jeyanandarajan D, Hansen C, et al. Intraoperative neurophysiological monitoring during spine surgery: A review. *Neurosurg Focus*, 2009, 27 (4): E6.

[29] Smith S J M. EEG in the diagnosis, classification, and management of patients with epilepsy. *Journal of Neurology, Neurosurgery and Psychiatry*, 2005, 76 (suppl2) : ii2- ii7.

[30] Murdoch-Eaton D, Darowski M, Livingston J. Cerebral function monitoring in paediatric intensive care: Useful

features for predicting outcome. *Developmental Medicine and Child Neurology*, 2001, 43 (2): 91-96.

[31] Saliba E, Marret S, Chavet-Queru M S, et al. Emergency electroencephalography during perinatal cerebral intensive care: Indications and results. *Neurophysiologie Clinigue*, 1998, 28 (2): 144-153.

[32] Taylor M J, Baldeweg T. Application of EEG, ERP and intracranial recordings to the investigation of cognitive functions in children. *Developmental Science*, 2002, 5 (3): 318-334.

[33] Wickelgren I. Neuroscience—Tapping the mind. *Science*, 2003, 299 (5606): 496-499.

[34] Inoue K, Hashimoto I, Nakamura S. High-frequency oscillations in human posterior tibial somatosensory evoked potentials are enhanced in patients with Parkinson's disease and multiple system atrophy. *Neuroscience Letters*, 2001, 297 (2): 89-92.

[35] Ozaki I, Suzuki C, Yaegashi Y, et al. High frequency oscillations in early cortical somatosensory evoked potentials. *Electroencephalography and Clinical Neurophysiology*, 1998, 108 (6): 536-542.

[36] Nunez P L, Srinivasan R. *Electric Fields of the Brain: the Neurophysics of EEG*. 2nd ed. Oxford, New York: Oxford University Press, 2006.

[37] Grech R, Cassar T, Muscat J, et al. Review on solving the inverse problem in EEG source analysis. *Journal of Neuroengineering and Rehabilitation*, 2008, 5 (1): 1419-1424.

[38] Schroeder C E, Steinschneider M, Javitt D C, et al. Localization of ERP generators and identification of underlying neural processes. *Electroencephalography and Clinical Neurophysiology Supplement*, 1995, 44: 55-75.

[39] Mecklinger A. Interfacing mind and brain: A neurocognitive model of recognition memory. *Psychophysiology*, 2000, 37 (5): 565-582.

[40] Sur S, Sinha V K. Event-related potential: An overview. *Industrial Psychiatry Journal*, 2009, 18 (1): 70-73.

[41] Winkler I, Karmos G , Naatanen R. Adaptive modeling of the unattended acoustic environment reflected in the mismatch negativity event-related potential. *Brain Research*, 1996, 742 (1-2): 239-252.

[42] Folstein J R, van Petten C. Influence of cognitive control and mismatch on the N2 component of the ERP: A review. *Psychophysiology*, 2008, 45 (1): 152-170.

[43] Schmitt B M, Munte T F, Kutas M. Electrophysiological estimates of the time course of semantic and phonological encoding during implicit picture naming. *Psychophysiology*, 2000, 37 (4): 473-484.

[44] Basar E, Basareroglu C, Rosen B, et al. A new approach to endogenous event-related potentials in man-relation between eeg and P300-wave. *International Journal of Neuroscience*, 1984, 24 (1): 1-21.

[45] Patrick C J, Bernat E M, Malone S M, et al. P300 amplitude as an indicator of externalizing in adolescent males. *Psychophysiology*, 2006, 43 (1): 84-92.

[46] Qian S, Chen D. Joint time-frequency analysis. *Signal Processing Magazine, IEEE*, 1999, 16 (2): 52-67.

[47] Fernandez T, Harmony T, Silva J, et al. Relationship of specific EEG frequencies at specific brain areas with performance. *Neuroreport*, 1998, 9 (16): 3681-3687.

[48] Klimesch W. EEG alpha and theta oscillations reflect cognitive and memory performance: A review and analysis. *Brain Research Reviews*, 1999, 29 (2-3): 169-195.

[49] Ray W J, Cole H W. EEG alpha activity reflects attentional demands, and beta activity reflects emotional and cognitive processes. *Science*, 1985, 228 (4700): 750-752.

[50] Boashash B. *Time Frequency Analysis*. Amsterdam: Gulf Professional Publishing, 2003.

[51] Jongsma M L A, Eichele T, van Rijn C M, et al. Tracking pattern learning with single-trial event-related potentials. *Clinical Neurophysiology*, 2006, 117 (9): 1957-1973.

[52] Quiroga R Q. Obtaining single stimulus evoked potentials with wavelet denoising. *Physica D-Nonlinear Phenomena*, 2000, 145 (3): 278-292.

[53] Quiroga R Q, Garcia H. Single-trial event-related potentials with wavelet denoising. *Clinical Neurophysiology*, 2003, 114 (2): 376-390.

[54] Mouraux A, Plaghki L. Single-trial detection of human brain responses evoked by laser activation of A delta-nociceptors using the wavelet transform of EEG epochs. *Neuroscience Letters*, 2004, 361 (1-3): 241-244.

[55] Zhang Z G, Hu L, Hung Y S, et al. Gamma-band oscillations in the primary somatosensory cortex—a direct and obligatory correlate of subjective pain intensity. *The Journal of Neuroscience*, 2012, 32 (22): 7429-7438.

[56] Lachaux J P, Rodriguez E, Martinerie J, et al. Measuring phase synchrony in brain signals. *Human Brain Mapping*, 1999, 8 (4): 194-208.

[57] Tang A. Applications of second order blind identification to high-density EEG-based brain imaging: A review. *Lecture Notes in Computer Science*, 2010, 6064: 368-377.

[58] Belouchrani A, Abed-Meraim K, Cardoso J-F, et al. A blind source separation technique using second-order statistics. *IEEE Transactions on Signal Processing*, 1997, 45 (2): 434-444.

[59] Tang A C, Sutherland M T, McKinney C J. Validation of SOBI components from high-density EEG. *Neuroimage*, 2005, 25 (2): 539-553.

[60] Makeig S, Debener S, Onton J, et al. Mining event-related brain dynamics. *Trends in Cognitive Sciences*, 2004, 8 (5): 204-210.

[61] Makeig S, Jung T P, Bell A J, et al. Blind separation of auditory event-related brain responses into independent components. *Proceedings of the National Academy of Sciences*, 1997, 94 (20): 10979-10984.

[62] Debener S, Ullsperger M, Siegel M, et al. Single-trial EEG-fMRI reveals the dynamics of cognitive function. *Trends in Cognitive Sciences*, 2006, 10 (12): 558-563.

[63] Jung T P, Makeig S, Westerfield M, et al. Analysis and visualization of single-trial event-related potentials. *Human Brain Mapping*, 2001, 14 (3): 166-185.

[64] Cepeda N J, Cave K R, Bichot N P, et al. Spatial selection via feature-driven inhibition of distractor locations. *Percept Psychophys*, 1998, 60 (5): 727-746.

[65] Chelazzi L, Miller E K, Duncan J, et al. A neural basis for visual search in inferior temporal cortex. *Nature*, 1993, 363 (6427): 345-347.

[66] Zhang D X, Zhou X L, Martens S. The impact of negative attentional set upon target processing in RSVP: An ERP study. *Neuropsychologia*, 2009, 47 (12): 2604-2614.

[67] Luo Y Y, Zhou X L. ERP evidence for the online processing of rhythmic pattern during Chinese sentence reading. *Neuroimage*, 2010, 49 (3): 2836-2849.

[68] Weiss S, Mueller H M. The contribution of EEG coherence to the investigation of language. *Brain and Language*, 2003, 85 (2): 325-343.

[69] Zhang Y, Tang A C, Zhou X L. Synchronized network activity as the origin of a P300 component in a facial attractiveness judgment task. *Psychophysiology*, 2014, 51 (3): 285-289.

Chapter 9 Detection, Processing, and Application of Neural Signals

Zhigong Wang, Xiaoying Lü, Boshuo Wang

9.1 Introduction to the Nervous System

Overall, the nervous system is a whole, whose morphology and functionality are indivisible. According to the location and function, it can be divided into the central nervous system and peripheral nervous system. Figure 9.1 is an overview of human's nervous system.

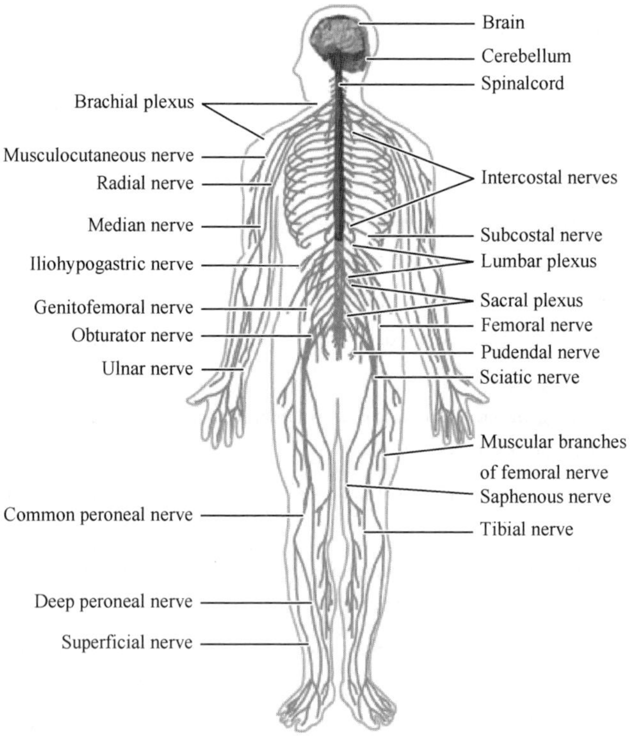

Figure 9.1 An overview of human's nervous system (dorsal view)

Sources: http: //www.wikiwand.com/zh/%E5%91%A8%E5%9B%B4%E7%A5%9E%E7%BB%8F%E7%B3%BB%E7%BB%9F

9.1.1 Central nervous system

The central nervous system (CNS) includes the brain and the spinal cord. The brain, located in the cranial cavity, is the enlarged rostral terminal of the central nervous system, and can be divided into six parts: telencephalon, diencephalon, midbrain, pons, cerebellum, and medulla oblongata. The midbrain, pons and medulla oblongata together are known as the brainstem. The medulla oblongata connects to the spinal cord through the foramen magnum. The lumens of the brain are known as ventricles, and contain cerebrospinal fluid. The telencephalon includes the left and right hemispheres of the brain, each covered by gray matter also known as the cerebral cortex. During the long course of evolution, the human's cerebral cortex has become highly developed. It is not only the high level center of human activities, but also the material basis of thought and awareness.

Located within the spinal canal, the spinal cord is of tubular shape, with the upper end connected to the medulla oblongata at the foramen magnum, and the lower end terminating at the level of the first lumbar vertebra. On the anterior and posterior side of the spinal cord are a series of very fine fibers called filaments. The filaments extend outwards and within certain ranges collect into bundles, forming the ventral and dorsal roots of the spinal cord. At their distal ends, the ventral root and the dorsal root join to form a mixed spinal nerve at the intervertebral foramen. Based on the level of exit on the vertebral column, the spinal cord is divided into thirty-one segments, i.e., eight cervical segments (C1-C8), twelve thoracic segments (T1-T12), five lumbar segments (L1-S5), five sacral segments (S1-S5), and one coccygeal segment (Co1). Figure 9.2 shows the vertebral segments corresponding to the spinal cord segments, and the functions of each spinal nerve.

9.1.2 Peripheral nervous system

The peripheral nervous system connects the central nervous system to the other systems and organs, and its majority consists of nerve fibers. Somatosensory nerves are distributed in the skin, skeletal muscle, tendons and joints, etc., and conduct afferent sensory signal from external or internal stimuli to the central system; visceral sensory nerves are distributed in organs, cardio vasculature, and glands, etc., and conduct sensory input from those internal organs; somatic motor nerves are distributed in skeletal muscle fibers and control their motion; and visceral motor nerve innervate smooth muscle, cardiac function, and regulate visceral glands.

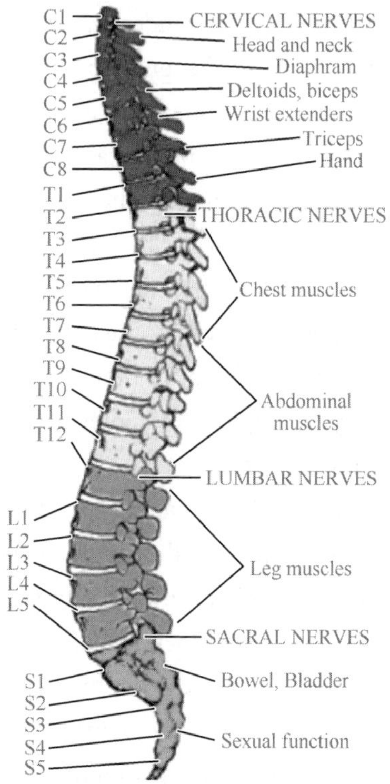

Figure 9.2 Vertebral segments corresponding to the spinal cord segments, and the functions of each related spinal nerve

Sources: https: //www.pinterest.com/pin/116038127875003394/

9.1.3 Neuronal structure

Neurons were confirmed as the individual building elements of the nervous system not until the end of the 19th century, when neuroanatomist Santiago Ramón y Cajal cleverly applied the silver staining method discovered by Camillo Golgi and established the "neuron doctrine". Cajal and Golgi shared the 1906 Nobel Prize in Physiology and Medicine for this contribution.

Highly specialized, neurons are the basic structural and functional units of the nervous system. They acquire stimulation and transmit the excitation. Neurons consist of soma and neurites (axons and dendrites). The soma contains a nucleus and cytoplasm.

Figure 9.3 (see the Color Inset, p. 24) shows the microscopic structure of a typical neuron, including the overall shape and characteristics of organelles.

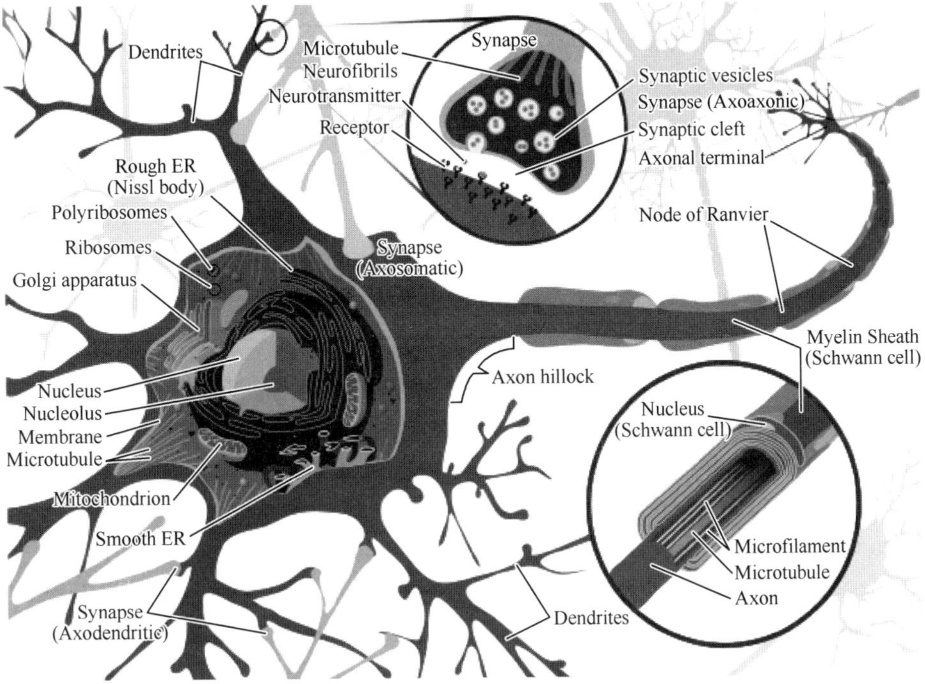

Figure 9.3 Microscopic structure of neurons

Sources: https: //commons.wikimedia.org/wiki/File: Complete_neuron_cell_diagram_en.svg

Apart from the organelles that most cells contain, such as mitochondria, endoplasmic reticulum etc., neuron's cytoplasm also contains unique structures such as neurofibrillary, Nissl bodies and so on. Compared to other cells, neurons have three unique structural components: axons, dendrites, and synapses. Dendrites are usually shorter in length but have more branches. They receive neural impulses, and conduct them to the soma. The dendritic tree of different neuron cell types varies in numbers and morphology. Unlike dendrites, the axon's function is to conduct the neural impulse generate from the soma outwards. Each neuron only has one axon, with length dependent on cell type. According to the numbers of neurites, neurons are classified as pseudo-monopolar neurons, bipolar neurons, and multipolar neurons. Neurons connect with each other mostly via synapses, and not by communication of the cytoplasm. Synapses are specialized compartments that form contact between neurons. Usually, the axon of one neuron connects to the dendrite or soma of another neuron via synapses, and this allows the neural impulses to be transmitted from one neuron to another.

According to the functions of neurons, they can be divided into sensory neurons, motor neurons and interneurons. Sensory neurons are known as afferent neurons, and are usually located in the outer periphery of the sensory ganglia. They are pseudo-unipolar or bipolar, accept a variety of stimuli from the external environment, and conduct neural impulse to the central nervous system. Motor neurons are also known as efferent neurons, and are usually located within the brain, spinal cord, peripheral motor nuclei or ganglia of the autonomous nervous system. They are mostly multipolar, and transmit impulses from the CNS to muscles, glands, or other effectors. Interneurons are also known as intermediate neurons. They are located between sensory and motor neurons, and play the role of connection and integration.

Neurons clustered together and form neural networks. In the network, neurons are connected by dendrites and axons. Via ions and chemical transmitters, neighboring neurons transmit neural signals. Neural network also processes the neural information, and related topics are under continuous research. The miraculous functions of the human nervous system are built on the basis of neurons and neuronal networks.

Besides neurons, the nervous system consists of large numbers of glial cells, which provide structural support, isolation, nutrition and protection to neurons. Glial cells have smaller soma, their projections are not divided into dendrites or axons, and they don't contain neurofibrillary nor Nissl body. Glial cells don't conduct neural excitations.

9.1.4 Neural electrophysiology and electrical transmission of neural information

Pioneer scientists have conduct numerous research regarding neurons and the signal transfer characteristics of nervous system in the following directions:

(1) The microscopic structure of specialized compartments of neurons;

(2) Signal transfer mechanisms within the neurons;

(3) Interconnection and information transmission between neurons;

(4) Relations between different forms of interconnection and behavior;

(5) Modification of neurons and neuronal networks due to experience, namely neural plasticity.

English physiologist Sherrington studied the knee-jerk reflex, and proposed that reflexes are a form of basic neural activity. Reflexes consist of both excitation and inhibition, two processes working in coordination. He discovered that the nerve bundle innervating muscles contain both motor and sensory nerve fibers, and put forward the concept of synapses and integration of the nervous system. The basic function of the nervous system relies on the inter-neuron communication, and it is the unique specialized structure of synapses that implements this function. Physiologist Edgar Adrian has specified the reflex theory, and by using a string galvanometer to record the neural impulses of a single nerve fiber, created a new era of modern electrophysiology study. Therefore, the two scientists shared the Nobel Prize in Physiology and Medicine in 1932.

The Nobel Prize of 1963 was awarded to the English physiologist Hodgkin and Huxley. They used microelectrodes, cathode-ray oscilloscope, and the voltage-clamp technique, and with microelectrodes of 1-μm size tapering tip detected for the first time action potentials and synaptic voltage around 10-mV amplitude on the squid's giant axons. They revealed the mechanism of potential changes, namely how the ion channels on the cell membrane behave, and also derived a classic mathematical model for describing nerve impulses. This opened up new methods for studying ECG and other clinical applications. Another recipient was the Australian physiologist Eccles, who studied under Sherrington the information transfer mechanism between neurons. He recorded the endplate potentials at the neuromuscular junctions, and further validated the existence of inhibitory synapses, which Sherrington has predicted in his later years.

With the rise of molecular biology in the 1970s, revolutionary breakthroughs occurred in the fields of life science including neurobiology. Direct study of the molecular mechanisms of the brain and neuronal activity became possible. German scientists Nehr and Sakmann first recorded the activity of single ion channel via patch clamp recording technique. The recording of electric current through a single ion channel on the cell membrane confirmed that the change of ion channels can affect intracellular ion concentrations and regulate cell function. For this bridging connection between electrophysiology and molecular biology, the two scientists were awarded the Nobel Prize in 1991.

9.1.5 Chemical transmission of neural information

In addition to the aforementioned neural signal transfer characteristics, Erlangely, Loewi, Dale, and other pharmacologists in the 1930s, have confirmed the existence of neurotransmitter and chemical transmission of nerve impulses at synapses. Several scientists have therefore been awarded Nobel Prize in Physiology and Medicine in 1936 and 1944 for research related to this milestone discovery. The Nobel Prize was awarded to Katz, Euler, and Axelrod in 1970 for their discovery in the storage, release, and inactivation processes of neurotransmitters.

In the 1950s, Harris first showed that the hypothalamus produces and releases hormones. Guillemin and Schally have utilized the radio immune assays developed by Yalow, and isolated TRF, LRF and other hypothalamic hormone releasing factor. A variety of neurotropic substances, neurotransmitters, and second messenger substances were identified despite their minute concentration. Guillemin, Schally and Yalow were therefore awarded the Nobel Prize in Physiology and Medicine in 1977. Moreover, Gilman and Rodbell were awarded the Nobel Prize in 1994 for their discovery G-proteins and the role of these proteins in intracellular signal transduction; Furchgott, Ignarro, and Murad were awarded the Nobel Pirze in 1998 for their discovery of nitric oxide (NO) as a signaling molecule; Carlsson discovered the importance role of neurotransmitter dopamine in movement control and mental activity and received the Nobel Prize in Physiology and Medicine in 2000 together with Greengard and Kandel for their contribution in studying signal transduction in the nervous system.

A major feature of the nervous system is that neural information transfers in both forms of electricity and chemistry, which must be considered when modeling and analyzing the nervous system.

9.1.6 Research on the functions of the nervous system

The studies of visual and auditory functions are the most prominent among the research of the sensory functions of the nervous system.

Gullstrand studied the refraction of light in the eye, and established laws of optical image in the eye. For this work, he received the Nobel Prize in Physiology and Medicine in 1911. Hartline, Wald, and Granit further contributed to the primary physiological and chemical visual processes in the eye, and received the Nobel Prize in Physiology and Medicine in 1967.

Barany studied the physiology and pathology of the vestibular organs and he was awarded the Nobel Prize in Physiology and Medicine in 1914 for his discovery of the functions of vestibular organs and diagnostic methods for related diseases. Békésy received the Nobel Prize in Physiology and Medicine in 1961, for his discovery of the physical mechanism for the cochlear to sense sound: the basal membrane vibrates in response to external acoustic input, and changes the cellular potential of the hair cells; the resulting stimulation of the auditory nerve is transmitted to the brain and perceived as hearing.

In addition to the sensory and motor functions, the nervous system serves other high level functions such as learning, memory, language, and sleep, etc. Studying the high level function of the nervous system, Pavlov first explained "conditioned reflex" from a physiological point of view, and received the first Nobel Prize in the area of neuroscience in 1904. Hubel, Wiesel, and Sperry were awarded the Nobel Prize in Physiology and Medicine in 1981. Sperry made important discoveries concerning the functional specialization of the cerebral hemispheres. Hubel and Wiesel collaborative researched the brain's visual cortex, revealing that visual columns are the functional unit in the cortex for processing visual information. Their study showed the significant plasticity in the development of the visual cortex, which demonstrated high importance for understanding the mechanism of how the cerebral cortex processes information.

In order to establish the theoretical basis for neural signal detection, functional electrical stimulation, and signal regeneration after spinal cord injury, the next section introduces the electrical characteristic of neurons and the transmission of action potentials.

9.2 Electrical Properties of Nerves and Propagation of Action Potentials

Information transmission within a neuron and among neurons is the most basic and important functions of nervous systems. This section describes the fundamental principles of how information is transmitted within the compartments of the neurons.

The transmission of nerve impulses is essentially flow of ionic current. However, they are reflected as electrical signals by the changes of transmembrane potentials. These electrical signals of neural impulse can be detected, while applying a suprathreshold electrical signal across the neuronal membrane can activate action potentials (AP). The APs encode neural activity and contain abundant information of behavior and functions, and enable the detection of neural signal and neural function stimulation via implantable microelectronics. Here, the passive and active electrical properties of neurons are introduced. The former is the foundation of transmission of the nerve signals in the cellular medium, and the latter is the basis for generating APs.

9.2.1 Passive electrical properties of neurons

Just like any other cell, neurons have a voltage difference across their cell membrane, known as the membrane potential. This is due to an uneven distribution of charge carrier ions across the membrane. Different ions cross the membrane at different rates, and the rates may differ for the same kind of ions under different condition of the neurons such as resting or during action potentials. The conductance of ions is regulated by ion channels, which are specific transmembrane protein that form aqueous channels allowing specific ions to flow through. The ionic current flow through different types of ion channels are responsible the transmission of electrical signal in neurons.

The membrane potential under the resting state, i.e. when not conducting nerve impulses, is called the resting potential and usually ranges between –90 mV to –40 mV. Typically, the membrane potential is said to be depolarized when it is less negative compared to the resting potential, and hyperpolarized when it is more negative than the resting potential.

In electrophysiological studies of neurons, the transmembrane potential can be measured via an intracellular electrode and potentiometer; while using the same electrode, cathodic or anodic current can be applied to the intracellular space to induce hyperpolarization or depolarization.

The membrane potential change under hyperpolarization or small depolarization current reflects the passive membrane properties of neurons, as shown in Figure 9.4. From the passive properties, the membrane shows behavior similar to a low-pass filter consisting of resistors and capacitors. Further studies may use an equivalent circuit to describe the electrical properties the neuronal bilipid layer membrane, as shown in Figure 9.5.

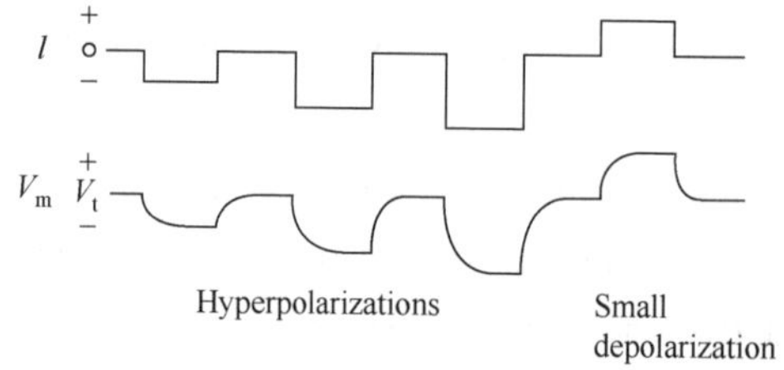

Figure 9.4 Passive membrane properties of neurons

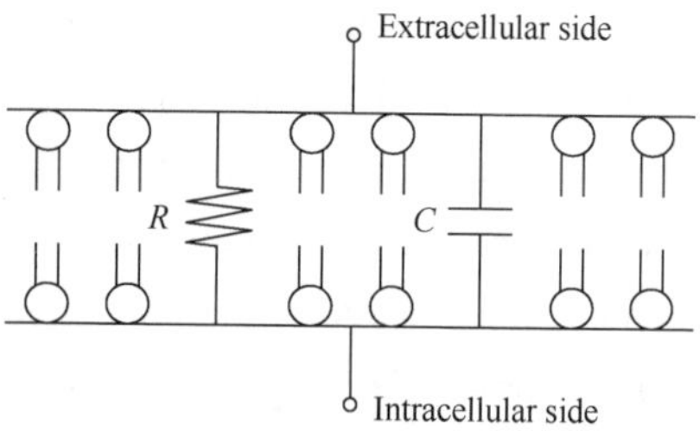

Figure 9.5 Equivalent circuit of neuronal membrane

With this RC equivalent circuit, the passive behavior could be accurately described. The membrane time constant can be expressed as $\tau = R_{\mathrm{m}}C_{\mathrm{m}}$, and the rate of change of the membrane potential is a function of the injected current and membrane capacitance $\mathrm{d}V_{\mathrm{m}} / \mathrm{d}t = I / C_{\mathrm{m}}$. The transmembrane currents can be divided into two parts, the displacement current by the capacitor and the ionic current via the conductance (ion channels), i.e.

$$I_{\mathrm{m}} = I_{\mathrm{c}} + I_{\mathrm{i}} = C_{\mathrm{m}} \cdot \mathrm{d}V_{\mathrm{m}} / \mathrm{d}t + V_{\mathrm{m}} / R_{\mathrm{m}} \tag{9.1}$$

From this equation, the time domain expression for the membrane potential subject to a current pulse input is given as

$$V_{\mathrm{m}}(t) = I_{\mathrm{s}} \cdot R_{\mathrm{m}} \cdot \left(1 - \mathrm{e}^{-t/\tau}\right) \tag{9.2}$$

9.2.2 Active electrical properties of neurons

Passive membrane properties of neurons described above apply to hyperpolarization and weak depolarization stimulation. However, when a large depolarization reaches the critical stimulus intensity, i.e. threshold, the neuron will exhibit active electrical properties. This is an active process because it involves the opening and closing of ion channels in response to changes in membrane potential. It is this active response that transmits information from one part of the neuron to another.

Neural information is encoded in the rapid changes of the membrane potential, known as action potentials, spikes, or nerve impulses. Specifically, when the depolarization exceeds the threshold, the membrane potential depolarizes quickly within factions of millisecond duration, and then returns to the resting level slightly slower, as shown in Figure 9.6. The action potential propagates along the axon from the cell body toward the distal part of the axon.

There are several important features of action potential generation:

(1) Threshold requirement. This ensures that the neuron doesn't generate an action potential in response to random small membrane depolarization and only to strong enough stimulus before they can pass information through the axon action potentials;

(2) The action potentials follow an all-or-none law, which is the basic characteristic of axonal transmission. Any suprathreshold stimulation will invoke an action potential of the same size, regardless of stimulus intensity. In other words, once the stimulus reaches threshold, the amplitude of the action potential doesn't reflect the amplitude of

the stimulation. Therefore information about the stimulus intensity is encoded in some other manner.

(3) Latency characteristics. The latency, defined as the time difference between the peak time of the action potential and the onset of the stimulus, is dependent on stimulation intensity. Within a certain limit, the stronger the stimulus, the sooner action potentials occur.

(4) Characteristics of refractory period. Within several milliseconds of the action potential, no action potential can be induced, no matter how strong the depolarizing stimuli is; This time is called the absolute refractory period. The absolute refractory period is followed by the relative refractory period, during which threshold of for stimulation has increased over its normal value at resting condition.

Given the relationship between stimulus intensity and latency and the existence of refractory period, action potentials can encode the stimulation intensity by the frequency of spiking. As long as the stimulus persists, the neuron will generate action potentials, which have to have an interval in between. After one of the action potentials, the threshold during refractory period must drop below the level of stimulation for the next action potential to occur. For large stimulation, the threshold does not need to drop much and the next spike occurs much faster compared to a small stimulus. Hence large stimulations have a higher frequency of action potentials, as shown in Figure 9.7.

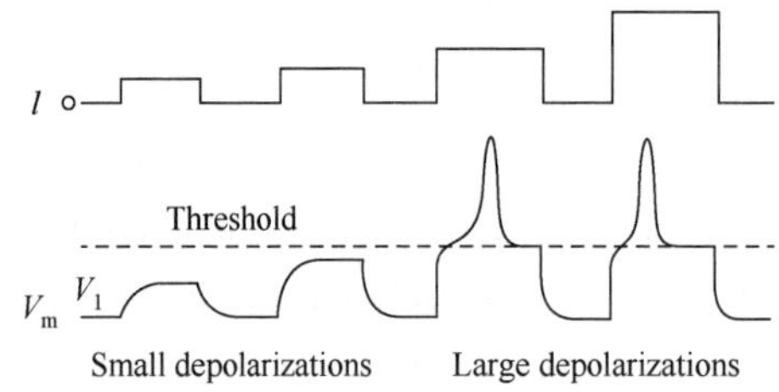

Figure 9.6　Depolarization exceeding a threshold induces a nerve action potential

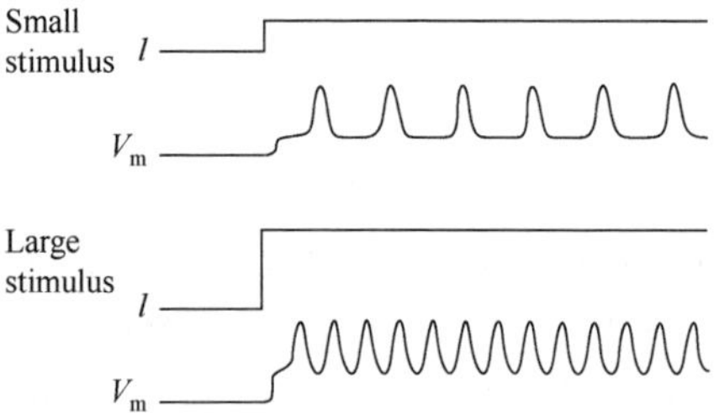

Figure 9.7　Action potential frequency modulated by stimulation intensity

As neurons control all kinds of different behavioral and physiological functions, neurons in different areas of the nervous system are diverse in their morphologies and electrophysiology. Neuronal electrical activities vary in many aspects, such as shape, magnitude, and duration of membrane potential activity. In addition, the modes of action potential generation are also diverse, with certain neurons not capable of firing action potentials automatically, while some may show regular periodic or bursting behavior.

9.2.3 Action potential propagation in axons

The transduction of action potential on the axon is close related with the passive diffusion properties of potentials on the axons. As shown in Figure 9.8, a few intracellular microelectrodes are placed at regular intervals on an axon. When a hyperpolarizing current is injected at the first electrode, a voltage response could be recorded at all electrodes. The amplitude is the largest at the first electrode, while decreasing exponentially with distance from the first one. This is due passive diffusion of the stimulation current.

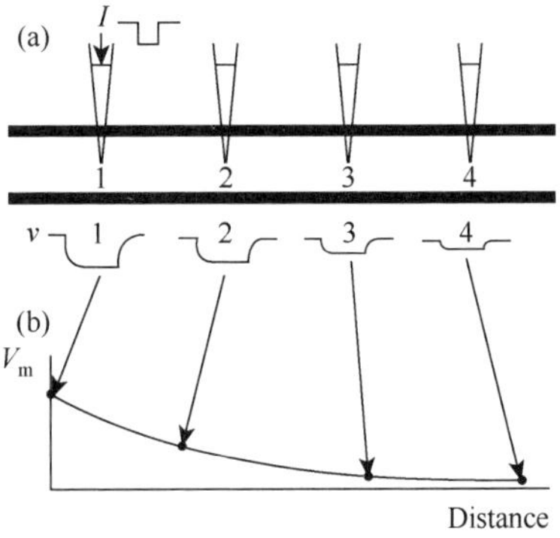

Figure 9.8 Passive diffusion

With the equivalent circuit model of neuron cell membrane, a mathematical expression for passive diffusion of membrane potential can be derived. Assuming non-myelinated axons, in shape of an ideal cylindrical shell of cell membrane, the transmembrane current and potential are axial symmetric. In the one-dimensional direction (Figure. 9.9), the membrane is modeled as a series of resistance r_m (Ω·m) and capacitance c_m (F/m) in parallel. The intracellular and extracellular medium are series of cascaded resistance per unit length of r_i (Ω/m) and r_e (Ω/m) respectively.

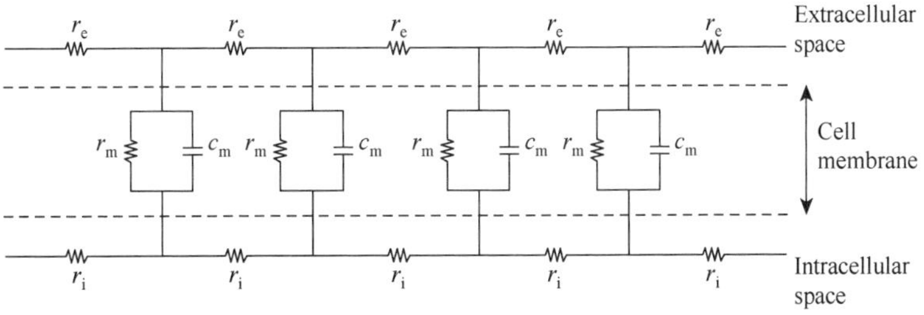

Figure 9.9 Equivalent circuit of the one-dimensional axon model

The following differential equation describes the membrane potential in the model of passive diffusion.

$$V_{\mathrm{m}}(x,t)=\varphi_{\mathrm{i}}(x,t)-\varphi_{\mathrm{e}}(x,t) \tag{9.3}$$

$$\frac{\partial \varphi_{\mathrm{i}}}{\partial x}=-r_{\mathrm{i}}i_{\mathrm{i}} \tag{9.4}$$

$$\frac{\partial \varphi_{\mathrm{e}}}{\partial x}=-r_{\mathrm{e}}i_{\mathrm{e}} \tag{9.5}$$

$$i_{\mathrm{m}}=-\frac{\partial \varphi_{\mathrm{i}}}{\partial x}=-\frac{\partial e_{\mathrm{i}}}{\partial x}=-V_{\mathrm{m}}+c_{\mathrm{m}}\frac{\partial V_{\mathrm{m}}}{\partial t} \tag{9.6}$$

Taking second order partial differential of x on both sides of (9.3), and combining the other equations, yields

$$\frac{\partial^2 V_{\mathrm{m}}}{\partial x^2}=\frac{\partial^2}{\partial x^2}(\varphi_{\mathrm{i}}-\varphi_{\mathrm{e}})=\frac{\partial}{\partial x}(r_{\mathrm{e}}i_{\mathrm{e}}-r_{\mathrm{i}}i_{\mathrm{i}})=\frac{r_{\mathrm{e}}+r_{\mathrm{i}}}{r_{\mathrm{m}}}V_{\mathrm{m}}+(r_{\mathrm{e}}+r_{\mathrm{i}})c_{\mathrm{m}}\frac{\partial V_{\mathrm{m}}}{\partial t} \tag{9.7}$$

Let $\lambda^2=r_{\mathrm{m}}/(r_{\mathrm{e}}+r_{\mathrm{i}})$, and the above equation can be simplified as

$$-\lambda^2\frac{\partial^2 V_{\mathrm{m}}}{\partial x^2}+\tau_{\mathrm{m}}\frac{\partial V_{\mathrm{m}}}{\partial t}+V_{\mathrm{m}}=0 \tag{9.8}$$

where $\tau=r_{\mathrm{m}}c_{\mathrm{m}}$ and λ are the time constant and spatial constant of the cell membrane, respectively.

If only a particular time is considered for analyzing the dependence of the membrane potential on location, then the second term, i.e. the partial derivative in regard with time t, is ignored in Equation (9.8), and an expression for the membrane potential as a function of location x is obtained:

$$-\lambda^2\frac{\partial^2 V_{\mathrm{m}}}{\partial x^2}+V_{\mathrm{m}}=0 \tag{9.9}$$

Applying the initial conditions $V_{\mathrm{m}}(0)=V_0$, the solution is $V_{\mathrm{m}}(x)=V_0\cdot \mathrm{e}^{-|x|/\lambda}$. The spatial constant is the distance at which the voltage decreases to $1/\mathrm{e}$ of its initial value.

Using the same method to ignore the location dependent term in (9.8), a solution can be obtained for the membrane potential change with time

$$\tau_{\mathrm{m}}\frac{\partial V_{\mathrm{m}}}{\partial t}+V_{\mathrm{m}}=0 \tag{9.10}$$

With initial condition $V_{\mathrm{m}}(0)=V_0$, the solution is $V_{\mathrm{m}}(t)=V_0\cdot \mathrm{e}^{-t/\tau_{\mathrm{m}}}$

The expressions for membrane potential under two simplified conditions has been obtained: time independent and distance independent. The result of the former is demonstrated in Figure 9.8. Solving for the membrane potential as a co-function of distance x and time t is more complex and therefore not further explored here.

The passive diffusion underlies the conduction of nerve signals on axons. When a location on the axon has been depolarized beyond the threshold, an action potential is generated locally. The action potential spreads out along the axon according to the passive properties of the neuron, with the amplitude attenuating in both directions. Due to the strong depolarization of the action potential, the neighboring area could still reach its threshold from the passive diffusion, therefore generating its own action potential. Each segment of the axon repeats this process in series to propagate the action potential to it downstream neighbor, allowing the action potential to travel down the entire axon (Figure 9.10). The whole process, in practice, there are many myelinated axons, which are wrapped around by glial cell processes, forming isolated axon segments. There is no extracellular space between the axon and the myelin and therefore no transmembrane current flow. The myelinated segments are separated by bare segments of the axons, known as nodes of Ranvier, where action potential could be generated. The depolarization of the action potential passively propagates between the myelinated segments, and exceed threshold at the nodes of Ranvier, thus allowing the action potentials be regenerated in a salsatory manner. The speed of action potential propagation is to a certain extent decided by the myelination of the axons.

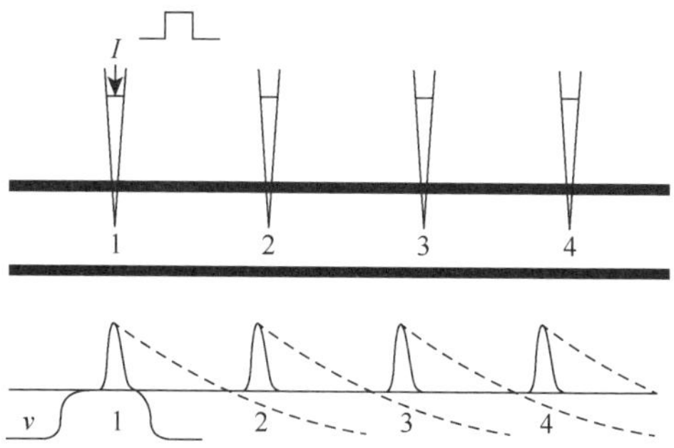

Figure 9.10 Active propagation

9.2.4 Action potential conduction in the extracellular space

When recording neural signal extracellularly, the electrode is located in the medium outside the epineurium. The relationship between the action potential and the signal recorded on the electrode should be obtained by solving the spatial equation of the electrical fields. By analyzing the biopotentials via volume conductor models, the analysis involves the geometric properties, the electrical conduction properties, and the current source characteristics of inside the neural tissue.

Volume conduction of resistive medium obeys the following relationship,

$$\nabla \cdot \sigma \nabla \varphi = \nabla \cdot J_{s} \tag{9.11}$$

where σ is the medium's conductivity; φ is the potential; J_s is the current density of internal sources; ∇ and $\nabla .$ are the gradient and divergence operator. For homogenous conductors, the equation can be simplified as the Poisson equation

$$\nabla^2 \varphi = \nabla \cdot J_{s} / \sigma \tag{9.12}$$

As the electrical field E is gradient of the potential $\nabla \varphi = -E$, and also by Ohm's law $1 = \sigma E$, the current density could be related to the internal sources by

$$\nabla \cdot J = -\nabla \cdot J_{s} \tag{9.13}$$

By Gauss's law, the integral of divergence of current density within a volume equals the net current via the high surface of this volume

$$\iiint_V \nabla \cdot J = \oiint_{\partial V} J \cdot n\mathrm{d}S \tag{9.14}$$

Combining 9.13 and 9.14 yields

$$\iiint_V \nabla \cdot J_{s}\mathrm{d}V = -\oiint_{\partial V} J \cdot n\mathrm{d}S \tag{9.15}$$

In which n is the unit normal vector on the surface of the volume. From a physical viewpoint Equation (9.15) states that the net current generated by sources inside the volume is related to the total current flux through the surface.

In neural signal detection, current sources like the electrode sites and nodes of Ranvier can be treated as point sources. If an infinitesimal region surrounding the point source is analyzed, Equation (9.15) can be rewritten as

$$\iiint_V \nabla \cdot J_{s}\mathrm{d}V = -I_{s} \tag{9.16}$$

And the potential can be obtained for a current source within an infinite homogenous medium. Combining Equation (9.15) and Equation (9.16),

$$\oiint_{\partial V} J \cdot n\mathrm{d}S = I_{s} \tag{9.17}$$

Taking a sphere centered at the point source with radius r, and substituting

$$J \cdot n = -\sigma \nabla \varphi \cdot n \tag{9.18}$$

and

$$\nabla \varphi \cdot n = \mathrm{d}\varphi / \mathrm{d}r \tag{9.19}$$

into Equation (9.20), yields the following on the boundary of the sphere

$$4\pi \cdot r^2 \sigma \mathrm{d}\varphi / \mathrm{d}r = I_{s} \tag{9.20}$$

Further, if an integral is taken from infinity to a radius of r, then

$$\varphi(r)=\frac{I_s}{4\pi\sigma r} \tag{9.21}$$

The above solution requires the conducting volume outside the nerve fiber to be infinite large and homogenous, while analytical solution for a realistic inhomogeneous and finite volume can be too complex and numerical methods have to be applied to solve specific situations.

Discussed above is a solution for a single point current source. Specifically to neural signal detection, the analysis is targeted at transmembrane ion current densities at nodes of Ranvier, and in the actual detection a number of nodes of Ranvier on the fibers have to be taken into account (Figure 9.11). While the same equation can be obtained for each node, the potential generated by all the nodes contribute to the recorded signal. An example is shown for a signal myelinated nerve fiber, with nodes of Ranvier distributed along the fiber. Each node can be treated as a current source.

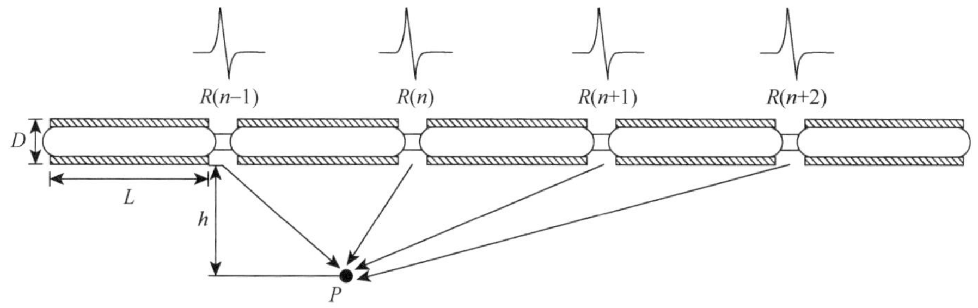

Figure 9.11 An illustration showing the contribution of many nodes of Ranvier from a myelinated nerve fiber to the potential at a point P

Assuming the transmembrane ionic current at node $R(n)$ to be $I_m(t)$, according to the all or none principle of action potentials, the current at any other node could be given as

$$I_{m,k}(t)=I_m(t-kT) \tag{9.22}$$

In which k is the number of nodes between the node and node $R(n)$, and T is the time for action potential to travel from one node to the next.

According to Equation (9.21), the potential at point P due to the current sources at all the nodes can be given as

$$\varphi(t)=\frac{1}{4\pi\sigma}\sum_{n}\frac{I_{\mathrm{m}}\left(t-\frac{x_n}{v}\right)}{\sqrt[n]{h^2+\left(x-x_n\right)^2}} \tag{9.23}$$

The compound action potential recorded at point P with an electrode is thus obtained. The exact transmembrane potential could be obtained via patch-clamp techniques, the distance from point P to the nodes is $\sqrt{h^2+\left(x-x_n\right)^2}$, and v is the transduction velocity of the nerve fiber. Equation (9.24) shows that the current due to the action potential at the nodes are summed with different weights and delays at point P. The weights at point R (n) is

$$w_n=\frac{1}{4\pi\sigma\sqrt{h^2+\left(x-x_n\right)^2}} \tag{9.24}$$

The parameters in Equation (9.23) could all be obtained, and therefore an accurate mathematical model describing the coupling of a single nerve fiber and an electrode could be obtained in homogenous conduction medium.

9.2.5 Propagation of action potential outside of the epineurium

The detection of neural signal is a more complicated issue when using cuff electrode to record from the surface of a nerve bundle.

As shown if Figure 9.12 (see the Color Inset, p. 25), the nervous system in the PNS takes mostly in the form of nerve bundles, which consists of several fascicles. Each fascicle in turn contains many axons and supporting tissue. Understanding the microscopic structure of the nerve bundles and their fascicles is the prerequisite to understand the principles and models of neural signal detection.

From a structural point of view, nerve bundles and fascicles are similar to a bunch of cables. Besides the vulnerable axons, there are durable and dense connective tissue that provides support, protection, and nutrition. The nerve bundle's outermost layer is the epineurium, the first layer of protection. Within, there are several nerve fascicles and blood vessels. The fascicles themselves are sub-systems with rather complicated structure, containing the bundled neuron axons and Schwann cells; the latter play an important role in the protection and regeneration of axons. The axon fibers branch out from the nerve bundle and connect to the motor and sensory organs at their terminal. Each nerve fiber is ensheathed by the endoneurium.

Because a nerve bundle could have multiple nerve fibers active simultaneously, it is very difficult to solve for the exact extracellular potential on a microelectrode site. Thus, the obtained neural signal is a compound signal, i.e. a combination of all the electrical signals from nerve fibers in the nerve bundle at the electrode site. This proves challenging for locating the detected signal. One viable solution is to place a multiple electrode array around the nerve bundle with sites at many azimuth angles to pinpoint the source of the signal via joint analysis on time and space of the multichannel recording system.

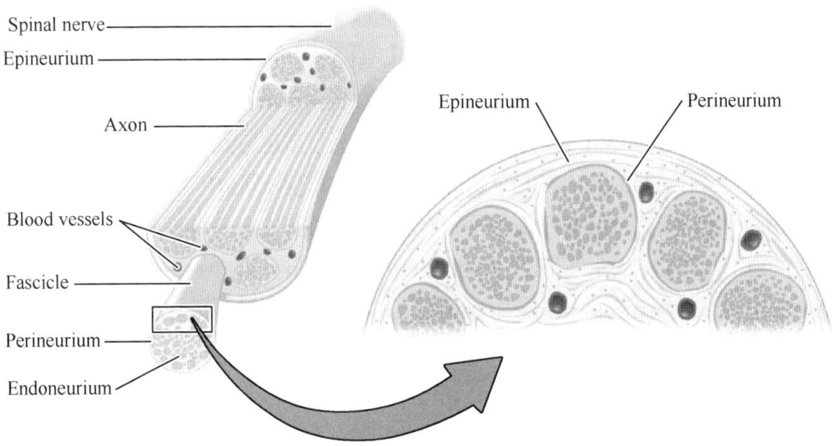

Figure 9.12 Internal microanatomy of nerve bundles

Sources: https: //en.wikipedia.org/wiki/Nerve#/media/File: 1319_Nerve_StructureN.jpg

9.3 Detection and Recording of Neural Signals

The recording and processing of neural signals from central and peripheral nervous systems has been the key aspect of neuroscience research. Because neural signals cannot be sensed by humans in a direct way, devices are designed to detect them. Usually, the process of converting a physical variable into a different one without affecting the signal itself is called sensing. A sensor is a device that accomplishes this conversion. Therefore, the detection of neural signals may also be considered as sensing of neural signals. As neural signals are usually long duration randomized pulse sequences, it is often necessary to record and analyze over a long time period to obtain useful information. That is why neural signals recording is a keyword in literature and a fundamental research topic of neuroscience. In the following discussion of the microelectronic neural bridge, however, the neural signals detected are processed in real time processing. Thus, signal recording become not necessary. Here we discuss only the neural signal detection problem.

9.3.1 Principle of neural signal detection

The bioelectrical phenomena enable the connection of the nervous system to an electronic information system. From the discussion in the previous sections, a biological potential exists across the membranes of neurons, nerve fibers, and nerve bundles. The so-called resting potential usually stays constant, and can be measured at any time across inside and the outside of the cell. When the potential on the nerve fiber exceeds a certain threshold at some point on the fiber due to stimulation, the ion permeability of the membrane will be changed, causing Na^+ inflow and K^+ outflow. This in turns causes the hyperpolarized membrane to depolarize, which then depolarizes adjacent cell membrane sequentially. The chain reaction spreads the potential change in a form of conduction of a nerve impulse. After the excitation, the cell membrane returns the potential back to its original resting state. Such a course from the resting to depolarization and again back to the resting constitutes a nerve's action potential (AP). The neural functions are transmitted throughout the body in the form of AP sequences that propagate along the nerve fibers. Due to the involvement of changing electrical parameters in this transmission process, neural signals are relatively easy to obtain. A pair of electrodes can be used to detect the potential changes from a nerve fiber, and thus detect the corresponding neural signal.

As the interface between biological information systems and electronic information systems, neural microelectrodes exchange information with nerve fibers via electrical coupling, with bioelectricity as the material basis to achieve this coupling. Microelectrodes implanted close to the nerve fibers can detect the weak electrical signal related to the potential changes on the neuronal membranes, and the subsequent circuitry implement signal processing including amplifying, filtering, analogue-todigital converting, etc.

9.3.2 Nerve signal detection electrode configuration structure

In the actual circuit design and animal experiments, there are a variety of electrode configurations for neural signal detection, including monopolar, bipolar, tripolar, and multi-polar settings. The number of poles is defined as the number of electrode sites for each channel of the system involved in terms of signal detection.

The structure of a monopolar configuration is relatively simple, and the physical meaning of the signal is relatively clear. However, for the neural signal detection circuit, the input signal is a potential difference between two electrodes. Even in a monopolar set-up, two electrodes are needed. One is used to obtain the signal, while the other serves as a reference electrode.

Monopolar systems are often very susceptible to noise and interference, so it is necessary to configure a multi-polar structure to solve this problem. The most important configurations are bipolar and tripolar electrodes set-up. A multipolar (including bipolar and tripolar) configuration means that the system exploits a plurality of electrode sites for each channel of the neural signal, and either for signal detection or stimulation. For detection, each channel's output is a combination of the signals from the electrodes. In general, the electrode sites are arranged in parallel along the nerve fibers, each of the contacts at the same distance from the nerve fibers, as shown in Figure 9.13.

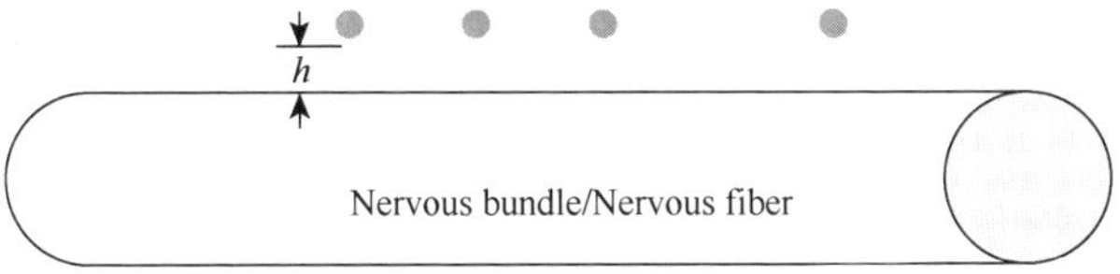

Figure 9.13 Polar electrode distribution over 9.12 nerve signal detection system

Because different electrode configurations have different characteristics for the detected signal, such as waveforms and signal to noise ratio, etc., single electrode, two electrode, and three electrode systems, i.e. mono-, bi-, and tripolar configurations, are discussed as follows.

1. Monopolar system

The neural signal detection system of monopolar configuration can be divided into two types: intracellular and extracellular. The intracellular probe is invasive, usually with a metal tip of of around 1-μm diameter piercing into the nerve to make a direct measurement of the potential of neurons. Figure 9.14 shows the diagram of a signal detection system with a single electrode.

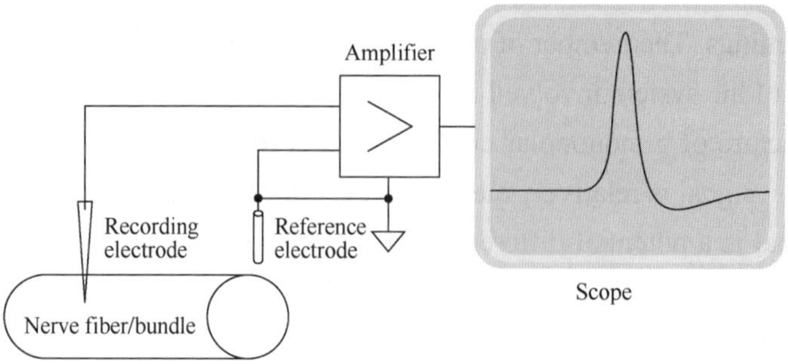

Figure 9.14　Intracellular neural signal detection system with a single electrode

For a signal detection system with a single electrode, the electrode potential is higher than the reference electrode when depolarization of the action potential waveform reaches the electrodes, resulting in a rising curve displayed on the oscilloscope. When the repolarization phase of waveform reaches the electrode, the electrode potential falls, and will continue to fall beyond the potential of the reference electrode, forming the overshooting hyperpolarization phase before slowly returning to the resting state. Eventually, a single-peaked waveform is shown on the oscilloscope. This typical waveform for an action potential can be used as the pattern recognition criteria of neural signal recorded from a single-electrode system.

Figure 9.15 is a typical monopolar system configuration of the signal detection from nerve bundles or nerve fibers. Here, for example, one electrode site of a cuff electrode is placed on (or close to) the outer membrane of the nerve bundle or fiber, and the reference electrode is located somewhere *in vivo* to provide a reference potential. The potential difference between two electrodes constitutes the detected voltage signal.

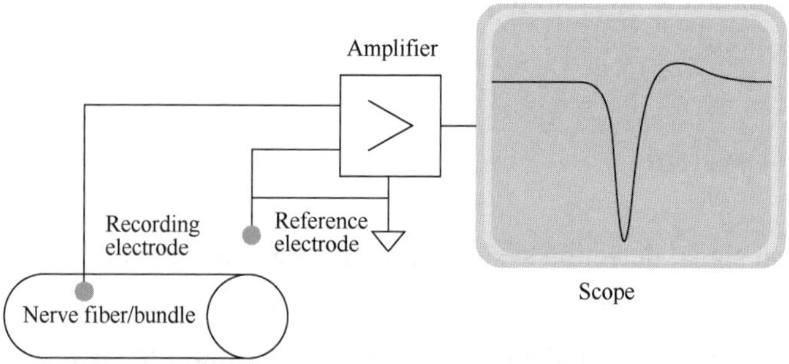

Figure 9.15　Extracellular signal detection system of a single electrode configuration consisting of a cuff electrode located on the outside of a nerve bundle

Since the electrode site is located on the outer membrane of a nerve bundle, when the depolarization phase of the action potential waveform reaches the electrode, the electrode's potential will be lower than that of the reference electrode, resulting in a downward curve on the oscilloscope. When the repolarization phase of the reaches the electrode, the electrode potential begins to rise, reaching at first the potential of the reference electrode and further rising to a higher potential during the hyperpolarized phase before slowly returning to the resting potential. Eventually, this gives a monopolar single-peaked potential, but with the opposite polarity to the signal waveform obtained by a intracellular electrode. This waveform can be used as a standard pattern for neural signal detection of a single electrode system outside of a nerve fiber bundle.

In a neural signal detection system with a single electrode configuration, clearly, the interference and the noise on both the detection electrode and the reference electrode will be passed to the next stage in the system. The reference electrode is also possibly affected by other biological signals. Once these interfering signals and noise appear, the detection amplifier circuit is unable to distinguish them from the signal to be detected. Therefore it would be best to place the reference electrode in a location with little signal or noise. However, as this is unrealistic to achieve, a monopolar configuration is very susceptible to noise and interference.

Another problem with the monopole configuration is that the signal-detecting electrode and the reference electrode are located in quite different environments, resulting in difference in their electrode-tissue interface impedances and possibly high DC offset voltages between two electrodes. For a high-gain amplifier, such a DC offset is already sufficient to saturate the amplifier, necessitating the removal of such offset voltage before the signal amplification.

2. Bipolar system

A bipolar system using two electrodes is one of the commonly-used configurations for neural signal detection systems. In this configuration, two electrode sites are used to detect the neural signal. These two electrodes may be both placed on the nerve fiber, or one on the nerve fiber and the other somewhere nearby. Nevertheless, these two electrodes must be located very closely, so that there environments are the same or almost identical. The electrode size, material, and other aspects must be also matched, as shown in Figure 9.16. When the action potential propagates down the nerve fibers, the waveform passes by the two electrodes successively. Although the signals directly detected by two electrodes are similar, the voltage signal output from the amplifier is a biphasic pulse as shown in Figure 9.16. This is because the two electrodes are connected to the positive and negative inputs of the differential amplifier and the waveforms arrive with a time difference, i.e. with a phase difference. When the action potential propagates in the opposite direction along the nerve fibers, the result is a biphasic waveform with the opposite polarity for each phase. Since the interval of the positive and negative peaks of the two phases depends on the velocity of the action potential and the distance between the two electrodes, there is no universal waveform that can be used as a standard or template, and the biphasic signal pattern for a bipolar electrode system has to be obtained from the experiment first, for example from artificial stimulation, before the later use in detection and recognition.

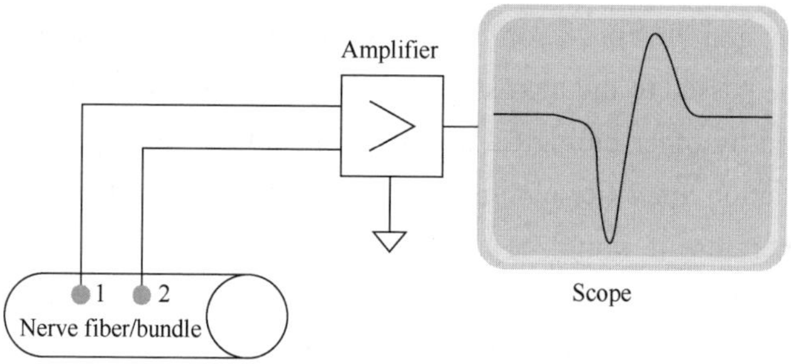

Figure 9.16 The two-electrode configuration of neural signal detection system

An important feature of the two-electrode system is that in most cases, the two electrodes are subjected to the same or highly correlated noise and interference. Utilizing the common-mode rejection of the differential amplifier, the noise and interference can be minimized. Therefore, the noise and interference suppression performance of the two-electrode configuration is much better than that of a single electrode configuration. However, not all of the two-electrode configurations have good interference suppression performance. The previous assumptions of similar interference signals are based on the equipotential lines being parallel to the nerve fibers and the electrode sites; if the equipotential lines are not parallel, then the interference signal can still cause a potential difference on the two electrodes, which is mixed together with the nerve signals and amplified by the detection amplifier.

3. Tripolar system

In a tripolar configuration, the system has three electrode sites involved in the detection of neural signals. The three-electrode configuration can further improve the signal-to-noise ratio of the system compared to a bipolar system. According to connection scheme, a three-electrode configuration can come in a "quasi-three-electrode" form or a form of "true three electrode", as shown in Figure 9.17 (a) and Figure 9.17 (b).

As seen from Figure 9.17, the output pulse patterns of both the quasi-three-electrode configuration and the true three-electrode configuration are characterized by bipolar waveform with three peaks. The spacing between the positive and negative peaks depends on the distance between the three electrodes and the propagation velocity of the action potentials. With stimulation elsewhere on the nerve fibers, the bipolar waveform with three peaks can obtained by the three-electrode system and then used as recognition pattern for neural signal detection later.

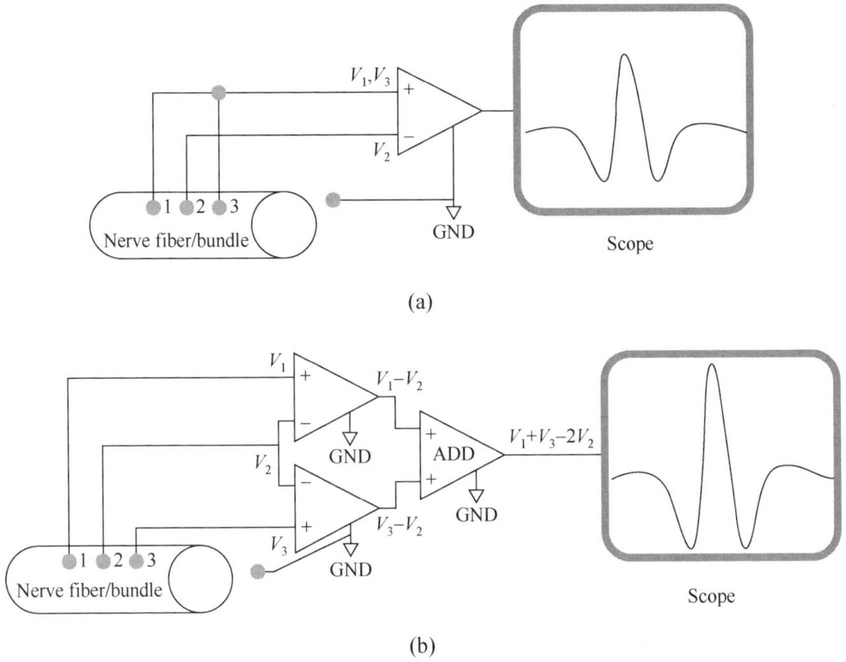

Figure 9.17 Two three-electrode systems for neural signal detection: (a) quasi-three-electrode configuration; (b) the true three-electrode configuration

The difference between the quasi-three-electrode and the true three-electrode configurations is reflected in two aspects:

(1) In the quasi-three-electrode configuration e, electrode 1 and electrode 3 are directly connected together, therefore there is a problem of signal shunting, while there

is no such a problem in the true three-electrode system and the output signal amplitude as well as the signal-to-noise ratio is higher in the latter;

(2) Since electrode 1 and electrode 3 in the quasi-three-pole configuration are connected together, the axial current along the nerve can be avoided, which helps to inhibit the common mode interference in implantation environment.

For interferences whose equipotential lines are parallel to the nerve fibers and electrode sites, the three electrode configuration has a suppression capability similar to that of the two-electrode configuration. But for interferences whose equipotential lines are not parallel to the nerve fibers and electrode sites, the three electrode configuration also has a good suppression capability.

The working principle of the three electrode configuration, with the quasi-three-electrode system for example, is explained as follows: Since the two outer electrode sites are short-circuited, the potential obtained from them is the mean value of each site. Because the interference signal sources in the body are generally far away from the electrodes (compared to the nerve fiber), the potential distribution of the source is monotonically decreasing, and the mean potential of both outer electrodes are basically equal to that of the middle electrode. Therefore, no matter how the equipotential lines of the interference source are aligned with the nerve fiber, a three-electrode set-up can always suppress the interference signals to a very small value by means of the common mode characteristics of the differential amplifier, as shown in Figure 9.18. The potential distribution due to the neural signal from in between the electrode sites, however, is not monotonically decreasing, and the short circuit of the outer two electrodes of the quasi-three-pole configuration does not affect signal detection. The advantageous interference suppression of the three-electrode configuration makes it the most effective electrode connection set-up for neural signal detection from nerve fibers.

The cuff electrode is the most common among the three-electrode configurations. After implanted on a nerve bundle, the cuff electrode naturally wraps around the nerve bundle. Thus the quality of the detected signal can be effectively improved as the electrode sites are on the inside and by isolated from the surrounding by the cuff. The nerve bundle is also separated from the extracellular fluid environment and the cuff forms a partially closed extracellular space inside its wrapping. The impedance of the extracellular medium increases and the detected signal amplitude becomes higher.

The volume conductor model, namely the potential equation derived in Section 9.2 is equally applicable to the cuff-type three-electrode configuration. Stein and Pearson, and Marks and Loeb, independently established the one-dimensional model for cuff electrode on nerve fibers with and without myelin, and proved that under applicable conditions very similar results can be obtained compared to the one-dimensional model[26]. Assuming the current signal source is located within the cuff, the potential equation can be obtained by one-dimensional model as

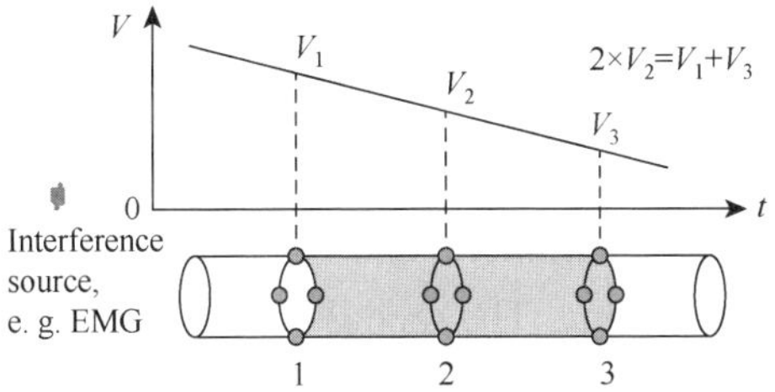

(a)

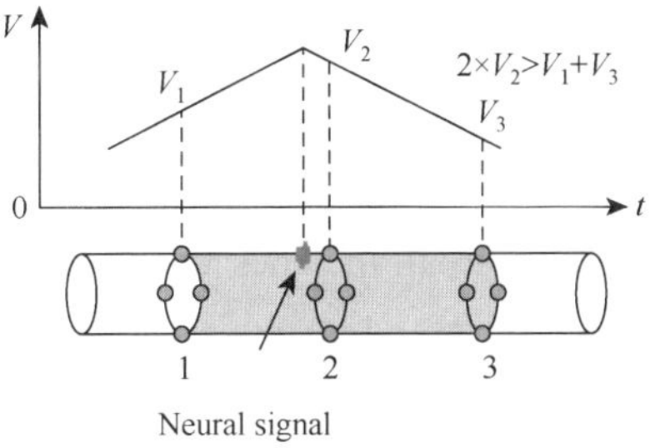

(b)

Figure 9.18 The voltage distribution of interference signal source and electrode nerve source within cuff electrode contacts

$$\varphi_e(t,x)=\frac{R_e}{R_e+R_a}\left[\left(1-\frac{x}{L_{cuff}}\right)V_m(t)-V_m\left(t-\frac{x}{v}\right)+\frac{x}{L_{cuff}}V_m\left(t-\frac{L_{cuff}}{v}\right)\right] \quad (9.25)$$

where υ is the propagation velocity of the neural signal, x is the location within in the cuff ($0 \leqslant x \leqslant L_{cuff}$), $V_m(t)$ is a transmembrane potential at the node of Ranvier, R_a is the internal resistance of the axon between two adjacent nodes of Ranvier, and R_e is the extracellular resistance with a value of

$$R_e=\left(\frac{2\pi}{L_{cuff}}\int_{r_{fiber}}^{r_{cuff}}\sigma\cdot r\mathrm{d}r\right)^{-1} \quad (9.26)$$

9.3.3 Neural signal detecting circuit

A neural signal source itself is a weak signal source with a high inner resistance. The voltage amplitude obtained by a detection electrode is on the level of microvolts.

Meanwhile, the high inner resistance has considerable noise and disturbance due to various *in vivo* physiological activities. This problem can be solved with the design of a high-sensitivity low-noise amplifier.

1. Neural signal detection circuit based on discrete components

At first, a four channel neural signal detection module is introduced, which is made of integrated operational amplifiers and other discrete components and interfaces with a 12-site cuff electrode array. The module is designed for *in vivo*, real-time, and parallel detection of compound potentials the spine cord or in peripheral nerve bundles.

Based on the analysis of a cuff electrode array, the 12-site electrode array is divided into four channels, each with three sites along the nerve bundle axis used to obtain one channel of neural signal. The electrodes are wired to detection module *ex vivo*, with metal shielding to suppress interferences.

Each channel of detection circuit includes an RC-coupling network, a pre-amplifier, an active filter, a 50 Hz-notch filter, and other circuit elements. According to the characteristics of neural signals, shield clamp circuit and a driven-right-leg circuit were designed to further suppress common mode interferences.

Figure 9.19 shows one schematic diagram of the single-channel detection circuit and connection diagram of the neural signal detection experiment in the rat spinal cord signal detection experiments. The single-channel circuit includes an RC-coupling network, a pre-amplification stage, an active band-pass filter, a 50-Hz notch filter, a shielding clamp circuit, and a driven-right-leg circuit unit.

The RC coupling network lies in the forefront of the circuit and consists of two C_as and four Rs. The main aim of using this RC-network is to achieve a nearly infinite CMRR, while a simple DC-coupling or AC-coupling network is not used here because:

(1) If a DC-coupling network was used, the high-gain preamplifier could easily be saturated, due to the presence of the polarization voltage. To prevent circuit saturation, the gain of a DC-coupled preamplifier can only be set within a few tens. The common-mode rejection ratio and anti-noise performance of the circuit is closely related to the gain of the first stage: the smaller the gain, the smaller the common mode rejection ratio, and the worse the noise performance. Therefore DC-coupling will reduce the overall performance of the circuit.

(2) Since there is an imbalance between the electrodes, an AC coupling easily transforms the common-mode interference into differential mode interference.

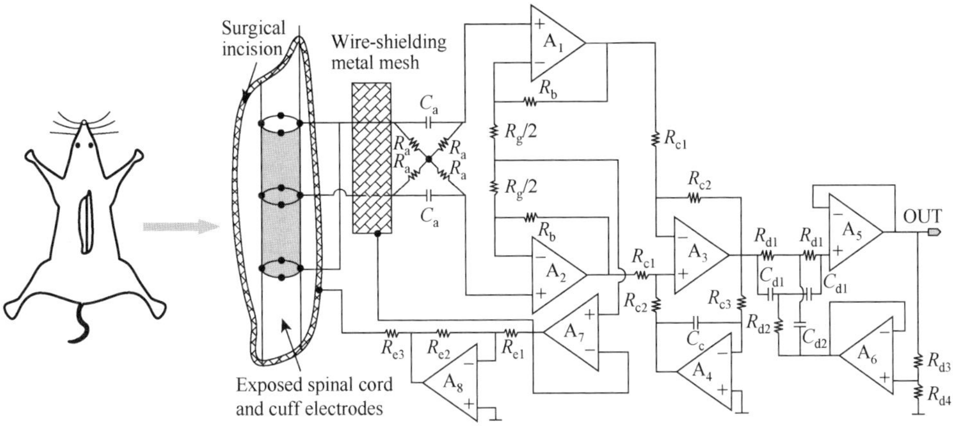

Figure 9.19 Schematic diagram of a single-channel detection circuit and connection diagram of the neural signal detection experiment in the rat spinal cord

Classic analog circuit configurations are used for all of the preamplifier, the bandpass filter, and the notch filter. The first-stage amplifier consists of OPAs A_1 and A_2 connected in parallel. The gain is determined by the feedback resistor network, i.e., R_b/R_g. In this module, the feedback resistor is controlled by a switch, and the gain can be adjusted in the range from 500 to 10000. The active filter consists of A_3 and A_4. The gain in the middle frequency range is determined by the negative feedback of A_3 (i.e. R_{c2}/R_{c1}), the lower corner frequency is determined by the negative feedback of A_4 (i.e. $R_{c3}C_c$), and the higher corner frequency is determined by the unity gain bandwidth of the OPA. The notch circuit, besides the traditional double-T structure, includes OPA A_6 with a positive feedback network to reduce the width of the off band and increase the depth of the notch. Corresponding to a 50-Hz AC power supply, R_{d1} is 47 kΩ and C_{d1} is 68 nF.

The shield clamp circuit and the driven-right-leg circuit are unique to biomedical electronic circuits. The shielding clamp circuit consists of the unity gain buffer A_7. Through A_7, the common mode signal outputted by the differential amplifier stage of the front-end amplifier is connected with the metal shield wire mesh, and clamps the shielding sheath and the inner core to the same potential, so that the radial displacement current caused by the wire's capacitance and the leakage can be reduced. In fact, it can be considered that the shielding cover is AC-grounded through the low-impedance output of A_7.

The driven-right-leg circuit consists of A_8 and its negative feedback loop. In bioelectric signal detection, the *in vivo* and *in vitro* common-mode interferences are very strong compared to the signal, and it is often necessary to ground the body via a metal clip attached to skin in order to reduce the interferences. Our design uses the alternative method, i.e. the more effective driven-right-leg (DRL) circuit, for which this negative feedback OPA circuit of biological grounding is commonly known as. The common mode signal is inverted and applied again to the body (to the right leg in ECG). The common mode interference can be strongly reduced by a deep negative feedback. In Figure 9.19, the common mode interference signals are reduced to $i_d \cdot R_{e3}/(1+R_{e2}/R_{e1})$. Another benefit is that the loop resistance to the ground is large, which ensures that the ground current does not exceed the range of biosafety.

The circuit module was fabricated on PCB, in which the Maxim's MAX4168 were used for the OPAs. The module's performance parameters listed in Table 9.1 are obtained by HSpice-simulation using a macro model of MAX4168.

Table 9.1 Performance parameters of a single-channel neural signal detection circuit

Voltage/V	$\pm$1.5	CMRR/dB @100 kHz	164
Current/mA	9.6	Power Supply Rejection Ratio/dB	120
Power/mW	28.8	Equivalent input noise/(nV/$\sqrt{\text{Hz}}$) @1 kHz	9.2
Gain range	500-10000	Voltage drift/mV	4.7
3 dB bandwidth/kHz	127-10.3	Notch depth/dB	45

The implemented PCB and the module photo are shown in Figure 9.20 (see the Color Inset, p. 25).

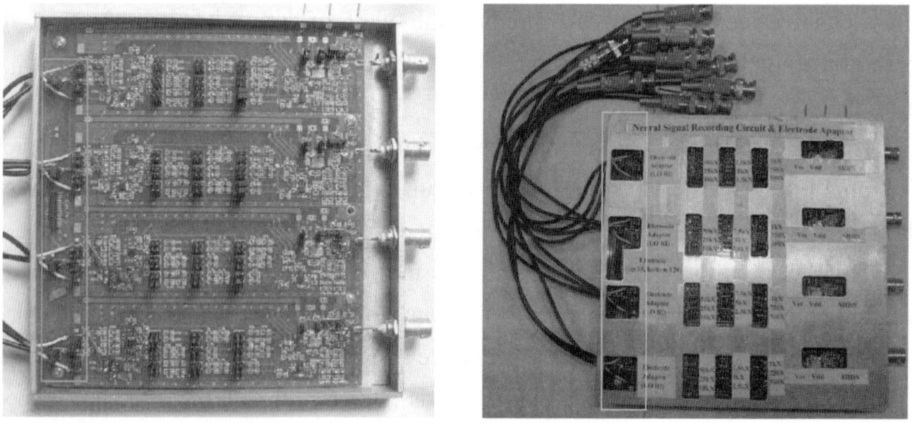

(a) (b)

Figure 9.20 (a) PCB photo of the four-channel neural signal detection module circuit board; and (b) the module photo. The orange frame shows the electrode switching units

The PCB includes the four-channel neural signal detection circuit and electrode lead outlets, and is connected with a 12-site cuff electrode array, enabling four-channel of neural signals to be detected simultaneously. The module also contains an electrode switching unit (marked by the frame in Figure 9.20, see the Color Inset, p. 25). The neural electrodes can therefore easily switch their roles as either detection electrodes or the stimulation electrodes. Therefore a signal detection channel can be flexibly combined with a functional electrical stimulation channel and further configured into a neural bridging system as described in the introduction section.

2. Nerve signal detection IC

An implantable neural signal detection device requires as high a degree of integration as possible and minimizes the use of off-chip components. Especially when the number of channels increases up to tens or even hundreds channels, this problem will become more prominent. Therefore, an ultimate goal for an implanted system is to achieve a fully integrated circuit. In this regard, we have implemented a variety of integrated circuits (ICs) for neural signal detection utilizing CMOS OPA. Figure 9.21 shows the circuit diagram of one of these ICs.

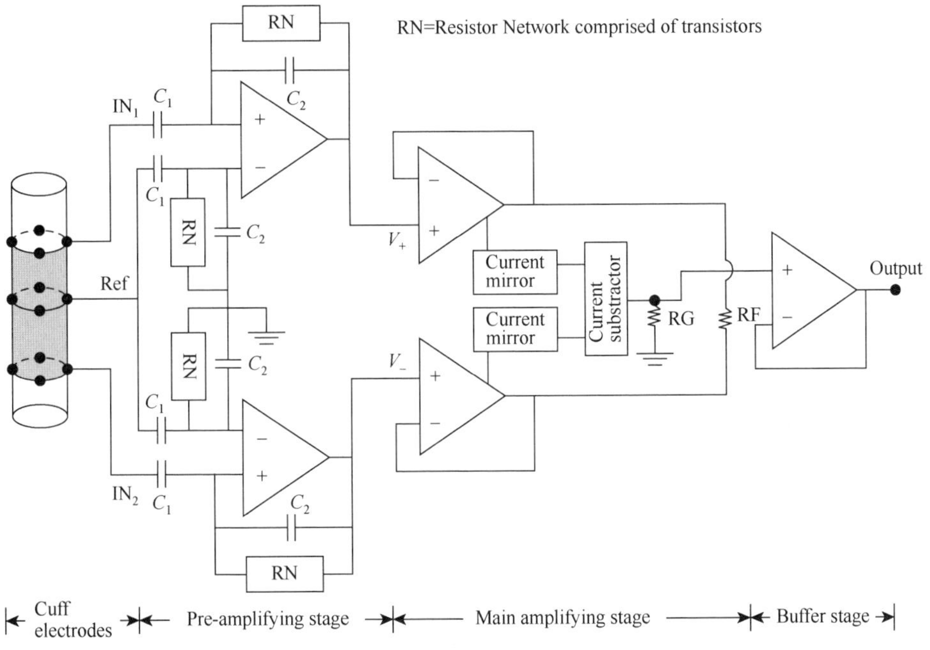

Figure 9.21 Circuit diagram of a monolithic integrated neural signal detection amplifier

The detecting amplifier includes a preamplifier and a current-mode instrumentation amplifier. The preamplifier stage uses capacitive feedback and MOSFET active resistors. Figure 9.22 (a) shows the MOS active resistor. If V_{GS} is negative, the transistor operates as a diode. When V_{GS} is positive, the parasitic "source-N well-drain" transistor comes into operation, and the transistor can be treated as a diode-connected BJT. The I-V characteristic curve of a MOS active resistor is shown in Figure 9.22 (b). As shown in the figure, in the voltage range of ΔV=–0.5–1 V, the current is almost unchanged, indicating that the equivalent resistance of a MOS transistor is very high. In fact, the MOS operates in its cutoff region, and its resistance characteristic is due to the channel's leakage. Another point illustrated by the curve is that the voltage across the active resistor should not exceed ΔV, otherwise the MOSFET is turned on and its equivalent resistance value will quickly decrease. Therefore the design should put MOSFETs in series properly to prevent the voltage between the source and drain of the MOSFETs to exceed the range when the circuits works with large signal work under dynamic conditions.

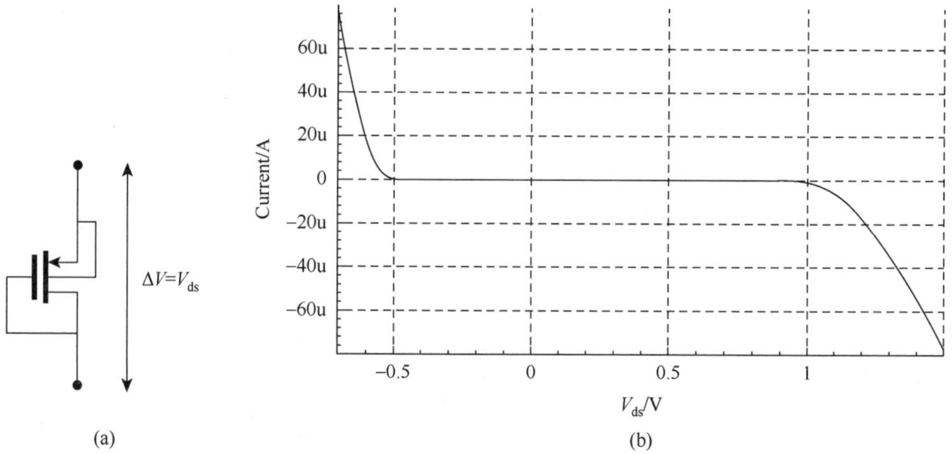

Figure 9.22 Implementation of (a) active resistance and its (b) *I-V* curve

The current-mode instrumentation amplifier consists of two OPAs connected with a unity gain, two current mirrors, one current subtractor, and two resistors (Figure 9.23). The main part of the amplifier stage is shown in Figure 9.23. Its operating principle is explained as follows: R_F is connected to two output stages of the OPAs, and the current flowing through R_F is $I_{RF}= (V_+-V_-) /R_F$; the currents of the OPAs' output stage are replicated by current mirrors, and the difference is outputted after current subtraction, thus, the output voltage obtained across R_G is $V_{out}=I_G \cdot R_G=2k (V_+-V_-)$ with $k=R_G/R_F$.

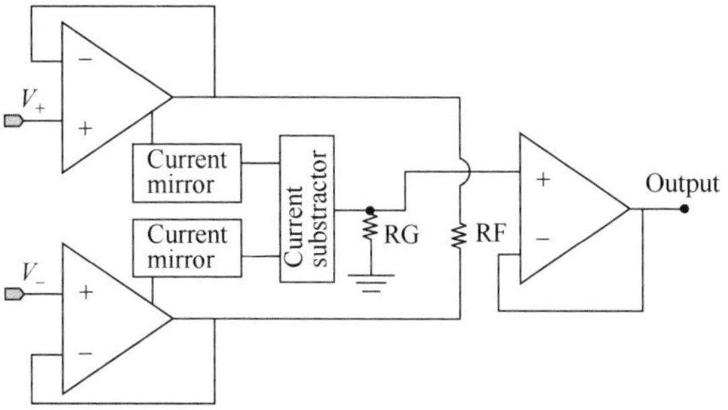

Figure 9.23 Schematic of the main current mode amplifier

The detailed circuit of the main amplifier is shown in Figure 9.24.

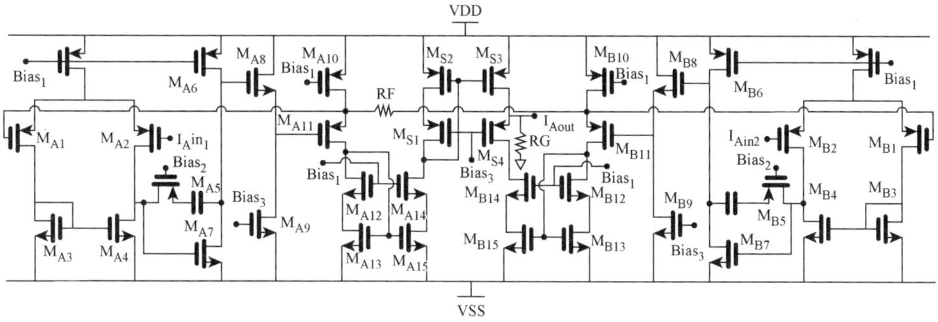

Figure 9.24 Circuit diagram of the current-mode main amplifier stage

The circuit was designed in standard 0.6-μm DPDM mixed-signal CMOS process (CSMC, Wuxi). After the simulation results met the design specifications, the layout was designed and validated using the EDA tools by Aether System (Empyrean, Beijing, China) and component library custom-developed by the Institute of RF-&OE-ICs, Southeast University. The IC was implemented through the MPW-program of Shanghai IC Center. The photos of the die and the chip-test module used for experiments are shown in Figure 9.25 (see the Color Inset, p. 25).

The chip function was validated by testing the response waveforms in time domain. The testing equipment includes: Cascade Microtech's chip probe station, Agilent's 33220A Arbitrary Waveform Generator and Tektronix's TDS5104 Oscilloscope. The test results show that the chip can work properly under positive-and-negative dual power supply and under single power supply, with good quality output waveforms obtained. The high-3 dB corner frequency is 10 kHz. Figure 9.26 shows the output signal waveform with an input of 2 kHz frequency and a square wave signal of 20 mVpp amplitude.

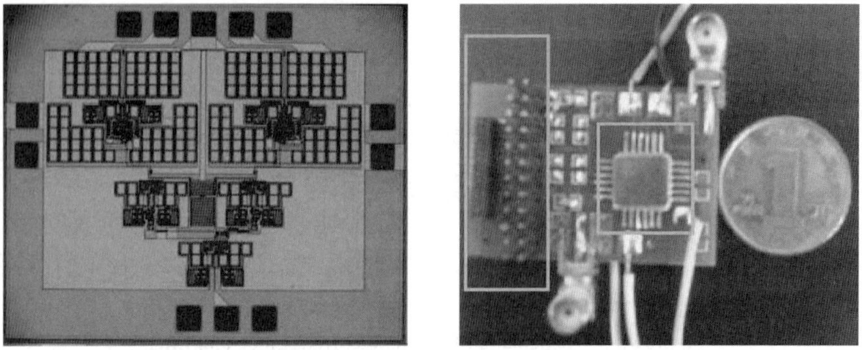

(a)　　(b)

Figure 9.25　Current mode main amplifier: (a) die photograph and (b) chip-test modules for experiments

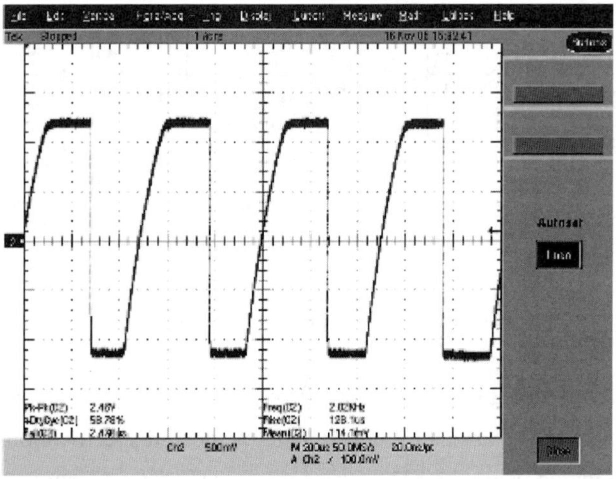

Figure 9.26　Output waveform with an input frequency of 2 kHz and a square-wave signal of 20 mVpp amplitude

9.3.4　Neural signal detecting experiments

A series of spontaneous neural signals have been obtained from the sciatic nerves and spinal cords of rats and rabbits, by using the 12-site cuff electrodes array (4 sites constitute a three-electrode channel), the neural signal detection module, the experimental nerve bridging systems, and the IC-packaged micro-module described in the previous sections.

Figure 9.27 shows the four channels of spontaneous neural signals which were detected from the sciatic nerve of a rat. The signal amplitude ranges from 100 μV to 1 mV and the duration of a single pulse is about 1 ms.

Figure 9.28 shows the four channels of spontaneous neural signals detected from the sciatic nerve of a semi-anesthetized rabbit when its head is gently tapped with an insulating rod. The signal amplitude ranges between 40-400 μV, and the duration of a single pulse is about 1 ms.

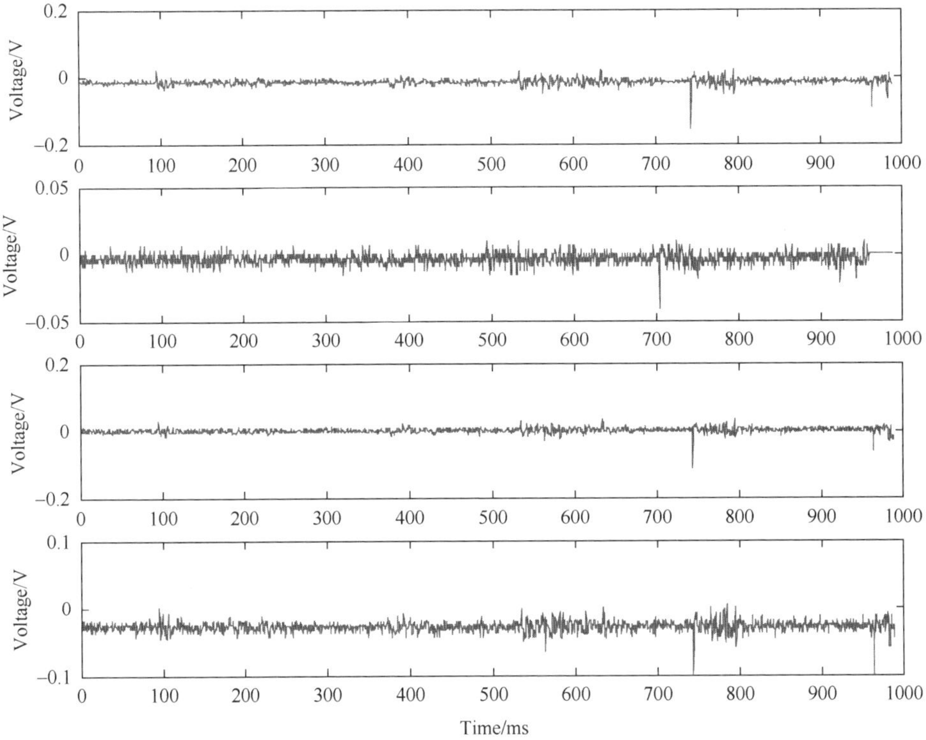

Figure 9.27 Four channels of spontaneous neural signals detected from the sciatic nerve of a rat

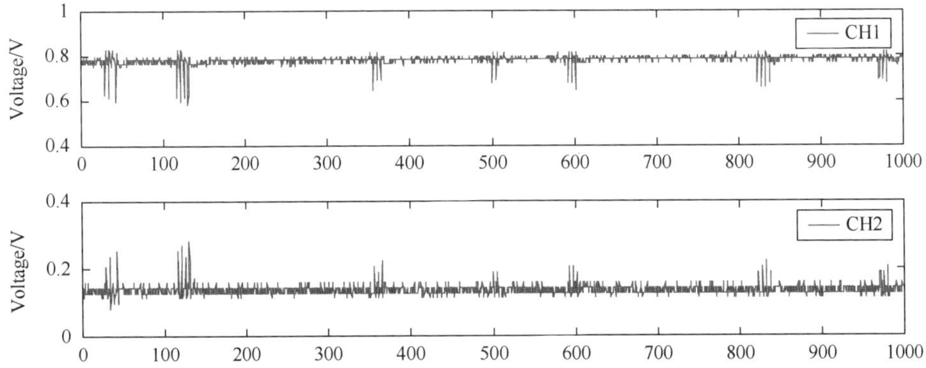

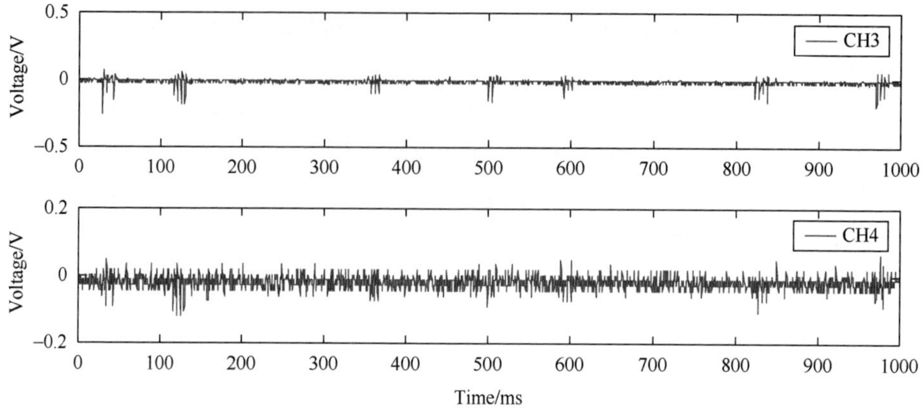

Figure 9.28　Four spontaneous nerve signals detected from the sciatic nerve of a semi-anesthetized rabbit when its head is gently tapped with an insulating rod

Figure 9.29 shows four spontaneous neural signals detected from the spinal cord of a rat without any external stimuli present. The burst of pulses appears in the duration of about 0.2 s. The peak amplitude of pulses is about 1 mV and the duration of a single pulse is 0.5-3 ms. The timing of the onset and the termination of the bursting is highly consistent between the four channels. From a viewpoint on the pulse amplitude and phase, the waveforms of the first and second channels are very similar, showing high correlation. The phase of the third channel is basically the same as those of the first and second one, but the amplitude is smaller. The fourth channel exhibits some special characteristics, and shows different trends compared to the other three channels.

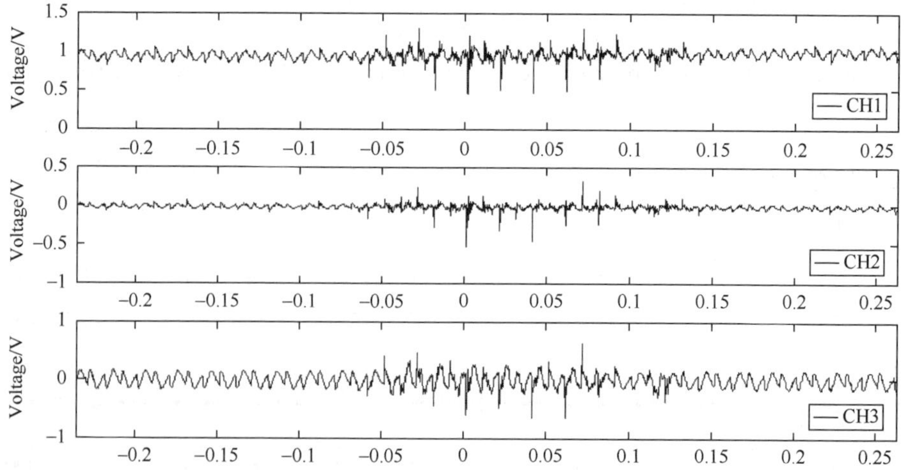

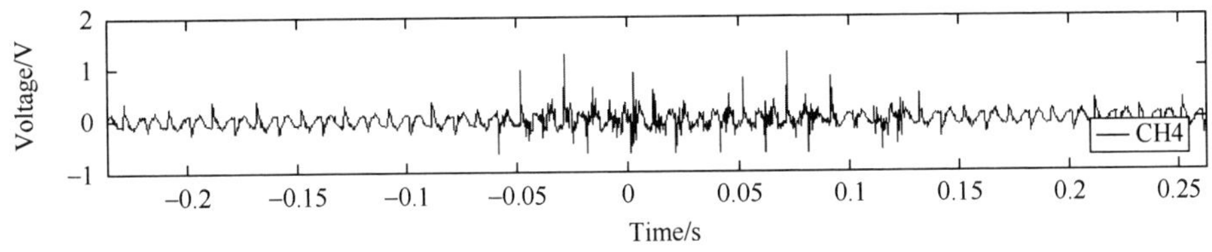

Figure 9.29 Four spontaneous neural signals detected from the spinal cord of a rat

9.4 Neural Signal Processing and Pattern Recognition

9.4.1 Neural signal processing

According to the literature and the recorded waveforms shown in Figure 9.27-Figure 9.29, it can be understood that the signals detected from the epineurium possess the following characteristics:

(1) The recorded signal consists of noise for most of the time, such as the initial 500 ms of the waveform shown in Figure 9.27;

(2) Spontaneous neural activity usually presents as highly random pulse sequence such as the waveforms of CH1-3 in Figure 9.28;

(3) Single pulses from the pulse sequence are spike-shaped and hence named spikes;

(4) The spikes are often subject to strong interference and background noise, such as the waveforms shown in CH4 of Figure 9.28, and the presence of neural signals within recorded waveforms cannot be determined by the naked eye.

The purpose of neural signal analysis is to extract the action potential spikes from the recorded signal with noise and interferences and characterize their properties; or conversely, to detect the spike sequence according to the characteristics of the neural spikes. An important aspect of this processing is to remove interference and noise.

The following are three areas of focus in neural signal denoising and interference removal.

1) Removing the electromagnetic interferences in the external environment

Common measures include:

(1) Electromagnetic shielding the entire detection system including the animal subject, to prevent wireless electromagnetic signals leaking into the experimental system;

(2) Suppressing high frequency induction currents passing into the system through the power cord from equipments such as oscilloscopes by using chokes on the power

input lines of the equipments;

(3) Using independent DC (battery) power supplies for the amplifiers, to eliminate the 50 Hz AC ripples otherwise unavoidable in AC-DC conversion.

2) Removal of electromagnetic interference inside the body

Measures include:

(1) Using 50-Hz AC notch filters to inhibit AC currents from passing into the body via electromagnetic induction or wires. Separate measures specific against AC interference is necessary due to its being the most frequent and strongest interference;

(2) Using low-pass or band-pass filters to suppress interference outside the frequency range of interest.

3) Removing background noise from the body

Measures include:

(1) Using low-pass or band-pass filters to suppress noise outside the frequency range of interest;

(2) Signal analysis and processing utilizing "a priori" knowledge of the neural spikes.

In order to use filters to extract neural spikes or filter out noise outside the signal spectrum, it is necessary to perform frequency domain and time domain analysis of neural signals.

9.4.2 Frequency domain analysis of neural signals

From a theoretical standpoint, the processing of recorded neural signals is just conventional electrical signal processing. Mathematically, any periodic time-varying signal could be analyzed via Fourier series to obtain its discrete spectrum; Fourier transform can be applied to non-periodic signals to obtain its continuous spectrum. In electrical circuits, the spectral components of these two types of signals can be selectively obtained by filters. Utilizing digital signal processing techniques, fast Fourier transform (FFT) can be applied for spectral analysis, time domain and convolution, and frequency domain analysis, and applying "a priori" knowledge of the signal (in communication, the characteristics of the signal to be transmitted are predefined by transmission protocols), its corresponding frequency components can be extracted and undesirable noise can be eliminated.

The frequency domain characteristics are an important part of any signals, and neural signals are no exception. A neural signal has a certain bandwidth, due to the fact that they contain spikes of relatively short durations (around 1 ms). Although there are different description of this bandwidth, it is generally agreed that a neural signal mostly contains frequency components of less than 4 kHz and are mainly concentrated in a range of 1-2 kHz. Therefore, the frequency domain analysis is an important criterion for determining whether a collected signal contains useful neural information, and provides critical design specification for the pass band of the amplifier and the hardware or software filter.

9.4.3 Time domain analysis and pattern recognition of neural signals

Individual neural signals are mostly temporal-coded. Therefore, a time domain analysis and a pattern recognition are the focus of neural signal processing.

The time-domain analysis of neural signals can be divided into two steps: The first step is to determine whether the signals contain information-carrying spikes, which is a pattern recognition problem; The second is to correlate the obtained spike encoding with specific biological functions or behaviors, which involves the studies of neurological function recovery, brain-computer interfaces, prosthetics control, and cognitive science etc. Below, we focus on the pattern recognition step only.

The response pattern in specific systems is necessary to implement pattern recognition of neural signals. The systems here refer to the configurations of electrodes. The output signal waveforms in response to a standard action potential were given for single-electrode, two-electrode, and three-electrode configurations in Section 9.3.2. These standard response waveforms are the neural signal patterns of the corresponding system. An action potential spike is identified whenever a recorded pulse has been matched to the corresponding neural signal pattern of the recording system.

A simple algorithm according to the discussion above could continuously shift and compare the standard neural response pattern to the recorded signal waveform, and a neural spike is identified by maxima in the similarity. Although feasible, this method is computational expensive, both in terms of time and memory space. Therefore, more efficient algorithms have to be developed.

Wavelet functions can very accurately capture any tiny changes of the pattern in signals by applying its temporal translation and scaling. Since wavelet analysis can decompose signals at multiple scales, and by utilizing the different characteristics of the signal and noise at different scales, a multi-scale wavelet transform can remove as much noise as possible from the original neural signal, therefore achieving noise reduction and pattern recognition.

Among many function families that accord with wavelet definition, the Daubechies wavelets which are named after their discoverer have special importance. For a given number of vanishing moments, Daubechies wavelets have a minimal support set, and therefore minimizes the number of wavelet coefficients. Daubechies wavelets are named numerically according to their parameters in ascending order, such as Daubechies-1 wavelet, Daubechies-2 wavelet, Daubechies-3 wavelet etc. Figure 9.30 shows an example of the Daubechies-3 wavelet.

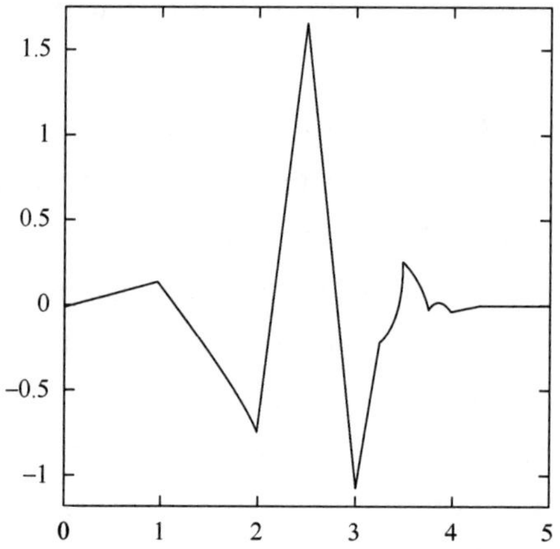

Figure 9.30　Waveform of Daubechies-3 wavelet

There is a high similarity between the Daubechies-3 wavelet and the neural signal waveform recorded from a three-electrode system as shown in Figure 9.17. Therefore, Daubechies-3 wavelet has been chosen for effective denoising and pattern recognition of neural signal recording from three-electrode systems.

Building on the above analysis, Chen and Lue proposed a neural pattern recognition algorithm resistant to potential drifts based on the QRS complex detection method for ECG.

Assuming the original digital signal is $x(n)$, in which n is the index number of the equally-spaced sampling time points. The original signal length (number of samples) is N. The Algorithm is as follows:

(1) Compute a first-order difference sequence of $x(n)$: $a(n-1)=x(n)-x(n-1)$, $2\leqslant n\leqslant N$. This yields the changes in the signal.

(2) Replace $a(n)$ by its sign: $a(n)=\mathrm{sign}[a(n)]$. That is: positive values are replaced by 1, negative values by −1, and zeros unchanged. This step gives a qualitative evaluation of how the signal changes.

(3) Compress data. Record the sign (1, −1, or 0) and start index of each block of continuous data with the same sign in array *b* (*m*) and *c* (*m*), where *m* is a positive integer whose maximum is the number *Z* of blocks continuous data with the same sign. This step records all turn points in the data.

(4) Define an array *d* (*m*) of size *Z* to store the smoothed and deburred signal. Setting a threshold value *T* for burrs, loop with index *i* in the range $1 \leqslant i \leqslant Z-1$; if $c(i+1) - c(i) < 2$, then *d* (*i*) is assigned the value of $\{x[c(i)] + x[c(i+1)]\}/2$; on the other hand, if $c(i+1) - c(i) \geqslant 2$, $w = \left|x[c(i)] - x[c(i+1)]\right|$ is compared with *T*: if $w \leqslant T$, then *d* (*i*) is assigned as $[\{x[c(i)]+x[c(i+1)]\}/2]$, with [*y*] representing the maximum positive integer not greater than *y*; If $w > T$, then *d* (*i*) is assigned as $x[c(i)]$. Finally *d* (*Z*) is assigned as *x* (*N*). The purpose of this step is to characterize the original signal by averaging the original data at the turning points and deburring.

(5) Apply a predefined recognition template (e.g., Daubechies-3 wavelet for a 3-electrode system) to the smoothed and deburred signal *d* (*m*). Detection is completed by searching in *d* (*m*) for the characteristics of the template to identify the position of each occurrence of the template.

The complexity of the algorithm is discussed as follows. The first step, which generates the difference sequence, requires $N-1$ subtractions in total. The second and third steps are comparison and assignment arithmetic operations. In the fourth step, considering that the pattern recognition is applied to the denoised signal, therefore the occurrence of $c(i+1)-c(i)<2$ is significantly less than $c(i+1)-c(i)\geqslant 2$. After $2\times(Z-1)$ necessary subtractions, the number of operations (denoted *L*) depends on how many patterns (spikes) this digital signal has. Usually, the sampling rates in experiments are relatively high (>4 kHz), thus, $L \ll N$. The assignments in the fourth step require only 2 *L* additions, and 2 *L* subtractions. In the fifth step there are only comparisons and no arithmetic operations. Hence, the entire algorithm requires $N-1$ subtractions, $2\times(Z-1)$ subtractions, 2 *L* additions, and 2 *L* subtractions. After denoising, $L<Z\ll N$, therefore the complexity of the algorithm is between *O* (*N*) and *O* (2*N*), and meets the requirement for real-time processing.

Applying this pattern recognition algorithm to three-electrode cuff electrode recording of spontaneous signals from rat cord signal, i.e. CH1 in Figure 9.31, the spike pattern recognition are shown in Figure 9.31. Panel (a) shows the raw recording signal, while panel (b) shows the raster image.

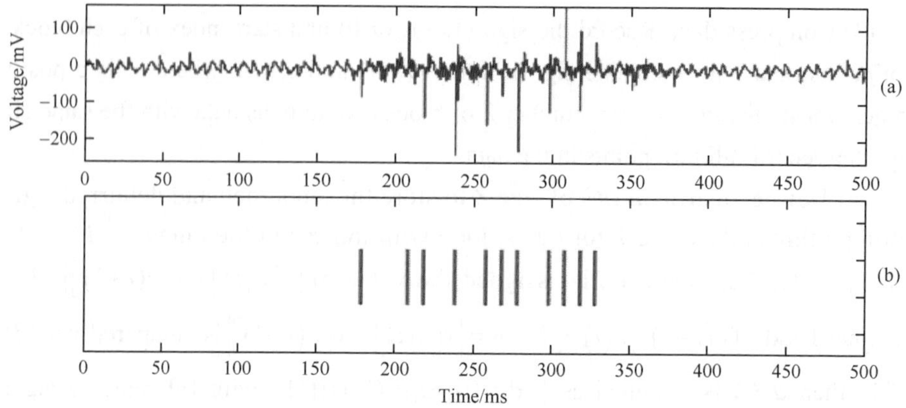

Figure 9.31 Applying the spike pattern recognition on three-electrode cuff electrode recording of spontaneous signal from rat spinal cord (a) yields the raster plot (b)

9.5 Functional Electrical Stimulation of Nerves

In neural function regeneration systems, functional electrical stimulation (FES) circuits generate electric signals of certain amplitudes and waveforms, and regenerate specific physiological functions by applying the signals to stimulating electrodes and invoke neural response. There are two types of stimulation circuits: voltage-driven and current driven. The principles of neural stimulation and the two types of stimulation circuits are introduced here.

9.5.1 Principles of functional electrical stimulation of nerves

Over the past half a century, experimental studies have used electronic devices to implement biological functions such as hearing, vision, touch, and movement via functional electrical stimulation (FES). This includes single-channel functional electrical stimulators, and implantable mixed-signal nerve stimulators.

The basic principle of electrically stimulating nerve is depolarizing the neuron membrane to reach threshold for firing action potentials. When an excitable neuron at resting state experiences external stimulation, the electrical field and current densities created by the electrodes causes cations to flow towards the cell membrane in the intracellular space and away from the membrane in the extracellular space, while the anions move in the opposite direction. Both ion movements reduce the depolarization of the cell membrane capacitance. When the depolarization reaches a threshold level, i.e. about 20 mV above resting potential, active voltage-gated ion channels open in response and the ionic current reverts the polarization of the membrane potential in a very short time from –60-–50 mV to +20-+40 mV. This constitutes the rising phase of the action potential. The membrane potential quickly returns from the peak to its resting value as delayed rectifying K channels repolarizes the cell membrane, forming the repolarizing phase. Functional regeneration is achieved when the invoked single or train of action potential (s) reaches the target organ, such as muscles, and generate physiological activities.

9.5.2 Neural stimulation circuits

Functional electrical stimulators need to output sufficient high voltages (several V) or high current (several mA). Due to uncertainty of the impedance of the electrode-tissue system, current stimulation is preferred to better maintain the stimulation level and balance the injected charge. However, voltage stimulation is also used due to its simplicity in implementation. As the frequency band of neural signal is centered below 10 kHz, the frequency requirement for neural stimulation circuits is relative low. Challenges in FES design are the high output current or voltage.

The choice of voltage driven or current driven stimulation in FES depends on the stimulation electrode and target nerve. Both types of stimulators exist in clinical applications. For example, products from ISOFLEX have maximum output of 90-V voltage or 10-mA current, given a 90-V supply voltage. On one hand, a sufficient large stimulus is required to achieve effective nerve activation. On the other hand, given the constraint of the implantable device and IC fabrication, the output voltage and current cannot be arbitrarily high; challenges such as high power consumption and tissue damage should be considered for an efficient and safe implant.

The load of functional electrical stimulators includes the electrodes and the neural tissue, which have complex impedance. The impedance not only depends on electrode geometry and location, and type of tissue, but also varies over time due to tissue reactions such as encapsulation. Under common conditions, the tissue-electrode impedance has a magnitude in the thousands of ohms range.

For a voltage driven stimulus, the device can be treated as a voltage source, and the output impedance should be as small as possible. Given the range of load, the output impedance of the circuit is preferably in the tens to hundreds of ohms range. For current driven stimulus, the device can be treated as a current source, and the output impedance of the circuit should be as large as possible. Given the range of load, the output impedance of the circuit is preferably in the hundreds of kΩ to several MΩ range.

1. Voltage stimulation circuit

Typically, the stimuli for activation nerves are rectangular pulses generated by a pulse generator, which is often of high voltage amplitude. Stimulators for neural function regeneration only need to generate voltage sufficient to induce neural activities.

Figure 9.32 is a typical voltage driver with 2 operational amplifiers (OpAmp). Stimuli signals of specified amplitude but opposite phase are output on two terminals via OpAmps. The opposite phases of two outputs can effectively activate nerves. A shared terminal between two output terminals is connected to the reference electrode in three-electrode system.

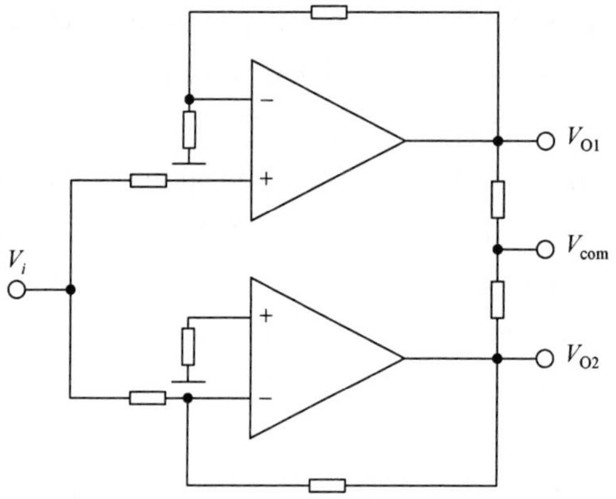

Figure 9.32 Voltage driver circuit implemented via OpAmps

2. Current stimulation circuit

A current stimulator using Zener diode and an OpAmp is shown in Figure 9.33.

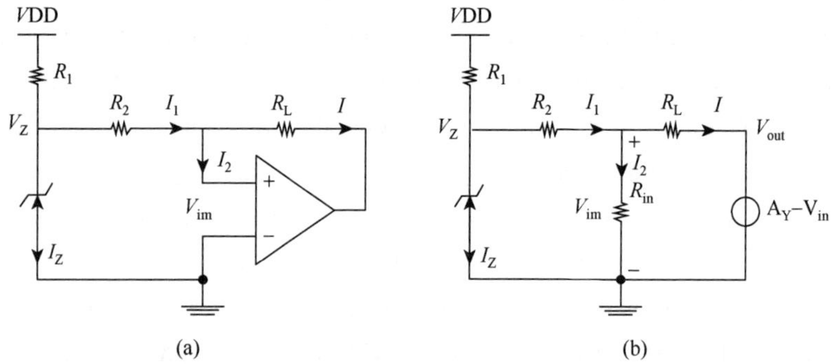

Figure 9.33 (a) Circuit diagram and (b) circuit model of a current source stimulator utilizing Zener diode

The voltage on the Zener diode V_Z is constant due to the operation characteristics of the diode, and the current I_1 through the resistor R_1 stays constant as well due to the virtual shorted-circuit characteristics of the OpAmp. I_1 is given as

$$I_1 = \frac{V_Z}{R_1} \tag{9.27}$$

For an ideal OpAmp, its input current I_2 is negligible due to the ultrahigh input resistance. And the current on the load R_L is

$$I = I_1 - I_2 \approx I_1 \tag{9.28}$$

As the highest negative value of V_{out} is $-V\mathrm{DD}$, therefore

$$R_L \leqslant \frac{V\mathrm{DD}}{I} = \frac{V\mathrm{DD}}{I_1} \tag{9.29}$$

and R_L has a maximum.

For the purpose of circuit analysis, the model of this circuit is given in Figure 9.15 (b), which shows that

$$V_{in} = V_Z - I_1 \cdot R_2$$

$$I_2 = V_{in} / R_{in}$$

$$I = I_1 - I_2 = \frac{V_{in} - V_{out}}{R_L} = \frac{V_{in}}{R_L}(1 - A)$$

The above three equations yields

$$I = \frac{I_1}{1 + \dfrac{R_L}{R_{in}(1 - A)}}$$

Typically,

$$\left| \frac{R_L}{R_{in}(1 - A)} \right| \ll 1$$

therefore

$$I \approx I_1$$

As I_1 is related to R_1, the current applied to the load R_L can be varied by adjusting the resistor R_1. The load and the current always follow

$$I \cdot R_L \leqslant V\mathrm{DD}$$

Due to the large power or voltage and current levels of this type of stimulation circuits, design consideration has to be taken regarding the characteristic of the circuit and load, which could be challenging.

9.5.3 Implementation of neural stimulation circuits

For *in vitro* use, discrete circuits composed of components such as OpAmp and resistors and capacitors are feasible designs, while for implantable devices, integrate circuits are the only viable option. Here, voltage stimulators are introduced, in the form of both discrete circuits and IC.

1. Discrete circuit neural stimulator

Figure 9.34 is the diagram of a voltage stimulation circuit of two stages. The first-stage amplifier consists of R_1, R_2, R_{f1}, and A_1; in the second-stage, R_3, R_4, R_{f2}, and A_2 constitute the non-inverting amplifier, and R_5, R_6, R_{f3}, A_3 constituting the inverting amplifier. By choosing the right resistance ratio, two voltage outputs of the equal amplitude and opposite polarity versus the reference electrode could be achieved. Figure 9.35 (see the Color Inset, p. 26) shows the circuit board of a four-channel neural stimulator for experimental use.

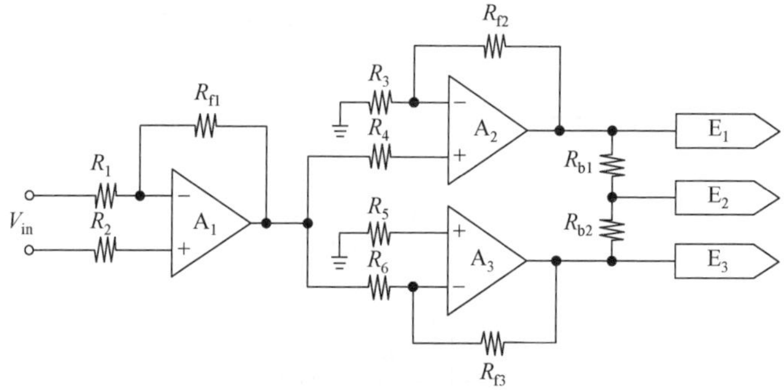

Figure 9.34 Circuit diagram of voltage stimulator with two outputs of opposite phase

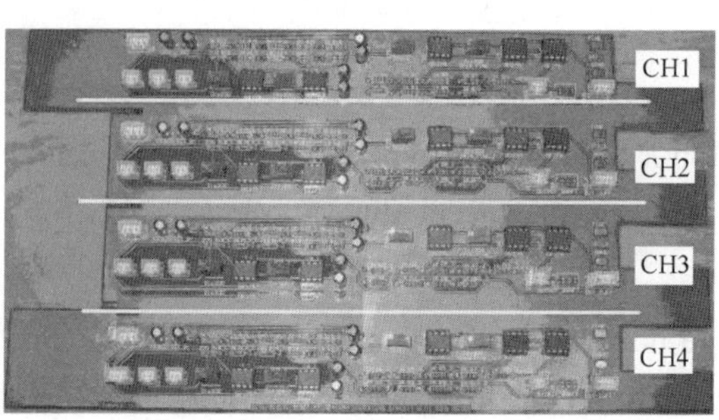

Figure 9.35 Circuit board of a 4-channel neural stimulator

2. Integrated circuit neural stimulator

A neural stimulation IC for neural signal regeneration consist of a pre-amplifier, a main amplifier, and an output stage. The pre-amplifier applies a constant gain to the differential signal from a neural signal recording circuit; and the main amplifier amplifies the signal with a high gain; while the output stage forms a buffer between the amplifiers and output load, and protects the output terminal of the circuits. A circuit of the preamplifier and main amplifier is shown in Figure 9.36.

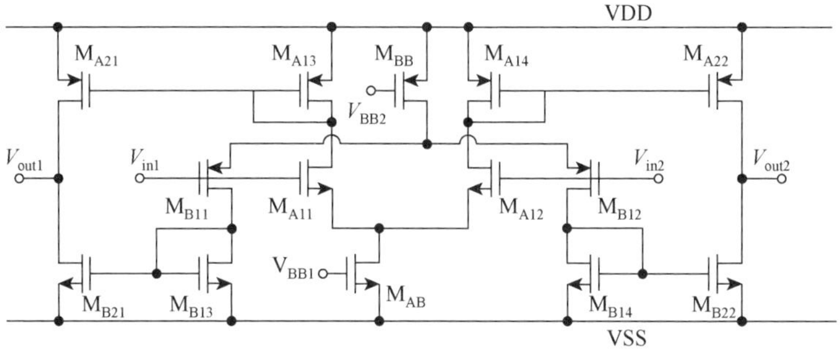

Figure 9.36 Preamplifier and main amplifier of a CMOS neural stimulation IC

The preamplifier consists of PMOS and NMOS input terminals, so that the full-swing voltage input could be taken. The NMOS input stage consists of transistors M_{A11}, M_{A12}, M_{AB}, M_{A13} and M_{A14}, in which M_{A11}, M_{A12} are the differential inputs, M_{AB} provides the voltage bias, and M_{A13} and M_{A14} operate in the saturation region in diode mode as the active load of the amplifier circuit. Similarly, the PMOS input stage consists of M_{B11}, M_{B12}, M_{BB}, M_{B13} and M_{B14}, with corresponding functionality as the transistors in the NMOS input stage. The gain of preamplifier stage is determined by the width to length ratio of the transistor M_{A11} (M_{A12}) and M_{A13} (M_{A14}). The main amplifier consists of common source amplifiers M_{A21} (M_{B21}), M_{A22} (M_{B22}), in which M_{A21} and M_{B21} are the load for each other.

The key components of the output stage circuit are MOS transistors M_{31}, M_{32} and amplifier A_1, A_2, as shown in Figure 9.37.

The width to length ratios of M_{31} and M_{32} are relatively large to improve the driving capability of the output transistors. For achieve full-swing output, M_{31} and M_{32} are connected in a common source mode. Additionally, the output resistance is reduced by the feedback of the amplifiers A_1 and A_2.

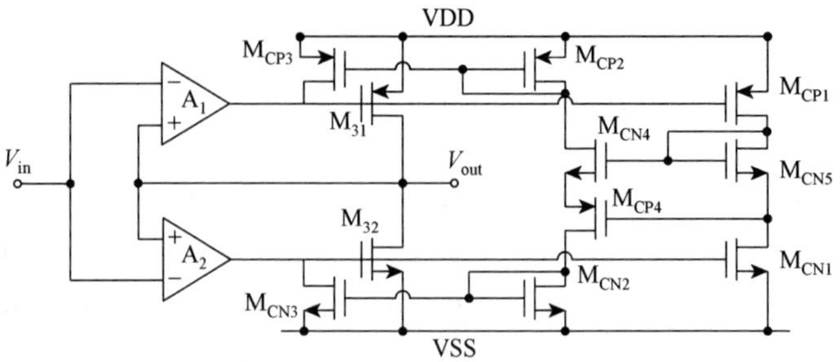

Figure 9.37　Output stage of a CMOS neural stimulation circuit

Due to full-swing operation of the neural stimulation circuit, an output protection circuit is included in the design. The protection circuit consists of M_{CP1} (M_{CN1}), M_{CP2} (M_{CN2}), M_{CP3} (M_{CN3}), M_{CP4} (M_{CN4}) and M_{CN5}. A small width to length ratio of M_{CN5} results in negligible current through M_{SC5} under normal output current of the circuit. Therefore, its VDS is small and M_{CP4} (M_{CN4}) are not conducting. Conversely, when the output current of the circuit becomes very large, the current through MCN5 increases, which increases its VDS as well. This causes M_{CP4} and M_{CN4} to turn on, and the feedback current is diverted via feedback loop through M_{CP2} (M_{CN2}), M_{CP3} (M_{CN3}) to M_{31} and M_{32}, therefore protecting the circuit.

This neural functional stimulation circuit was simulated and fabricated in a 0.6-μm CMOS process by CSMC (Wuxi, China). Figure 9.38 (a) (see the Color Inset, p. 26) and Figure 9.38 (b) (see the Color Inset, p. 26) show the die photo of the voltage stimulator IC and the PCB on which the packaged IC was mounted. Figure 9.38 (c) (see the Color Inset, p. 26) shows the output waveform under a test input of 500 mVp-p.

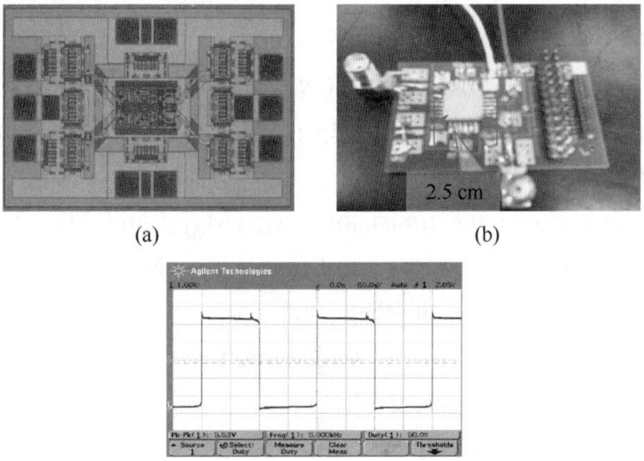

(a)　　(b)

(c)

Figure 9.38　Die photo (a), PCB circuit (b), and output waveform under 500 mV p-p test input (c) of a voltage stimulator for functional neural stimulation

9.6 Microelectronic Neural Signal Regeneration

9.6.1 Nerve injury and regeneration

Pathological changes and neural function degenerations are major kinds of illness of neural systems. Among them, spinal cord injury is (SPI) one of the most destructive trauma and lesion. With the development of modern transportation, the occurrence of SPI has been increasing over the years. In the US, there are 11 thousands of new SPI patients annually, with a total of 253 thousand in the year 2006. According to a preliminary estimation of the Chinese National Production Safety Committee, there are 120 to 140 thousand SPI patients due to accidents at work and traffic accidents every year. The December 1st, 2012 report by the 2nd National Disability Sampling Survey Leadership Group and the National Bureau of Statistics of China shows that there are currently 24 million people with limb disability. Past statistics reveal that SPI patients account for more than 20%, which gives an estimate of about 3 million in China. Among these patients, about half have complete loss of volitional control and sensory functions below the lesion site due to complete spinal cord injury. They usually cannot take care of themselves in everyday life, which causes personal suffering and burdens their family and the society.

Function rebuilding after SPI has been an unresolved challenge in the field of health science. The limb dysfunctions after lesion to the central nervous system (CNS) are due the lack of drive and control from the injured brain and/or spinal cord. Normally, the neurons of the CNS send signals in the form of bioelectricity conducted to the corresponding muscles through neural pathways, which results in coordinated muscle contractions and joint movements. The nervous system spontaneously lay out these pathways during development, and recovers some of the connections after lesion. For example, axons in the arm can grow back after nerve damage and regain functions. There are some axon regeneration and function recovery in the CNS in frogs, fish, or invertebrates. If the optic nerve of frogs or fish is cut, the nerve fibers can regrow to the brain and recover vision. However, the CNS of adult mammals does not have this capability. For a century, biologists, neuroscientist, and neurologists have devoted to a biological method to reconstruct neural functions.

Nerve regeneration in the true sense reconnect neural pathways via proliferation, growths of neurons through the original pathway of the disrupted nerve. For a long time, people are exploring a range of biological methods to achieve repair of damaged spinal nerves.

The first is the use of cell and tissue transplantation. Neural tissue and cell transplantation can replace various lost neurons, thus restoring levels of neurotransmitters, hormones, and neurotrophic factors. Nerve tissue and cell grafts can also provide new neurons as neural transduction transits, thereby restoring the access to the distal lesioned spinal cord, or bridging the two terminals and guide axonal regeneration.

In September 2005, a research team at UC Irvine led by Prof. Anderson announced breakthrough in spinal cord injury repair in rats via human neural stem cells. However, the experiment prompts the following questions:

(1) Human neural stem cells were injected in mice after nine days of spinal cord injury. How long after the spinal cord injury would stem cells therapy remain effective-no one knows how much time does nine days in a rat life corresponds to in human life.

(2) Rats recovered some form of motion after four months of treatment. Would this time be too long?

(3) The researchers suppressed the immune system of the rats during the therapy. If this applies for the treatment in human, the suppressed immune system will bring significant issues.

Other studies showed that stem cell therapies are not effective for neural injury for more than six months. In reality, more than 90% of the paraplegic patient has a history of more than six months. Therefore, even if the stem cells therapy repairs relative new spinal cord injuries, to achieve recovery of paraplegic patients on a broader base still proves a difficult course.

Given the challenges over the past decades in neural regeneration via biological methods, our group proposed the idea using microelectronic circuit to bridge neural pathways and regenerate neural signals in 2003. A project titled "implantable central nervous system function regeneration SOC and related biological experiments" has been funded by National Natural Science Foundation of China as major scientific program on basic research of semiconductor integrated chip systems "Preliminary results have been obtained after many years of theoretical research, circuit designs and fabrication, and animal experiments". The relevant principles, system design and animal experiments are introduced here.

9.6.2 Principles of neural signal regeneration

The basic idea of microelectronic systems for neural signal regeneration is to use a microelectronic devise implanted on the disrupted spinal cord to channel bridging and signal regeneration of the injured spinal cord, and regenerate function actively via signal amplification and processing. Figure 9.39 (see the Color Inset, p. 26) is an illustration of the signal bridging in the central nervous system.

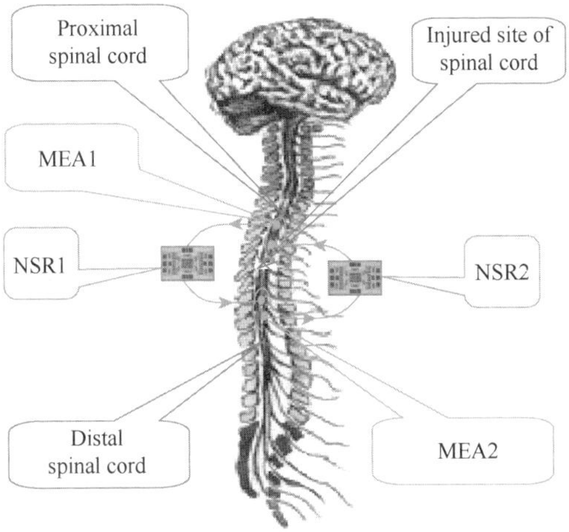

Figure 9.39 Illustration of bridging the spinal cord via electronics

The system as shown in the figure detects motor control signal from the brain via a microelectrode array (MEA1) from the proximal end of the injured spinal cord nerve bundles, processes them in the neural signal regeneration chip (NRS1), and then stimulates the distal end of the injury site via another microelectrode array (MEA1), therefore achieving neural signal pathway bridging and motor signal regeneration.

Similarly, another neural signal regeneration chip (NRS2) and a pair of microelectrode arrays can be used to bridge sensory neural pathways, and achieve regeneration of sensory signals.

The four sets of electrode arrays and the two neural signal regeneration chips will work towards rehabilitation of paralysis patients due to spinal cord injury.

9.6.3 Microelectronic systems for neural signal regeneration

Given that each nerve bundles contain tens to hundreds of motor and sensory neural pathways, the required electronic systems are extremely complex to completely or just partially reconstruct the function of damaged nerve bundles. Not only the regeneration of the signal is required, but the bridging and switching of pathways needs to be considered. Figure 9.40 (see the Color Inset, p. 27) shows the complete structure of microelectronic system for neural pathway bridging and neural function regeneration, including the microelectrode arrays, channel switching units, microelectronic signal processing units, the control unit and a monitoring unit.

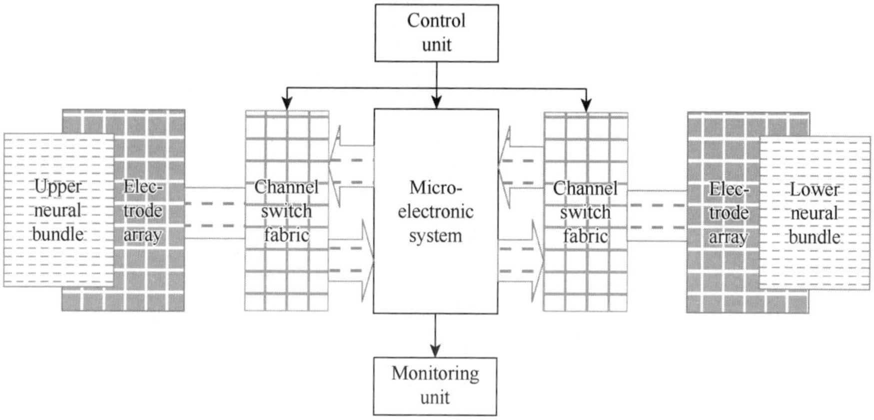

Figure 9.40　Block diagram of a microelectronic systems for neural pathway bridging and neural function regeneration

A functional block diagram of a single unidirectional neural signal regeneration is shown in Figure 9.41 (see the Color Inset, p. 27). Neural signal detection electrodes record from the proximal end of the injured nerve fibers, and the weak signal are amplified to a certain voltage level. Interference and noise are removed in the signal processor and then sent to control FES pulse generation generator. Finally, the FES pulse is applied to the excitation electrodes, to reproduce the neural signal on the distal end of the nerve fiber with high fidelity.

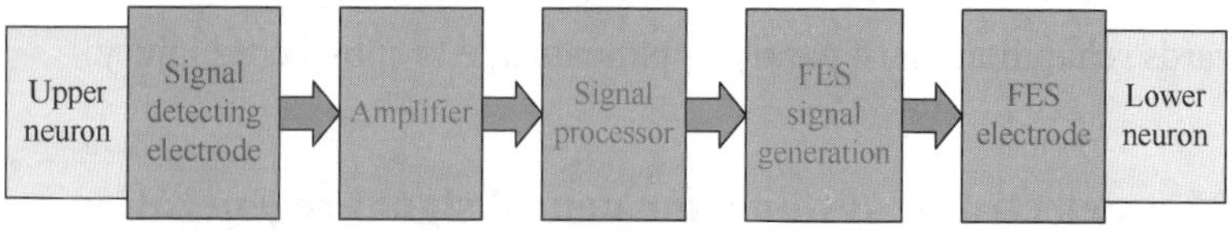

Figure 9.41　A functional block diagram of a single unidirectional neural signal active regeneration

In fact, if a signal recording devices such as an oscilloscope replaces the FES parts, the system is a neural signal recording system. If an artificial neural signal generator is used instead of the parts before the neural signal processor, the system is an artificial FES system. If the neural signal recording and stimulation electrodes are replaced with receiving and transmission antenna or a wireless transceiver, the neural signal regeneration system is a typical communication channel regeneration system in principle

Packaging the neural signal detector and amplifier in Section 9.4, and the neural function stimulator in Section 9.6 into a multi-chip module (MCM) package, i.e. to encapsulate the detection and stimulation ICs into one interconnected entity, forms a complete neural signal regeneration circuit. Such an MCM neural signal regeneration devise and experimental PCB are shown in Figure 9.42 (a), see the Color Inset, p. 27 and Figure 9.42 (b) , see the Color Inset, p. 27. The size of the IC package is 9 mm×9 mm, and the size of the PCB is 60 mm×30 mm. According to the figure, the chip area of each channel of neural signal generation circuit is about 0.15 mm^2, and the capacity of the package is 27 mm^2, allowing multiple channels to be included in one package.

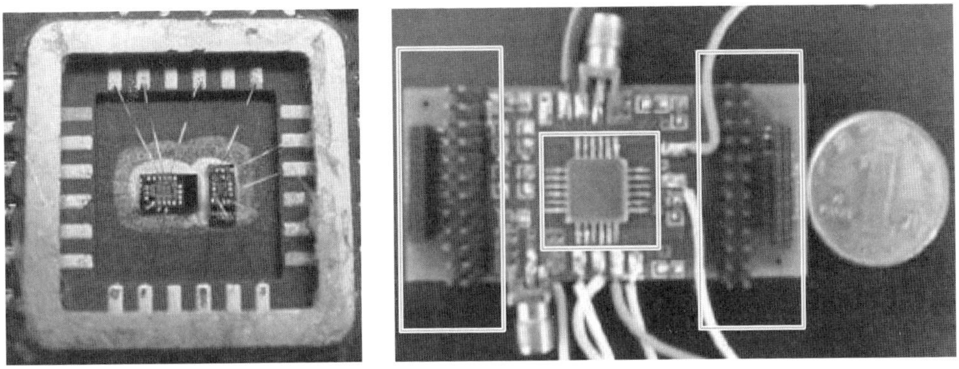

(a) (b)

Figure 9.42 (a) Photo of micro components of neural signal regeneration system and (b) photo of experimental PCB

Similarly, integrating the neural signal detection and stimulation circuits on a single chip yields a monolithic neural signal regeneration integrated circuit. Figure 9.43 shows the circuit diagram of one channel, in which the signal detection part consists of OpAmps A_1, A_2 and A_3, and the stimulation part consists of A_4, A_5 and A_6. The output is in the form of a differential voltage.

Figure 9.44 (a) (see the Color Inset, p. 27) is the die photo of a 6-channel neural signal regeneration circuit. The size is 2.82 mm×2.00 mm. Figure (b) (see the Color Inset, p. 27) shows the test result of one channel.

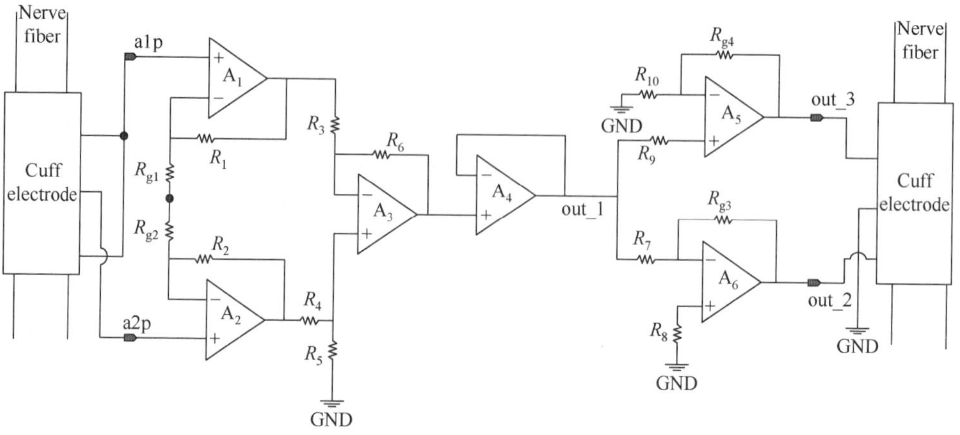

Figure 9.43 Circuit diagram of voltage stimulation neural signal regeneration system

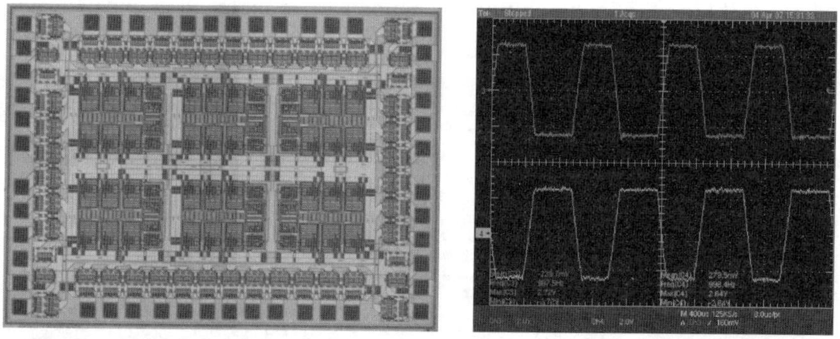

(a) (b)

Figure 9.44 (a) The die photo of a 6-channel neural signal regeneration circuit and (b) the test result of one of the channels

9.6.4 Animal experiment of microelectronic neural signal regeneration

Figure 9.45 shows the regeneration of neural signal from left sciatic nerve to the distal end of the interrupted right sciatic nerve in rat using microelectronic neural signal regeneration circuit. A stimulator applies extern neural activation on electrode A_1, which wraps around the left sciatic nerve of the rat and provides FES. A second electrode A_2 on the same nerve detects the activation, and sends the signal into the neural regeneration circuit, where the signal is amplified and processed to form a new stimulation voltage. This stimulus is then applied to the electrode B_1 wrapped around the distal stump of the interrupted right sciatic nerve, and the invoked neural signal is detected by electrode B_2, which is more distal on the same nerve stump, and sent to the oscilloscope.

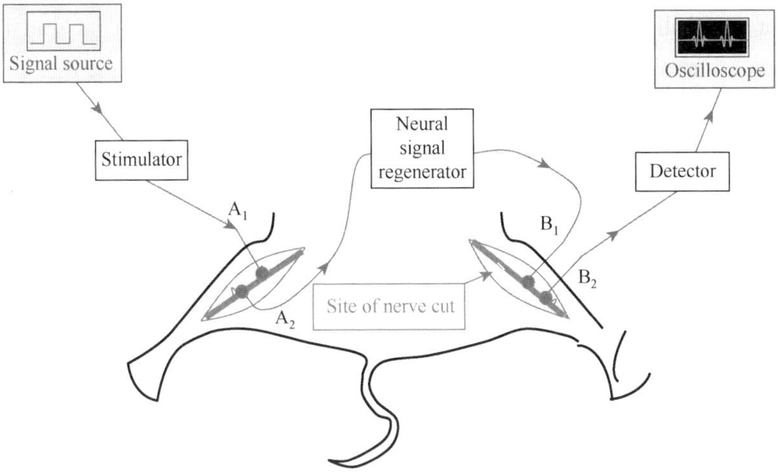

Figure 9.45 Illustration of neural signal regeneration in rat sciatic nerve

Figure 9.46 shows the recorded signal waveforms from the four electrodes/channels when external stimuli were applied to the rat under awaken state. In the red ellipses, the external excitation signal, induced neural signal on the left nerve bundle, the regeneration stimulus signal on the right nerve bundle, and the induced neural signal on the right bundle after regeneration. Comparing the four signals demonstrates that the externally-invoked neural signal on the left sciatic nerve has been synchronously regenerated on the right sciatic nerve.

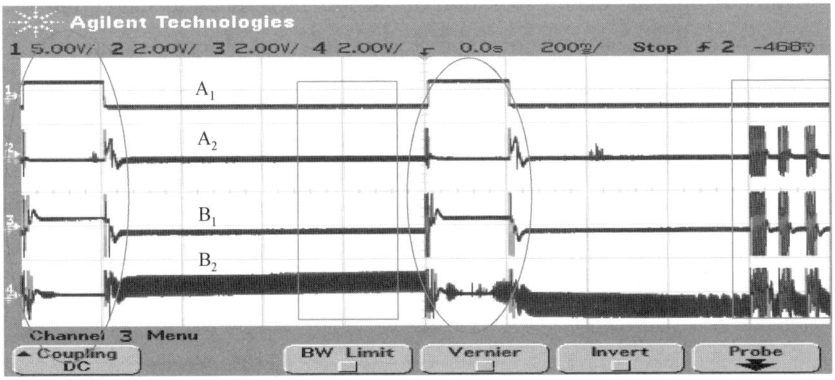

Figure 9.46 Neural signal regeneration from left to right sciatic. A_1: Stimulation source signal on the left sciatic nerve; A_2: Induced neural activity detected on the left sciatic nerve; B_1: Regenerated stimulation applied on the right sciatic nerve; B_2: Neural activity induced by the regeneration signal detected from the right sciatic nerve.This four-signal waveform described on the left sciatic nerve on the right side of the applied signal in synchronism with sciatic nerve regeneration obtained

Figure 9.47 illustrates bridging and neural signal regeneration in rat spinal cord. A signal source is applied through a stimulation circuit to electrode A_1 on the proximal end of the disrupted spinal cord, forming FES; the induced neural signal (red trace) is detected by electrode A_2 nearby on the same nerve bundle and feed to the neural signal regeneration circuit, where it is amplified and processed to form new stimulation voltages. The stimulus is applied to electrode B_1 placed on the distal end of the disrupted spinal cord, and the invoked neural signals are detected from electrode B_2 nearby and sent to the oscilloscope.

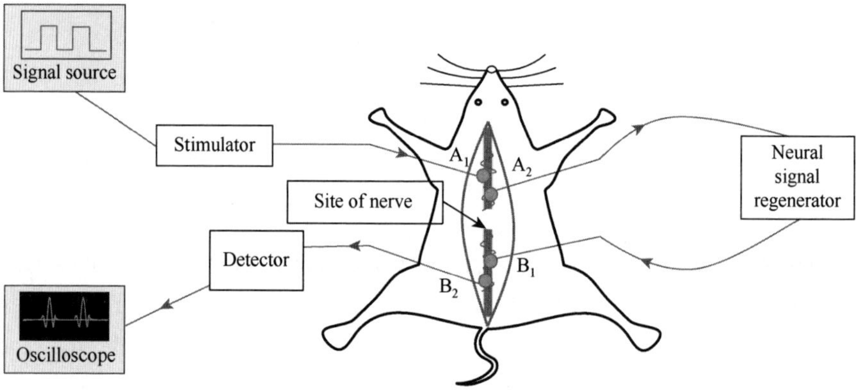

Figure 9.47 Illustration of neural pathways bridging and signal regeneration in rat spinal cord

Figure 9.48 shows the signal from channel A_1 and B_2-B_4 (each cuff electrode array has four channels), respectively, which, together with the physical reaction observed from the rat, demonstrate the regeneration of the spinal cord neural signal in the rat.

Figure 9.48 of the signal waveform and physical reactions observed in rats demonstrated that rat spinal nerve signals has been regenerated.

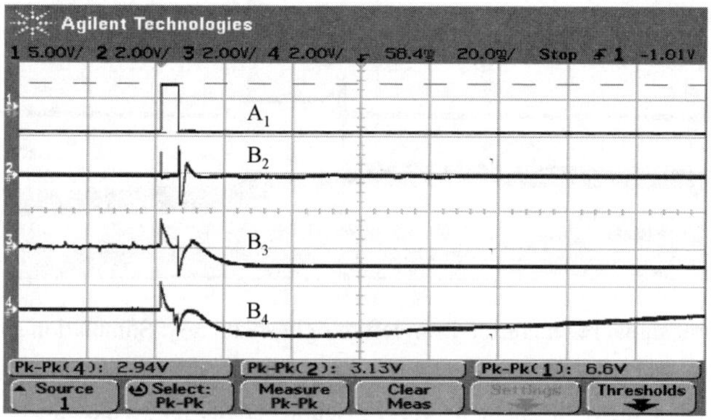

Figure 9.48 The four signal waveforms from channel A_1 and B_1-B_3 in the experiment of bridging neural pathways in rat spinal cord

9.7 Summary

The nervous system of vertebrates consists of the central nervous system and the peripheral nervous system, with neurons as the basic element. The signal transmission in a neuron is an electrophysiological process reflected by the change in its membrane potential. There are passive and active electrical components in the axon of neurons: the former is responsible for the propagation of neural signal in the cytoplasm, and the latter is the basis for action potentials. Regeneration and function rebuilding after lesion to neurons in the CNS is an unsolved challenge. Exploration of functional rebuilding after CNS lesion via implanted chips combines microelectronics and neurobiology, and shows significance in the research of trans-system IC between biological information system and electronic information system and in the field of brain-computer interface. These researches involve a myriad of projects covering neural signal recording, selection and fabrication of stimulation electrodes, IC design for neural signal amplification, processing, and functional electrical stimulation, biocompatibility of implantable devices, techniques for implant surgery, animal experiments, and development of evaluation criteria etc. Exploration in these fields reveals the feasibility of neural signal regeneration and function rebuilding through microelectronics.

Chapter 10 Cerebral Glioma Grading Using Bayesian Network with Features Extracted from Multi-modal MRI

Jisu Hu, Suiren Wan, Yu Sun, Bing Zhang

10.1 Introduction

Glioma is a most common type of brain tumor. The grading of glioma is the premise of treatment planning. Gliomas of different grades will vary in the therapies and prognoses, so it is of clinical significance to determine the grade of a glioma. Up to now, although being still the "golden standard" for grading of brain tumors, histopathologic grading has some inherent limitations. On one hand, the biopsy in some case is physically impossible (brain stem, for example). Even the biopsy is physically possible, there are complications and risks due to its invasiveness. On the other hand, due to the heterogeneous growing of a tumor tissue, the biopsy can only sample very limited region, which may not reflect the true malignancy of the whole tumor.

Considering the above limitations of histopathologic grading, radiological examinations, especially MR imaging techniques, are of great help and importance in evaluating the grade of a glioma. With the development of many state-of-the-art MR imaging sequences, some of them are applied in the clinical diagnosis. Hence, radiologists lay hopes on these tools to make preoperative glioma grading more accurate. This goal, however, is still somewhat difficult to achieve in the current circumstances.

Conventional MR imaging modalities, including T1-weighted imaging with and without gadolinium-based contrast agents, T2-weighted imaging and fluid-attenuated inversion-recovery imaging (FLAIR), have provided a large number of characteristics like tumor size, signal intensity, mass effect, peritumoral edema, hemorrhage, necrosis, contrast enhancement, etc. Among them, Dean et al.[1] pointed out that mass effect, necrosis and contrast enhancement are closely associated with tumor grade and they can be regarded as evidence for determining the grade of a glioma. Particularly, gliomas with contrast agent enhancement, meaning blood-brain barrier breakdown, is often indicative of higher grade[2]. Information from conventional MR imaging, however, is insufficient to make an accurate grading. For example, gliomas of different grades can present similar signal characteristics, and another example is that a small proportion of high-grade gliomas can show no contrast enhancement while some low-grade ones can exhibit contrast enhancement conversely. All these cases remind us that more useful information should be added to improve the accuracy of grading.

In recent years, more advanced MR imaging techniques, including perfusion weighted imaging (PWI), MR spectroscopic imaging (MRSI), diffusion tensor imaging (DTI), susceptibility weighted imaging (SWI), have been proposed and become routinely used not only in research but also in clinical settings, which give us a more comprehensive understanding of brain tumors.

Increased tumor vascularity is an essential point in tumor diagnosis, for it can directly reflects the growth and malignancy of a tumor. PWI is the ideal tool to show this characteristic and many good results have been reported in tumor grading using regional cerebral blood volume (rCBV) maps extracted from PWI.

The *in vivo* brain multi-voxel MR spectroscopy (also named as MR spectroscopic imaging, MRSI) is another powerful tool that can greatly benefit tumor grading. Due to its advantages in diagnosis and treatment monitoring, it has drawn a lot of interests and has been studied extensively in recent years[3, 4]. With MRSI, we can non-invasively observe metabolite alterations in the tumoral and peritumoral areas[5, 6]. For cerebral gliomas, many metabolites can be used in grading. Specifically, increased choline-containing (Cho) compounds along with decreased N-acetyl aspartate (NAA) indicate the presence and aggressiveness of a tumor[3], and quantities of lipid and lactate metabolites generally reflect the extent of tumoral necrosis which is also associated with tumor grade[4]. MRSI is superior to single-voxel spectroscopy technique in that its volume-of-interest (VOI) covers both tumoral and peritumoral regions and it can present metabolite variations between voxels. At present, the main difficulty of using MRSI clinically lies in the quantitation of metabolites. LCModel[7, 8] is a software that objectively uses basis metabolites to fit a spectrum to get all metabolite concentrations and this book proposed a method to extract spectroscopy features based on LCModel results.

In clinical settings, accurate diagnosis would typically require more than one type of information; this is particularly true for tumor grading. Galanaud D et al.[9] proposed grading methods based on multi-modal imaging. However, due to the difficulty in quantifying MR imaging features, these category of methods are only used to aid the clinical decision rather providing a diagnosis tool. Tsolaki E et al.[10] proposed machine learning based grading systems, but the performance is limited by the accuracy of spectroscopy features extraction and not satisfactory for clinical use. Bayesian Network (BN)[11] is a powerful tool in the field of artificial intelligence and has many applications in the fields like medical diagnosis[12-15] and fault diagnosis. In a BN, every feature is regarded as a random variable and the BN structure is the delineation of casual relationship between variables. After parameter learning[16] with enough data, the posterior probability of the new patient's tumor grade given the imaging evidence can be computed through the BN. The predictive grade is assigned to the one that has the highest probability, which is also a measure of belief that it truly belongs to this grade.

In this book, we proposed to build a grading system for preoperative cerebral gliomas using BN that integrates features of tumor volume, mass effect, contrast enhancement, perfusion features and MRSI. We evaluated the grading performance using real clinical data; and the result showed that the proposed system is promising in glioma grading. We also discuss contributions of every feature used in the grading system and possible future improvements.

10.2 Methods

10.2.1 Patients

All clinical cases involved in this study came from the Drum Tower Hospital in Nanjing, China, consisting of histopathologically confirmed 30 with high grade gliomas (WHO Ⅲ-Ⅳ) and 22 with low grade (WHO Ⅰ - Ⅱ) from January 2010 to May 2014. All of them underwent conventional MR imaging, 40 had PWI and 23 had MRSI, preoperatively. The patient's ages ranged from 20 to 83 years with a mean of 49 years and there were 29 males and 23 females.

10.2.2 MR examination protocols

All the imaging data were collected using a clinical whole body 3.0-Tesla scanner (Philips Achieva 3.0 T TX MR; Best, the Netherlands) with a phased-array 8-channel sensitivity encoding head coil. Conventional sequences (axial T1-weighted, axial T2-weighted and coronal FLAIR) were acquired first and contrast agent enhanced T1-weighted imaging in all axial, coronal and sagittal directions was performed after the acquisition of PWI data.

Two-dimensional-MRSI (2D-MRSI) data were acquired using Point-RESolved

Spectroscopy (PRESS) pulse sequence with automatic shimming and water suppression. Measurement parameters used in 2D-MRSI were 2000/144 milliseconds (TR/TE), (28×28) phase encoding steps, 8 mm section thickness, and the field-of-view (FOV) size was adjusted to each patient's brain anatomy. The positioning of the whole voxel was placed by experienced neuroradiologists on typical slices of either T2-weighted imaging or T1-weighted imaging with contrast agents. The VOI was set to cover all tumoral, peritumoral and contralateral normal regions.

10.2.3 Feature extraction

From conventional MR images, brain tumors can be characterized structurally by their locations, sizes, shapes, boundaries and their relationships with adjacent tissues which all can make contributions to grading. Clinically, the descriptions of location, shape and boundary are often subjective and ambiguous. Hence, in this research, tumor volume and mid-sagittal line displacement was applied to evaluate tumor size and mass effect respectively.

In measuring tumor volume, tumors were approximately treated as ellipses. We first located the axial T2W image that exhibits the largest size and measured the major axis a and minor axis b. Then we located the coronal T2W-FLAIR image that presents the most of the tumor to measure another length c from the head-end to the foot-end of the tumor. With these three lengths, the volume of the tumor can be calculated by the formula $V = \pi abc/6$ (Figure 10.1, see the Color Inset, p.28).

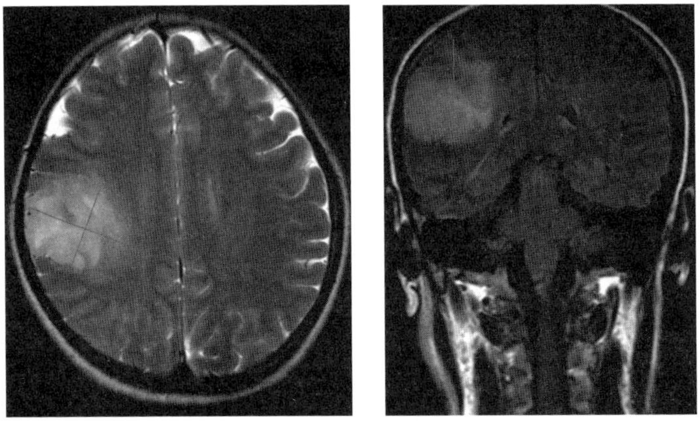

Figure 10.1 Measurement of tumor volume. Major and minor axes a and b from the T2W image (left), and head-foot length c from the T2W-FLAIR image (right), are shown in red

The extent of mass effect was measured by the maximum displacement of the mid-sagittal line. Specifically, we first located the axial T2-weighted image with the heaviest mass effect, and then drew a line that went through the anterior and posterior point of the longitudinal fissure. This line should be the mid-sagittal line before mass effect occurred. Finally, we drew another line which was perpendicular to the first line from the point with the maximum displacement and the length of the line segment (blue in Figure 10.2, see the Color Inset, p.28) was the measure of mass effect. We named the line segment as mid-sagittal line displacement (MLD).

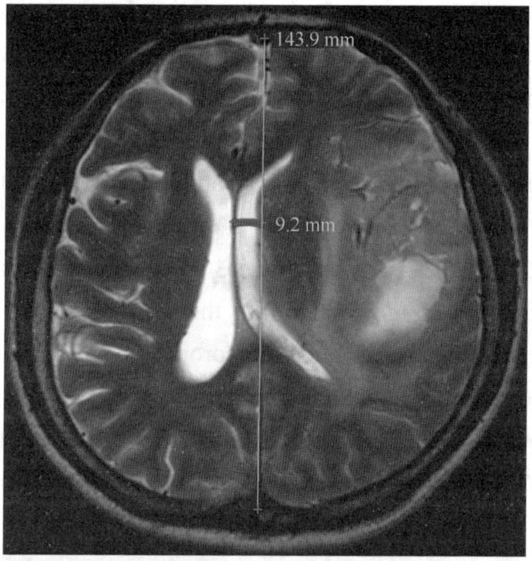

Figure 10.2 Measurement of mid-line displacement

The contrast enhancement features were extracted by comparing T2-weighted images against contrast enhanced T1-weighted images. We define three categories of contrast enhancement, namely negative contrast enhancement, minor contrast enhancement and apparent contrast enhancement. Negative contrast enhancement was recognized while minor and apparent ones were measured by signal intensities and enhancing areas. Apparent contrast enhancing was defined as the cases that either present high signal intensities or have large enhancing areas. Conversely, minor contrast enhancing is difficult to recognize, for its signal intensity is much lower and its area is small (less than 5 mm in diameter). Usually, it needs all axial, sagittal and coronal enhancing images to determine whether the contrast enhancement is minor or negative. Images below show examples of the three kinds (Figure 10.3).

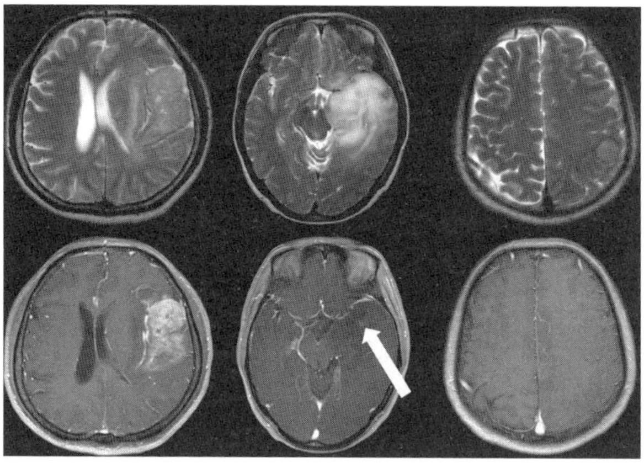

Figure 10.3 Examples of different contrast enhancement. The left column is a case of glioblastoma with apparent enhancement, the middle a case of high grade glioma with minor enhancement (white arrow) and the right a low grade glioma with no enhancement. Upper row Axial T2-weighted images. Lower row Axial T1-weighted images with contrast agents

For perfusion imaging, the following parameters are recorded clinically: regional cerebral blood volume (rCBV), mean transit time in millimeter (MTT), regional cerebral blood flow (rCBF), time of arrival of the contrast agent in the slice after injection (T0) and time corresponding to the maximum contrast variation (time to peak, TTP). Generally, studies related to perfusion imaging often use rCBV alone. As rCBV value is the product of rCBF and MTT, we used rCBF and MTT instead as a two-dimensional vector, which maybe more accurate in grading. For each case, we recorded the ratio of rCBV and MTT from tumoral region to those from contralateral normal region respectively as the perfusion feature(Figure 10.4, see the Color Inset, p.28).

Metabolites in MRSI were first quantitated using LCModel[7, 8] with the processing chemical shift ranging from 4.0 ppm to 1.0 ppm. In each voxel within the VOI, all metabolites concentrations were obtained along with their corresponding standard deviations (SD) in percentage, which are measures of confidence in the quantitated concentrations and are associated with the metabolites' true concentrations and signal-to-noise ratios in each voxel. The LCModel manual recommended that quantitated values with the SDs greater than 20% should be discarded and the ratios relative to creatine (Cr) are of more clinical importance. Cho and NAA are two major biomarkers of a tumor's aggressiveness. In both tumoral and peritumoral regions, the voxel with the highest Cho/Cr is regarded as the place where the tumor is highly active in growing and infiltration while the one with the lowest NAA/Cr is where neurons have been damaged the most. Hence, we searched for the voxel with the highest Cho/NAA with all the SDs of Cho, Cr and NAA meeting the 20% criteria. In this way, we used the maximum of Cho/NAA ratio across the ROI as the spectroscopy feature for each case (Figure 10.5, see the Color Inset, p.28).

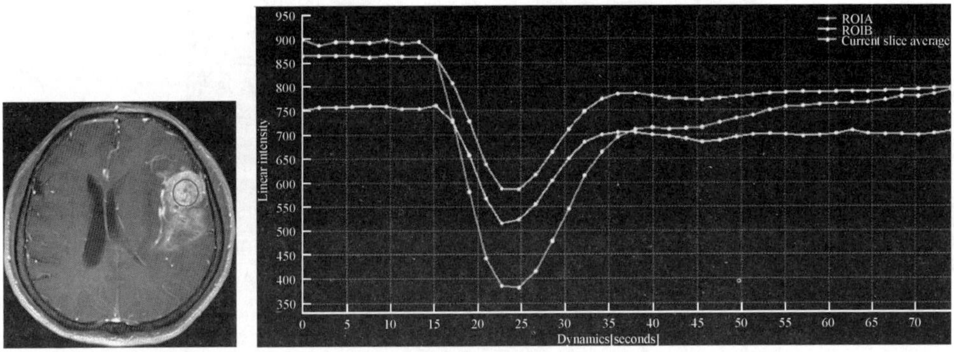

(a) (b)

Figure 10.4 (a) Tumoral ROI on the right and contralateral normal ROI; (b) Signal curves of tumoral ROI and contralateral normal ROI are in pink and white and the average signal curve in yellow

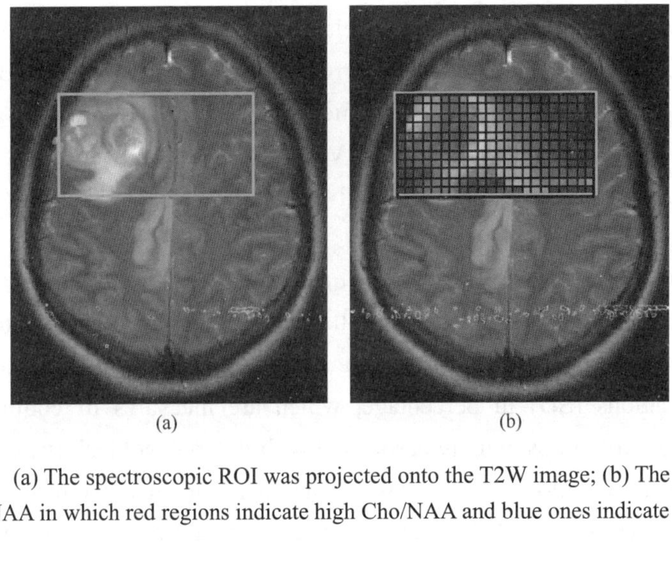

(a) (b)

Figure 10.5 (a) The spectroscopic ROI was projected onto the T2W image; (b) The pseudo-color map of Cho/NAA in which red regions indicate high Cho/NAA and blue ones indicate low Cho/NAA

10.2.4 Statistical analysis

The JMP software was used to perform Kruskal-Wallis test to see if there was statistical significance between high grade and low grade groups for each of the features extracted in the above methods. Additionally, Logistic regression analysis was performed on perfusion features.

10.2.5 Building Bayesian network for grading

There are two ways to determine the structure of a BN. One is to construct manually by domain expert, the other is to learn the structure through data with structure learning algorithms. The BN used in this study was built in cooperation of and confirmed by neuroradiologists. The network structure (Figure 10.6) and detailed description of each node or random variable in the network (Table 10.1) are shown below. Rectangles denote discrete nodes and ellipses denote continuous ones.

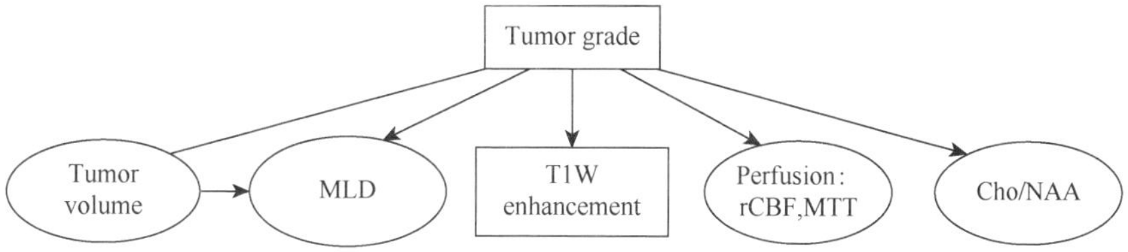

Figure 10.6 Network Structure

Table 10.1 Detailed description of nodes in the network

Node name	Node type	Values
Tumor grade	Discrete binomial	High grade=1, low grade=2
Tumor volume	Continuous Gaussian	In the unit of cm^3
MLD	Continuous Gaussian	In the unit of mm
T1W enhancement	Discrete multinomial	Apparent=1, minor=2, negative=3
Perfusion feature	2D Continuous Gaussian	2D vector (MTT ratio, rCBF ratio)
Cho/NAA	Continuous Gaussian	Scalar

Parameters of distributions for each node were learned after network construction. In the case of complete data, we can get the parameters using maximum likelihood estimation. In this study, however, there were a number of patients who did not have

PWI or MRSI scans, which made the dataset incomplete. In this situation, we used the Expectation-Maximization (EM) algorithm to iteratively estimate the parameters. In each step of iteration, the algorithm contains two steps: the E-step computes a posterior distribution for each case using a chosen BN inference engine and the M-step gets the parameters that maximize the log-likelihood[11, 17]. Parameters were randomly initialized and the algorithm is running iteratively until either the parameters converge according to the criteria or the maximal number of iteration. In this study, the convergence threshold was set to be 0.001 and maximum number of iteration to be 20.

With Bayesian network of determined structure and parameters, we can make inferences on new clinical cases. For a new case of preoperative cerebral giloma, if we observe one or more of the four features, the posterior probability of tumor grade can be computed. The grade with the highest probability was selected as the predictive grade of this case. In general, probabilistic inference can be divided into exact inference and approximate inference. Due to the relative simplicity of the network structure in this research, we used one of the most common exact inference methods of Junction Tree[11].

Considering that the dataset size was comparatively small, we adopted the leave-one-out analysis to evaluate the grading performance. In other words, we each time left one case out for testing and used the rest of data to learn the parameters.

All the work in this study was implemented in Matlab programming environment (MathWorks, Natick, MA) with the help of Bayes Net Toolbox which is a powerful Matlab programming toolbox and supports most of the tasks of network building, parameter learning and inference[18].

10.3 Results

10.3.1 Statistical analysis results

The box plots of tumor volume, MLD and Cho/NAA and the mosaic plot of T1W enhancement between high and low grade groups are shown as below (Figure 10.7, see the Color Inset, p.29). Table 10.2 lists the ranges, means, SDs and *P* values for all the features except T1W enhancement that can only take discrete values. The *P* values indicate that all the features have statistically significant difference between low grade and high grade tumors. The logistic regression analysis also showed that there exists significant difference in the perfusion features between low and high grade groups (Figure 10.8, see the Color Inset, p.29).

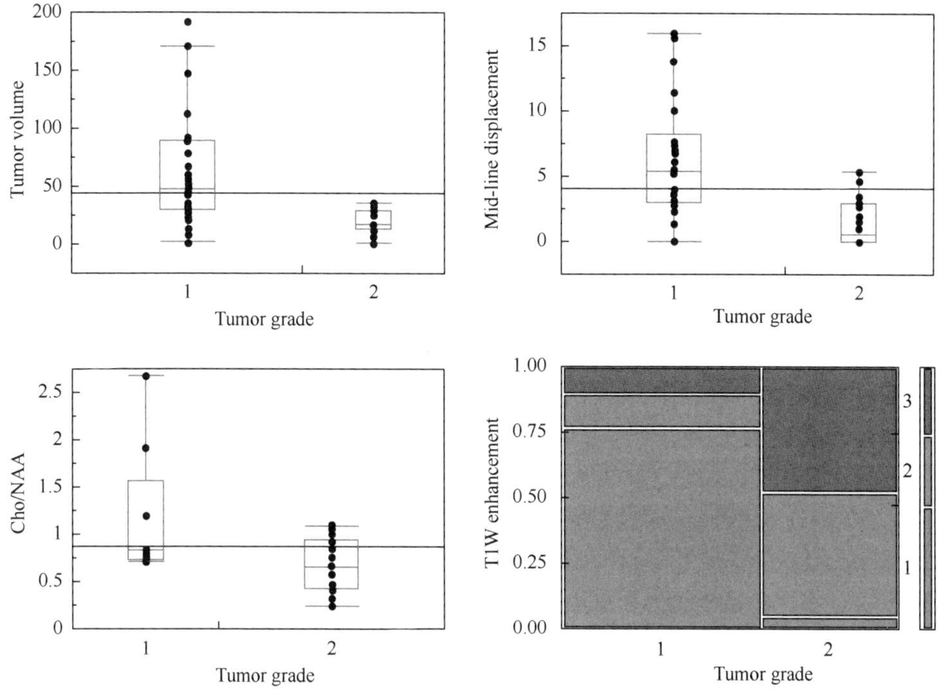

Figure 10.7 The box plots of tumor volume, MLD and Cho/NAA and the mosaic plot of contrast enhancement in terms of tumor grade

Table 10.2 Statistical analysis results

Kruskal-Wallis test Features	High grade cerebral gliomas			Low grade cerebral gliomas			*P* value
	Range	Mean	SD	Range	Mean	SD	
Tumor volume/cm^3	1.74-192.1	61.63	46.92	0.88-36.19	20.04	10.25	<0.0001
T1W enhancement							<0.0001
MLD/mm	0-16	6.00	4.36	0-5.4	1.44	1.77	<0.0001
Cho/NAA	0.70-2.69	1.20	0.68	0.24-1.09	0.67	0.28	0.0275
Logistic regression							
				Accuracy/%			*P* value
Perfusion: rCBF, MTT				82.5			<0.0001

10.3.2 Grading performance

We calculated the posterior probabilities of tumor grade given one, two, three, four and all observed features respectively and figured out the accuracies in each case with the leave-one-out analysis. In the case of one observed feature, knowing tumor volume, MLD, T1W enhancement, perfusion and Cho/NAA reach accuracies of 74.47%, 76%, 88.24%, 87.5% and 87.5%, respectively. When knowing all the features, the accuracy reaches 88.24%. The detailed information is listed below (Table 10.3).

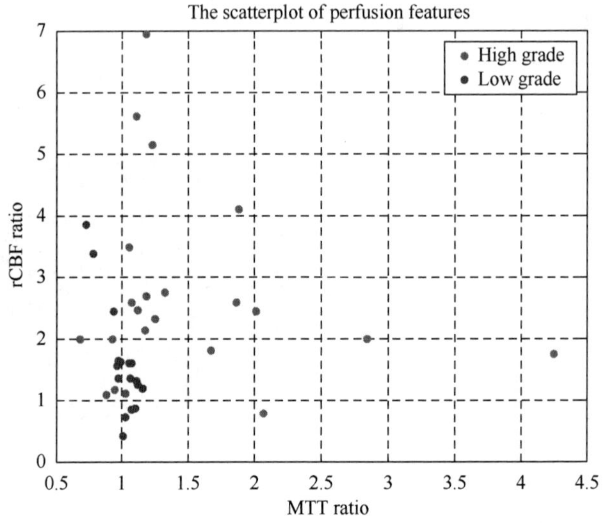

Figure 10.8 The scatterplot of perfusion features in high grade and low grade cerebral gliomas. This scattering plot shows that there is noticeable overlapping between high and low grade tumor rCBF ratio

Table 10.3 Grading accuracies when different combinations of features are observed

Observed nodes		Accuracy/%
Nodes observed	Tumor volume	74.47
	MLD	76
	T1W enhancement	88.24
	Perfusion features	87.5
	Cho/NAA	87.5
Nodes observed	Tumor volume, T1W enhancement	86.27
	MLD, perfusion features	86.27
	Others	88.24
Nodes observed	Tumor volume, MLD, perfusion features	86.27
	Others	88.24
Nodes observed	MLD, T1W enhancement, perfusion features, Cho/NAA	86.27
	Others	88.24
Nodes observed		88.24

10.4 Discussion

In this work, we measured MRI biomarkers for glioma including tumor volume, T1W enhancement, MLD and perfusion with MTT ratio, rCBF ratio in tumor parenchyma, the highest Cho/NAA within FOV in patients with high grade and low grade gliomas. And then constructed the grading system by using Bayesian Network (BN). From the grading results of different observed biomarker features, we found that overall the grading accuracies were generally higher with combined observed features. This agrees with the expectation that it is often unreliable to make a diagnosis based on only one or a few biomarkers. We also found that although there were statistically significant differences ($P<0.05$) between features of low grade and high grade gliomas, all the five biomarker varied in the ability to differentiate low grade from high grade groups and there are still cases which each individual feature cannot discriminate. Concretely, with single observed feature, the feature of T1W enhancement came the first in the grading accuracy, followed by perfusion features, Cho/NAA, MLD and tumor volume in that order.

We found that tumor volume and MLD are weaker in tumor grading and this is possibly due to following reasons. The low grade tumor is smaller than high grade tumor. However, the volume variance for high grade tumor is large. In other words, the volumes of many high grade tumors can also be small, making the data overlapped heavily with that of low grade ones. It is the same case in the data of MLD. Additionally, as tumors come in various shapes, the approximate measurement in this research can introduce errors in the data. The much higher accuracies of the other three features indicate that they can possibly present more intrinsic characteristics of brain tumors and remind us that T1W contrast enhancement, PWI and metabolite concentrations from MRSI are of great value in tumor grading.

Cho/NAA achieves comparably high grading accuracy in this study and there are two major reasons for this. One is that we used a different method to extract MR spectroscopy features. At the beginning, measuring the peak height is a way to quantitate metabolites in a MR spectrum, but it is highly subjective and can introduce unexpected errors. Generally, the spectrum of a single metabolite consists of not only one prominent peak but also some other non-prominent ones. LCModel, however, treats a spectrum as a linear combination of all the basis metabolites detectable in a certain echo time and estimates an optimal set of coefficients to fit the spectrum and minimize the residual signal[7, 8]. Another one is that the highest value of Cho/Cr generally appears in the region where cellular proliferation is most active while more neuron damaging often results in lower value of NAA/Cr. Compared to low grade ones, we may find high grade tumors growing much more rapidly with more neurons damaged, making the Cho/NAA ratio in many high grade group higher than that in low grade group. It is worth noting that the dataset of MR spectroscopy is relatively small and more data are needed to validate the grading ability of Cho/NAA.

Perfusion features used in the study is also different from most others in which only rCBV data are measured. Being the product of rCBF and MTT, rCBV can be replaced by a two-dimensional vector of rCBF and MTT which extend the grading ability of rCBV due to its higher dimension. Another logistic regression analysis on rCBV ratio alone was performed and the results (Table 10.4) validated that using rCBF and MTT instead of rCBV alone did enhance the grading ability of perfusion data. Additionally, T0 and TTP can also be added to form a higher dimensional vector to represent perfusion feature, but the performance of doing so needs to be further investigated.

Table 10.4　Grading accuracies of different perfusion features in logistic regression

Features	Accuracy/%	*P* value
rCBV ratio	77.5	<0.0001
rCBF and MTT ratio	82.5	<0.0001

The grading system based on BN has many advantages over other methods as follows. Firstly, the domain prior knowledge of medical diagnosis can be well applied in the construction of the network owing to BN's flexibility. Each type of data can be described using different distributions and encoded as random variables in the network. This gives BN the ability of integrating different types of data that other methods do not have. More importantly, the network structure represents the probabilistic casual relations and conditional dependencies in the nodes, which will make the results more interpretable than methods like artificial neuro network and support vector machine. Secondly, BN is a tool for systematic inference and data analysis under uncertainty. Law et al.[19] tend to set thresholds for features to discriminate one type from another. These thresholds, however, are somewhat subjective and cannot tell us how confident the predictive type of a new case is. When conflicts occur among different features, it is hard to give a correct prediction. The BN in this book, however, can tackle this problem neatly through calculating the posterior probability objectively regardless of the specific values of features. Thus, it not only tells clinicians which grade the tumor is but also the confidence of that prediction as well. Thirdly, it is inevitable to encounter situations where data are incomplete due to data loss or other reasons. Discarding the incomplete cases, especially some typical ones, will have the grading results adversely affected or in other words inaccurate to some extent. With EM algorithm, BN can deal with incomplete data, which is very important in clinical settings.

It is inspiring to demonstrate the advantages the BN presented above in tumor grading using real clinical data. Nevertheless, there are some limitations in this study and some future studies are needed to improve or extend the grading system. Some important and helpful features, including textural features in the image and other metabolites except Cho and NAA, are not used in the network. Neuroradiologists find that tumors with more heterogeneity in the image have more chance of being in the higher grade. Hence, further study should be sought to explore the usefulness of signal intensities and texture patterns in conventional MR images. There are some literature that extracted textural features[20-22] via co-occurrence matrix[23] or Wavelet transform[24] and achieved a good classification performance. This implies that appropriate textural features can also be added in BN to see if they can make a difference in tumor grading. In regard to MRSI, metabolites associated with tumor grade are not limited to Cho and NAA. Lipids and lactate, which are the reflection of necrosis, are also indicators of tumor grade. In future research, the methods in this study should not be limited to the grading of preoperative cerebral gliomas, but can be extended to classification involving more types of brain tumors, thus making a big step in building the expert system for brain tumors. Some literature[10], for example, investigated the performance of Naïve Bayes classification in differentiating glioblastomas mutiforme from metastases. With no doubt, it requires more clinical expertise and more features to construct networks of more complexity. In this study, we only used features from conventional MR imaging, PWI and MRSI. Future networks can include features from DTI, SWI and some other imaging sequences.

10.5 Conclusions

For clinical practice, cerebral glioma grading preoperatively is the foundation of a patient's treatment. Considering that any single feature is insufficient to make an accurate prediction of the tumor grade, we propose a Bayesian Network for tumor grading using multiple features of tumor volume, T1W enhancement, MLD, Cho/NAA and perfusion features. With single observed feature, T1W enhancement achieved the highest accuracy of 88.24% in tumor grading, followed by Cho/NAA, perfusion features, MLD and tumor volume in that order. The overall grading accuracy reaches 88.24%, higher than or at least equal to all those with single features observed, which to some degree validated that knowing more features will improve the grading performance. Future efforts should be put in adding more features and dealing with more complicated classification tasks and thus more clinical cases should be collected for further uses.

10.6 Acknowledgement

This work was supported by the National Science Foundation of China, (2014-2016, 81300925, B.Z.), the Provincial Natural Science Foundation of Jiangsu (2014-2016, BK20131085, B.Z.), and Jiangsu Province Medical Key talent people and "the 12th five years plan for China development" (2011-2-16, RC2011013, B.Z.).

References

[1] Dean B L, Drayer B P, Bird C R, et al. Gliomas: Classification with MR imaging. *Radiology*, 1990, 174 (2): 411-415.

[2] Arvinda H R, Kesavadas C, Sarma P S, et al. Retracted article: Glioma grading: sensitivity, specificity, positive and negative predictive values of diffusion and perfusion imaging. *Journal of Neuro-Oncology*, 2009, 94 (1): 87-96.

[3] Callot V, Galanaud D, le Fur Y, et al. 1H MR spectroscopy of human brain tumours: A practical approach. *European Journal of Radiology*, 2008, 67 (2): 268-274.

[4] Sibtain N A, Howe F A, Saunders D E. The clinical value of proton magnetic resonance spectroscopy in adult brain tumours. *Clinical Radiology*, 2007, 62 (2): 109-119.

[5] Posse S, Otazo R, Dager S, et al. MR spectroscopic imaging: Principles and recent advances. *Journal of Magnetic Resonance Imaging*, 2013, 37 (6): 1301-1325.

[6] Stadlbauer A, Moser E, Gruber S, et al. Improved delineation of brain tumors: An automated method for segmentation based on pathologic changes of 1H-MRSI metabolites in gliomas. *Neuroimage*, 2004, 23 (2): 454-461.

[7] Provencher S W. Estimation of metabolite concentrations from localized *in-vivo* proton NMR-spectra. *Magnetic Resonance in Medicine*, 1993, 30 (6): 672-679.

[8] Provencher S W. Automatic quantitation of localized *in vivo* 1H spectra with LCModel. *NMR in Biomedicine*, 2001, 14 (4): 260-264.

[9] Galanaud D, Nicoli F, Chinot O, et al. Noninvasive diagnostic assessment of brain tumors using combined *in vivo* MR imaging and spectroscopy. *Magnetic Resonance in Medicine*, 2006, 55 (6): 1236-1245.

[10] Tsolaki E, Svolos P, Kousi E, et al. Automated differentiation of glioblastomas from intracranial metastases using 3T MR spectroscopic and perfusion data. *International Journal of Computer Assisted Radiology and Surgery*, 2013, 8 (5): 751-761.

[11] Koller D, Friedman N. *Probabilistic Graphical Models: Principles and Techniques*. Massachusetts: MIT Press, 2009.

[12] Cruz-Ramirez N, Acosta-Mesa H G, Carrillo-Calvet H, et al. Diagnosis of breast cancer using Bayesian networks: A case study. *Computers in Biology and Medicine*, 2007, 37 (11): 1553-1564.

[13] Mani S, Valtorta M, McDermott S. Building Bayesian network models in medicine: The MENTOR experience. *Applied Intelligence*, 2005, 22 (2): 93-108.

[14] Wang X H, Zheng B, Walter F, et al. Computer-assisted diagnosis of breast cancer using a data-driven Bayesian belief network. *International Journal of Medical Informatics*, 1999, 54 (2): 115-126.

[15] Reynolds G M, Peet A C, Arvanitis T N. Generating prior probabilities for classifiers of brain tumours using belief networks. *BMC Medical Informatics and Decision Making*, 2007, 7(21): 1-6.

[16] Daly R, Shen Q, Aitken S. Learning Bayesian networks: Approaches and issues. *Knowledge Engineering Review*, 2011, 26 (2): 99-157.

[17] Seixas F L, Zadrozny B, Laks J, et al. A Bayesian network decision model for supporting the diagnosis of dementia, Alzheimer's disease and mild cognitive impairment. *Computers in Biology and Medicine*, 2014, 51: 140-158.

[18] Murphy K. The Bayes net toolbox for MATLAB. *Computing Science and Statistics*, 2001, 33: 2001.

[19] Law M, Yang S, Wang H, et al. Glioma grading: Sensitivity, specificity, and predictive values of perfusion MR imaging and proton MR spectroscopic imaging compared with conventional MR imaging. *American Journal of Neuroradiology*, 2003, 24 (10): 1989-1998.

[20] Castellano G, Bonil ha L, Li L M, et al. Texture analysis of medical images. *Clinical Radiology*, 2004, 59 (12): 1061-1069.

[21] Kassner A, Thornhill R E. Texture analysis: A review of neurologic MR imaging applications. *American Journal of Neuroradiology*, 2010, 31 (5): 809-816.

[22] Mahmoud-Ghoneim D, Toussaint G, Constans J M, et al. Three dimensional texture analysis in MRI: A preliminary evaluation in gliomas. *Magnetic Resonance Imaging*, 2003, 21 (9): 983-987.

[23] Asselin M C, O'Connor J P B, Boellaard R, et al. Quantifying heterogeneity in human tumours using MRI and PET. *European Journal of Cancer*, 2012, 48 (4): 447-455.

[24] Liu Y, Muftah M, Das T, et al. Classification of MR tumor images based on gabor wavelet analysis. *Journal of Medical and Biological Engineering*, 2012, 32 (1): 22-28.

Chapter 11 Histotripsy: Image-guided, Non-invasive Ultrasound Surgery for Cardiovascular and Cancer Therapy

Zhen Xu

11.1 Introduction

Wouldn't it be great to perform a surgery without incision or bleeding? "Histotripsy" is a non-invasive ultrasound tissue ablation technique that fractionates and removes tissue using focused, high pressure (>10 MPa), short duration (<20 μs) ultrasound pulses at low duty cycles (<1%) [1-3]. "Histo" means soft tissue in Greek, and tripsy means "breakdown". Unlike ultrasound thermal therapy that uses focused ultrasound energy to heat and kill tissue[4-6], histotripsy uses ultrasound to control cavitation for tissue fractionation. Using ultrasound pulses applied from outside the body and focused to the diseased tissue, histotripsy produces a cluster of energetic microbubbles from the endogenous nanometer sized cavitation nuclei in the target tissue. The microbubbles rapidly expand and collapse, producing very high strain and stress to surrounding cells and resulting in fractionation of the cell structures. The generation, expansion, and collapse of the microbubbles are part of a physical phenomenon termed cavitation[7-9]. With sufficiently number of ultrasound pulses, histotripsy can completely fractionate soft tissue into an acellular liquid homogenate, which can be reabsorbed by the body and resulting in effective tissue removal (Figure 11.1, see the Color Inset, p.30)[10, 11]. Histotripsy is currently being studied for many clinical applications where non-invasive tissue removal is desired including benign prostatic hyperplasia (BPH)[12], thrombolysis[13, 14], perforation of the atrial septum in the treatment of congenital heart diseases[15, 16], tumor ablation[17], kidney stones[18], and fetal interventions[19, 20]. This chapter covers the physics, image guidance, and cardiovascular and cancer applications of histotripsy.

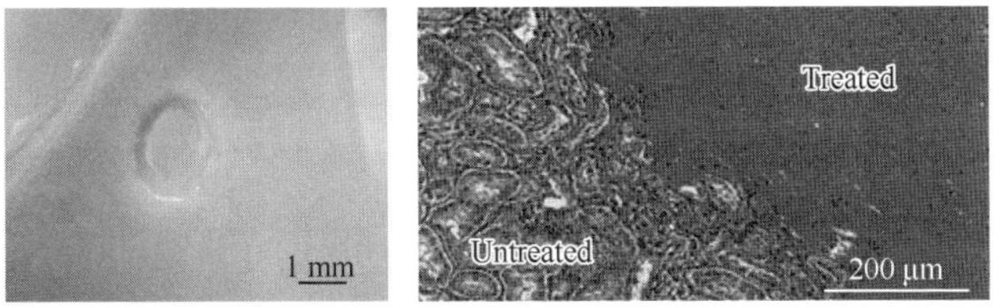

(a) (b)

Figure 11.1 (a) A perforation generated using histotripsy in the porcine atrial wall tissue; (b) Histology of histotripsy generated tissue fractionation in the porcine liver, with no cellular structures in the treated region and intact tissue outside the treated region. The left image is quoted from papers of Xu et al.[21]

11.2 Physics

11.2.1 Cavitation initiation mechanisms

Histotripsy depends on the initiation and maintenance of a dense cavitation bubble cloud to produce mechanical tissue fractionation[8, 9, 11]. Histotripsy utilizes endogenous cavitation nuclei in the tissue to generate cavitation, thus no external agents are needed. There are two mechanisms in which a histotripsy cavitation cloud can be initiated.

In the first mechanism, termed the "shock scattering mechanism", a dense bubble cloud is formed from a single multi-cycle histotripsy pulse (e.g., 3-20 cycles) using shock scattering from single sparse bubbles formed and expanded during the initial cycles of the pulse[8]. In this process, initial single sparse bubbles, which we term as "incidental bubbles" are potentially formed from large heterogeneous nuclei in the focus in the target tissue or due to growing of cavitation nuclei over multiple cycles. These incidental bubbles are formed significantly below the intrinsic threshold, which is termed as the threshold to generate a dense cavitation cloud directly from the incident negative pressure phase of a single cycle pulse. These incidental bubbles act as pressure release surfaces wherein the following positive pressure shock fronts are inverted and superimposed on the incident negative pressure phase to form negative pressures that produce a dense cavitation cloud growing back toward the transducer (Figure 11.2). Using the shock scattering mechanism, bubble clouds are initiated at negative pressures ranging from 10-28 MPa. Because of the complexity of the shock scattering process, multiple factors determine whether a cloud initiated, including the distribution of heterogeneous nuclei in the focal region, the size and shape of initial single bubbles, the number of cycles, the shock rise time, the positive pressure amplitudes, and the tissue stiffness[8, 22].

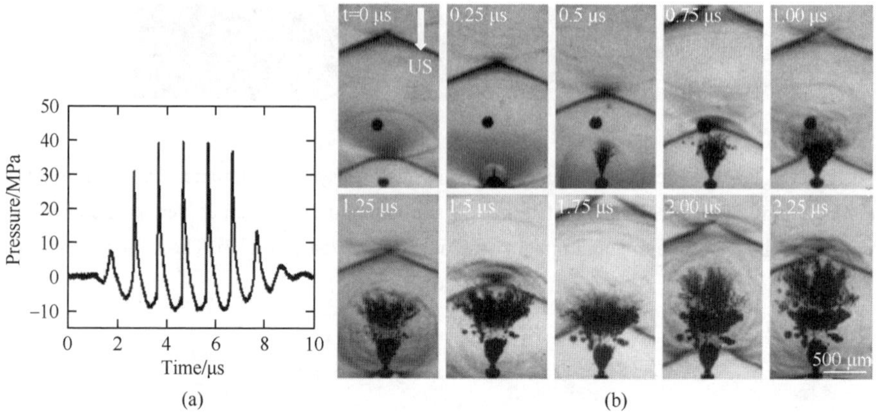

Figure 11.2 Ultrasound waveform of a 5-cycle histotripsy pulse at 1 MHz (a) and the high speed images of a bubble cloud initiated via the shockscattering mechanism (b). Ultrasound propagation is from top to bottom. At t=0, a single bubble is present at the bottom of the frame. After the shock impinges on the bubble (t=0.25 μs), a spherical wave is visible, apparently scattered by the bubble. Over the next cycle, a cloud of bubbles stems from the center of the single bubble behind this scattered wave. A second cycle produces another section of the cloud. These images are quoted from papers of Maxwell et al.[8]

The second mechanism for histotripsy cavitation cloud formation is termed the "intrinsic threshold mechanism", in which a 1-2 cycle pulse with a single dominant negative pressure phase is used to generate a bubble cloud directly from the negative pressure of the incident wave[9, 23]. In the previous study by Maxwell et al.[9], an intrinsic threshold of approximately 26-30 MPa was observed for water based soft tissues and tissue phantoms, while the threshold for tissue composed primarily of lipids was significantly lower (15.4 MPa for adipose tissue). With these short pulses, cavitation initiation depends solely on the negative pressure when it exceeds a distinct threshold intrinsic to the medium, without the contributions from shock scattering (Figure 11.3, see the Color Inset, p.30). Using this intrinsic threshold mechanism and a very short acoustic pulse (<2 cycles) to prevent shock-scattering, the volume of the histotripsy-induced cavitation cloud and lesion corresponded well to the volume of the focal regions above the intrinsic cavitation threshold. Because the supra-threshold portion of the negative half cycle can be precisely controlled, lesions considerably smaller than a wavelength are easily produced, therefore the intrinsic threshold approach is also termed as Microtripsy (Figure 11.4, see the Color Inset, p.30)[23].

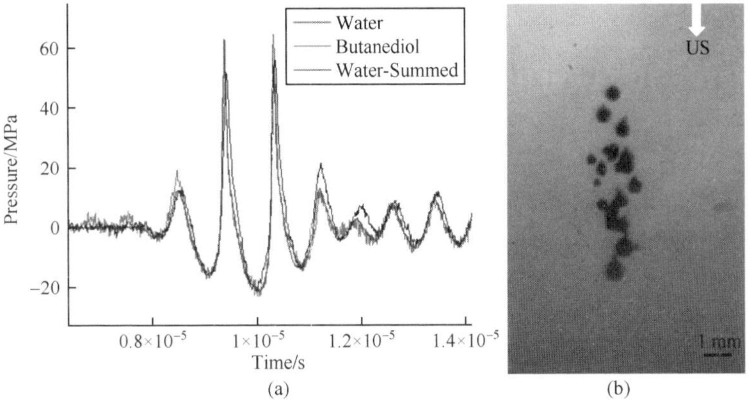

Figure 11.3 (a) Ultrasound waveform of a 2-cycle histotripsy pulse at 1 MHz and (b) the high speed images of a bubble cloud initiated via the intrinsic threshold mechanism. Ultrasound propagation is from top to bottom. The bubble cloud is generated along the ultrasound propagation direction when the negative pressure exceeds the intrinsic threshold directly. At the high amplitude, the ultrasound waveform could not be directly measured in the water due to instantaneous cavitation. The waveform was measured in Butanediol (cavitation suppression fluid) and also estimated by linear summation of the waveform from subapertures. This figure is referenced from papers of Maxwell et al.[9]

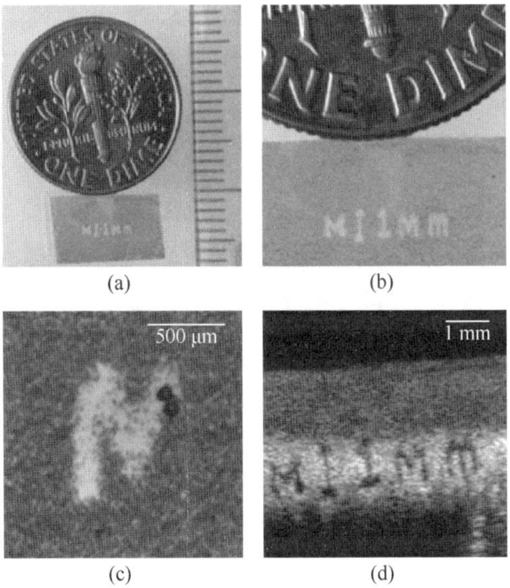

Figure 11.4 A representative lesion in RBC phantom generated by a 3-MHz transducer using microtripsy. The lesion reads "M 'vertical scale bar' 1 mm" placed below a US dime. (a) An overlook of the lesion along with a US dime and a ruler with millimeter tick marks; (b) A magnified view of the lesion; (c) A high-speed image showing the bubble cloud generated during treatment; (d) An ultrasound B-mode image of the lesion after treatment. The focal zone of the 3-MHz transducer is 1.4 mm×0.5 mm×0.5 mm. This figure is quoted from papers of Lin et al.[23]

11.2.2 Tissue fractionation mechanism

Using one of the above mechanisms, histotripsy can generate a millimeter-sized cloud of microbubbles in the target tissue using ultrasound pulses applied outside the body. The cavitation nuclei rapidly grow from nanometer size to microbubbles exceeding 50 μm in diameter to induce high stain and high stress to adjacent cells, which results in fractionation of the cells in the target tissue (Figure 11.5)[24, 25].

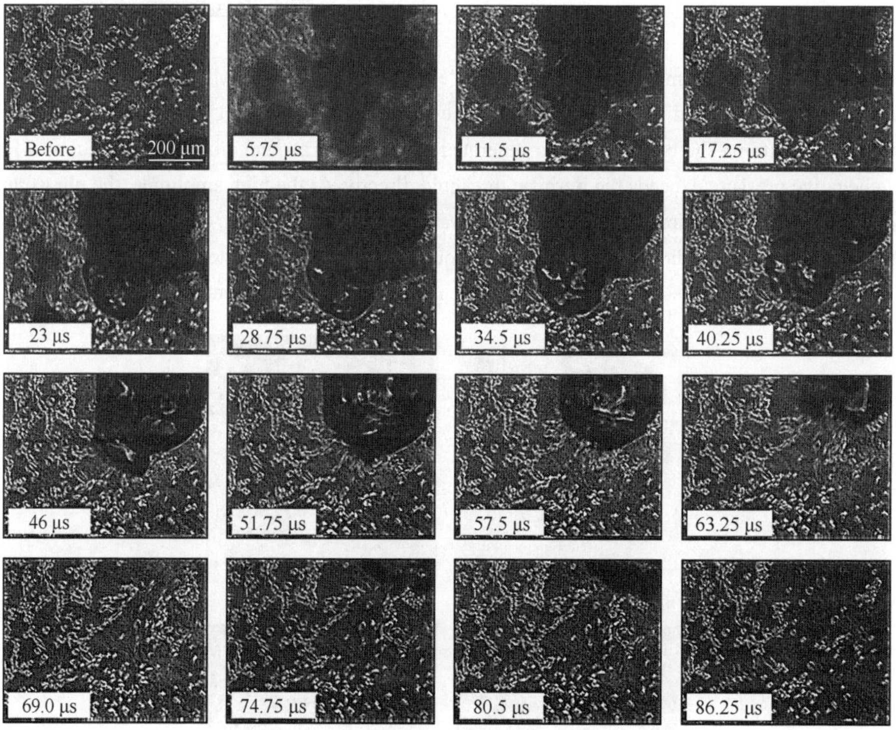

Figure 11.5 Cancer cells exposed to a single histotripsy pulse were significantly deformed during both bubble expansion and collapse. The dark void at the top right corner is a cavitating bubble cloud

Each histotripsy pulse fractionates a fraction of the tissue matrix and cellular structure. Increasing the number of pulses increases the level of fractionation in the tissue (Figure 11.6, see the Color Inset, p.30)[26, 27]. With sufficient number of pulses, histotripsy eventually completely fractionate the target tissue into liquid acellular homogenate[1]. TEM micrographs of the tissue treated by histotripsy show no recognizable cellular features and little recognizable subcellular structures in the histotripsy-treated tissue regions (Figure 11.7, see the Color Inset, p.31)[28]. A boundary, or transition zone, of a few microns separated the affected and unaffected areas. The size of the acellular homogenate debris particles within the treated region were measured with>99.9% smaller than 6 μm, and no particles greater than 100 μm[29]. Demonstrated in the *in vivo* kidney and liver, the acellular homogenate resulted from histotripsy treatment is reabsorbed by the body entirely within a month, leaving minimal fibrous tissue at the treatment site[10, 20].

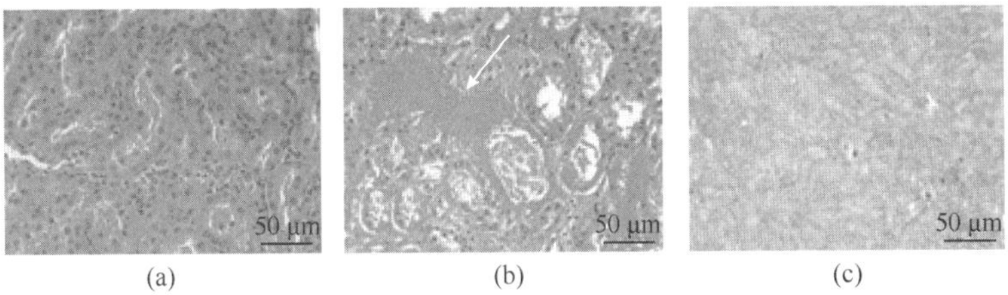

Figure 11.6 Histotripsy fractionated porcine kidney tissue *in vitro* (a: control), creating partial fractionation in the tissue structure (b: arrow). Increasing doses created higher level of fractionation until complete homogenate (c). This figure is quoted from papers of Wang et al.[26]

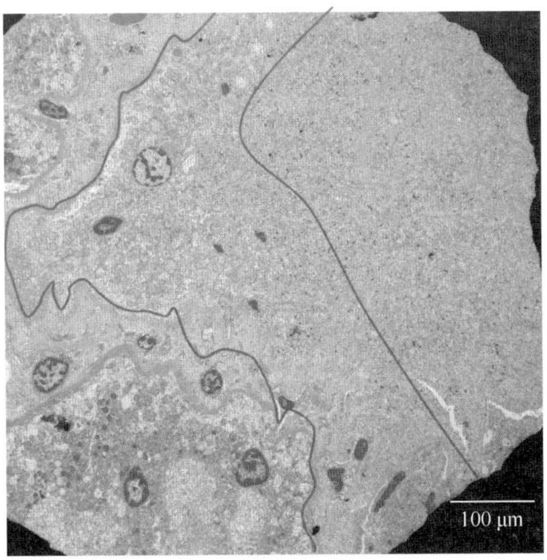

Figure 11.7 TEM micrograph of the kidney cortex treated by histotripsy at 1450X magnification. The red lines represent the approximate boundary between the treated and untreated regions and a transition in the morphology of the subcellular material is clearly evident: there is a clear absence of membranes and other subcellular features in the transition zone. Nuclei appear pyknotic and there is less presence of mitochondria (the two more robust organelles). To the right of the transition zone is the treated region, with less recognizable subcellular features. To the left, the cellular material is more clearly defined. This figure is quoted from papers of Winterroth et al.[28]

11.2.3 Effects of tissue stiffness on histotripsy process

Histotripsy has been demonstrated to successfully fractionate many different tissues, though stiffer tissues such as cartilage or tendon (Young's moduli > 1 MPa) are more resistant to histotripsy-induced damage than softer tissues such as liver (Young's moduli ~9 kPa)[30]. The effects of tissue mechanical properties on various aspects of the histotripsy process was investigated, including the pressure threshold to generate cavitation, the bubble dynamics, and the stress-strain applied to tissue structures[25]. We found that the intrinsic threshold to initiate cavitation microbubbles is independent of tissue stiffness while the bubble expansion is suppressed in stiffer tissues, leading to a decrease in strain to surrounding tissue and an increase in damage resistance. Based on these results, a histotripsy selective tissue ablation approach has been developed, in which a target tissue (i.e. liver, tumor) is completely fractionated while stiffer vital tissues within the focal region (i.e. blood vessels, bile ducts) are preserved[31]. This strategy has been tested effective in *in vivo* animal models and is detailed in Section 11.6.

11.3 Image Guidance

For a non-invasive tissue ablation technique, the operating clinician cannot directly see the target and operation progress. An image guidance that can target, monitor the treatment progress, and evaluate the treatment completion in real-time would be extremely valuable and essential to the treatment accuracy and efficacy of the non-invasive ablation technique. Ultrasound imaging provides such guidance for the histotripsy therapy. Typically an ultrasound imaging probe is placed co-axially with the therapy transducer such that the imaging plane can visualize the therapy plane.

Histotripsy generates and uses a cavitation bubble cloud as its effective scalpel for non-invasive surgery. As microbubbles are natural contrast agents for ultrasound imaging, the cavitation cloud can be visualized on ultrasound imaging, which is used to guide pre-treatment targeting and treatment monitoring during the treatment. For a typical histotripsy experiment, prior to the treatment, histotripsy pulses are applied to an empty water bath to generate a cavitation cloud, which shows as a hyperechoic (bright) zone on the ultrasound image and marked as the therapy focus [Figure 11.8 (a)][16, 32]. The therapy transducer is then placed on the subject and moved by a motorized positioning system until the therapy focus marker is aligned with the target.

To ensure precise targeting, a few pulses are applied to confirm the hyperechoic cavitation zone is on the target tissue. After targeting confirmation, histotripsy treatment is applied to the target tissue. During the treatment, a temporally changing hyperechoic (twinkling) zone is a good indication that the treatment is progressing normally [Figure 11.8 (b) and Figure 11.8 (c)][16, 32].

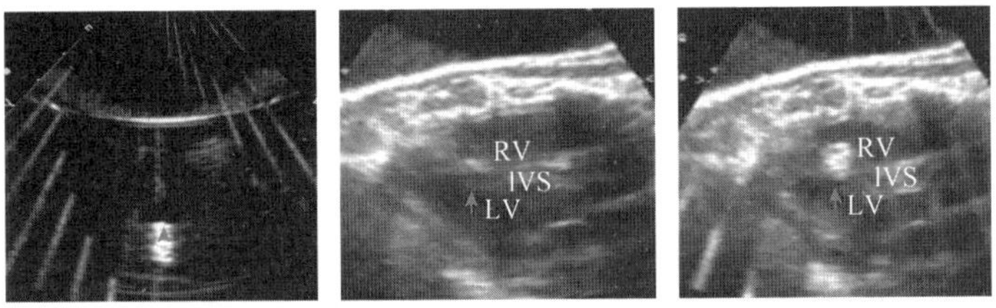

(a) (b) (c)

Figure 11.8 (a) Prior to treatment, a cavitation bubble cloud is formed in degassed water and shown on ultrasound image as a hyperechoic zone to mark the therapy focus point, seen by inserted arrow. (b) The therapy transducer is then moved to position the marked point over the target tissue, in this case, the interventricular septum (IVS) in the heart of a neonatal pig. Histotripsy was delivered from outside the chest wall from the top to the bottom. (c) Histotripsy is started and a hyperechoic bubble cloud is formed on the IVS. RV (right ventricle); LV (left ventricle). This figure is quoted from a papers of Owens et al.[32]

After treatment, as histotripsy physically breaks up tissue structures, the resulted tissue fractionation displays as a hypoechoic zone (dark zone) on ultrasound imaging, due to the reduced number and size of effective sound scatters within the treatment region (Figure 11.9)[33]. Our previous study showed that the normalized echogenicity (i.e., backscatter intensity) reduction was linearly correlated with the reduction of the percentage of normal appearing cell nuclei in the treated region[26]. This suggests that the backscatter intensity change in the treatment region on ultrasound image can be used to evaluate the level of tissue fractionation generated by histotripsy. Ultrasound elastography has also been used to monitor histotripsy-induced tissue fractionation, as tissue becomes increasingly soft with increasing level of fractionation[34, 35]. The ultrasound elastography approach provides higher sensitivity than the backscatter intensity reduction method. It can detect fractionation at an early stage, while substantial tissue fractionation needs to be generated for the lesion to be detectable through backscatter intensity reduction method. Although ultrasound elastogrhapy is difficult to implement in real-time during the treatment due to the interference from cavitation and the processing time required. Recently, we have developed a new ultrasound imaging based technique to monitor histotripsy-induced tissue fractionation, termed bubble-induced color Doppler (BCD)[36]. BCD detects the motion of the cavitation residual nuclei in the treatment region. As the tissue is increasingly fractionated and completely homogenized, the temporal profile of this motion expands and saturates. This trend persists regardless of tissue type and mechanical properties of the tissue. In addition to ultrasound feedback, histotripsy-generated cavitation and lesions show clearly on magnetic resonant imaging (MRI)[37]. Even though currently MRI is slower and more expensive than ultrasound imaging, it can be a great alternative in conditions where ultrasound imaging may not be suitable, e.g., tumor and brain applications.

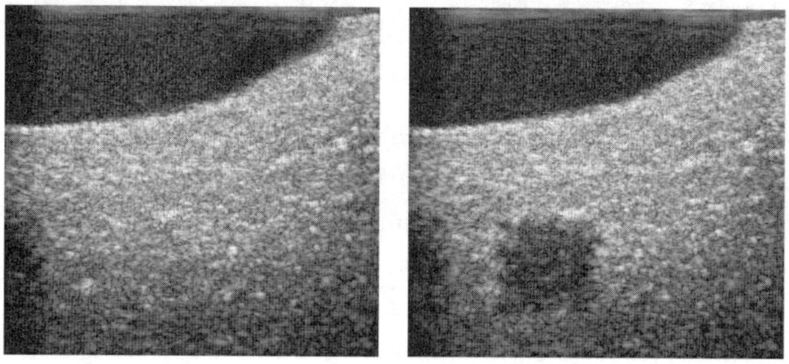

(a) (b)

Figure 11.9 Ultrasound images of porcine liver tissue before (a) and after (b) histotripsy treatment. The histotripsy-generated lesion shows as a hypoechoic (dark) zone on ultrasound image. This figure is quoted from papers of Hall et al.[30]

11.4 Congenital Heart Disease Application

11.4.1 Introduction

Hypoplastic Left Heart Syndrome (HLHS) is a rare and complex congenital heart disease[38]. Infants born with Hypoplastic Left Heart Syndrome (HLHS) have very small or absent left ventricles which are unable to pump enough blood to sustain life. If left untreated mortality approaches 95% by the first month of life[39]. The only successful treatment, other than a heart transplant, is reconstructive heart surgery performed in 3 stages through the first 3 years of life. To extend the patients' survival until the reconstructive surgery and optimize their physiological situation, a flow channel between through the atrial septum (a membrane separating the left and right atria) is needed to allow the only functioning ventricle (right ventricle) to provide the blood circulation for the entire body. Clinically, ASD is created by variations of catheter-based septostomies, which carries morbidity and mortality rates as high as 50%. Thus, an alternative non-invasive method such as high intensity focused ultrasound (HIFU) for creation of ASDs in this patient population may prove a beneficial innovation.

HIFU can ablate internal organs non-invasively from an extracorporeal position[40]. For cardiac applications, researchers have investigated the feasibility of using extracardiac HIFU to ablate cardiac tissue to treat arrhythmias[41-43]. Lesions were created inside the beating heart without damaging the intervening heart wall. The main mechanism of these HIFU studies is that continued or long ultrasound pulses can cause heating in the target tissue which results in thermal necrosis. However, ASD creation requires tissue removal that does not occur with HIFU thermal therapy.

Clearly demarcated perforations have previously been created by histotripsy in excised porcine cardiac tissue[21]. In human neonates, a subcostal acoustic window to access the atrial septum exists without bone and lung obstruction[16]. However, in quadruped animal models, the acoustic and thus therapeutic window for the atrial septum is suboptimal due to the different orientation of the heart and position within the thorax. Since there is a workable acoustic window to access the ventricular septum in neonatal pigs, the *in vivo* feasibility and safety of using histotripsy for ASD creation has been evaluated using ASD creation in an open chest canine model[44] and perforation of the ventricular septum (i.e. ventricular septal defect, VSD) in an intact neonatal porcine model[32, 45]. The histotripsy ASD or VSD treatment was guided by ultrasound imaging in real-time. In the intact neonatal porcine model, the treated pig were survived for a month and monitored for any clinically significant complications. These *in vivo* studies have been previous reported[16, 32, 45] and summarized in the following.

11.4.2 Methods

1. Canine open chest model

All animal procedures described in this chapter were reviewed and approved by the University Committee on Use and Care of Animals at the University of Michigan. In ten canines, a sternotomy incision provided direct access to the heart. The atrial septum was exposed to histotripsy by an ultrasound transducer positioned outside the heart. Ultrasound pulses of 6 μs duration at a peak negative pressure of 15 MPa, a peak negative pressure of 61 MPa, and a pulse repetition frequency of 3.3 kHz were generated by a 1 MHz focused transducer. The therapy transducer has a geometric focal length of 90 mm, an outer diameter of 100 mm, and a 40 mm inner hole (Imasonic, S.A., Besançon, France). The procedure was guided and monitored by a 5 MHz phased array ultrasound imaging probe (GE VingMed System FiVe, GE, United Kingdom) inserted into the central hole of the therapy transducer. Prior to the procedure, histotripsy pulses were applied to a water bath, resulting in a bright (hyperechoic) cavitation zone on a two-dimensional ultrasound image and marked as the focal position. The therapeutic transducer was then moved by a 3-axis positioning system to align the focus marker on the atrial septum surface in the right atrium. Histotripsy ultrasound exposures of 2 min were applied to the atrial septum at one time. Repetitive 2 min exposures were delivered to the atrial septum until an ASD was generated, which was identified as blood flow through the atrial septum detected using Doppler color flow.

2. Intact porcine neonatal model

A total of fifteen 3-5 kg neonatal pigs were utilized and histotripsy was applied to the ventricular septa. VSDs were created because of the poor acoustic window to the atrial septum in this neonatal model. During the histotripsy procedure, the pig was submerged neck deep in a semi-upright position in degassed water. Histotripsy treatment was delivered by the same 1 MHz ultrasound transducer used in the canine study. Ultrasound imaging guidance was provided by an 8 MHz phased array imaging probe (S8, SONOS 7500, Philips Healthcare, Andover, MA). The peak negative and positive pressures used in the following experiment were 16 MPa and 32 MPa, respectively. Taking into account the attenuation caused by the approximately 1.5 cm thick chest wall tissue in the pathway, the peak negative pressure reaching the ventricular septum was estimated to be 13 MPa. The ultrasound pulses were 5 μs in duration (5 cycles at 1 MHz) and separated by 1 ms (i.e., pulse repetition frequency=1 kHz). The targeting and treatment procedure were carried out following the same protocol used in the canine study with the ventricular septum as the target instead. Six animals were sacrificed immediately, three at 2-3 days post procedure, and six animals survived for one month. Brain MRIs were performed on long-term survivors to evaluate any ischemia brain injury associated with potential thrombo-embolic events produced by histotripsy. After euthanasia, the heart and lungs were extracted for pathological examination to evaluate the VSD creation, damage to surrounding and intervening tissue, and any potential hazardous emboli lodged in the lung.

11.4.3 Results

1. Canine open chest model

In nine of ten canines, an ASD was successfully generated in 6-16 min of histotripsy treatment. During the procedure, generally a ~2-4 mm wide bubble cloud was created at the focus on the atrial septum. After treatment, shunting across the atrial septum was clearly visualized using ultrasound imaging and color flow Doppler. The width of the color jet through the ASD was measured on Doppler images [(3.4±0.9) mm]. Gross morphology shows that the location of the ASD was consistent with that indicated on ultrasound image. The diameter of the dark colorization area (including the ASD and surrounding hemorrhage) was larger on the right atrial side (ultrasound entrance side) and smaller on the left atrial side. Pathology of the hearts showed ASDs all the way through the atrial septum with acellular tissue debris and red blood cells in the central damage zone. The width of the acellular zone measured on H&E cross-section slides were (5.0±1.1) mm on the right atrial surface (ultrasound entrance side) and (2.6±0.4) mm on the left. The hemorrhage zone surrounding the acellular zone was (2.6±1.1) mm on the right atrial surface and (2.2±0.3) mm on the left. There was no damage was found on the epicardial surface of the heart or other structures. All animals survived the immediate procedure and there were no complications such as pericardial effusions or sustained arrhythmias.

2. Intact neonatal pig model

Histotripsy therapy propagated through multiple tissues layers including skin, bone, and portions of lung, and created a VSD in 15 animals within 20 seconds to 13 minutes of therapy (mean 3.1 minutes, ±3.2). VSDs confirmed by color flow Doppler (Figure 11.10, see the Color Inset, p.31) ranged in size from 2 mm to 6.5 mm (mean 3.7 mm, SD±1.6). All VSDs persisted until scheduled euthanasia except for two initially small VSDs (2 mm and 2.5 mm) in the intermediate-term group, which spontaneously closed by one month. In all fifteen neonatal pigs studied, gross morphology showed no damage to intervening tissues and the epicardial surfaces appeared intact. Cardiac dissection revealed demarcated lesion in the ventricular septum in all animals. Flanking injury and hemorrhage seen acutely were resolved by one month with tissue remodeling present. In acute animals (sacrificed within two hours) the VSD was filled with acellular debris and platelet/fibrin deposits with flanking hemorrhage and myocyte injury spanning approximately (3.3±1.1) mm from the VSD border [Figure 11.11 (a), see the Color Inset, p.31]. By 2-3 days (sub-acute), the lumen of the VSD was clear and the extent of flanking injury was less [(1.4±1.2) mm] with the infiltration of inflammatory/remodeling cells such as fibroblasts and macrophages [Figure 11.11 (b), see the Color Inset, p.31]. By one month (intermediate-term) the VSD was completely endothelized and surrounded by fibroblastic scar tissue immediately adjacent to the defect and then flanked by normal myocardium [Figure 11.11 (c), see the Color Inset, p.31]. In the two animals where the VSD spontaneously closed, fibrotic tissue was visualized at the previous defect without surrounding injury. There were no fatalities and all recovered animals thrived with normal weight gain. Brain MRI and lung pathology revealed no evidence of thrombo-embolic events. No damage to intervening tissue was seen.

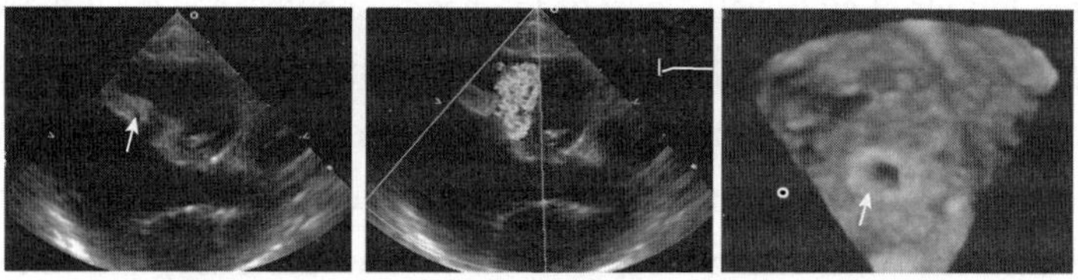

(a)　(b)　(c)

Figure 11.10　(a), (b) Parasternal long axis view of VSD 2 days after creation in the intact neonatal porcine model in 2D (arrow indicates VSD location) and with color flow Doppler; (c) 3-dimensional representation of same VSD shown in a modified apical four chamber view showing persistence of the VSD. Note the hyperechoic (bright) rim surrounding the VSD marked by white arrows. This figure is quoted from papers of Owens et al.[45]

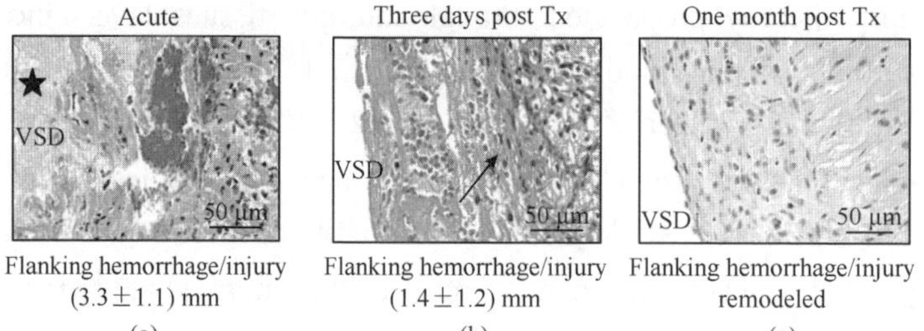

Figure 11.11　Histology of histotripsy-created VSDs in the intact neonatal porcine model. Acutely (a) the VSD is consumed with acellular fractionated fibrin debris (star) with flanking hemorrhage and myocyte injury. By three days (b) the VSD proper is clear and cellular remodeling has begun with the infiltration of fibroblasts (arrow) and an improvement in flanking hemorrhage. After 1 month (c) an endothelial border exists surrounded by fibroblastic scar tissue immediately adjacent to normal appearing myocardium. This figure is quoted from papers of Owens et al.[45]

11.4.4 Discussion

The ability to create or augment ASDs within the heart of an infant born with congenital heart disease such as HLHS, avoiding open-heart surgery or invasive cardiac catheterization, could significantly advance the field of pediatric cardiology. The *in vivo* safety and feasibility to create ASDs in the heart non-invasively using histotripsy was investigated in an open chest canine model and an intact neonatal porcine model. In nine of ten canines, ASDs were successfully created in an acute study. Due to the heart orientation difference between human and four-legged animals, while the acoustic window to the atrial septum exists in human neonates without bone and very little lung in the pathway, the acoustic window to access the atrial septum in neonatal pigs is poor with significant blockage of rib and lung. Instead, the acoustic window to access the ventricular septum in the neonatal pigs is closer to the acoustic window to reach the atrial septum in the human neonates. Therefore, using the intact neonatal porcine model, the ventricular septum was targeted. In all fifteen pigs, VSDs were successfully created, and all animals survived the procedure. The six animals sacrificed after one month had no clinical symptoms and gained weight appropriately. In addition, ventricular systolic function was preserved and no evidence of thrombo-embolic events was observed based on lung pathology and MRI of the brain. The lack of embolic events is consistent with *in vitro* data revealing that most particles created by histotripsy induced tissue fractionation are less than six microns (size of a red blood cell) in size[46]. These data suggest that histotripsy cardiac therapy is safe and that effects are limited to the targeted region, similar to other reports using this technology in different regions of the body[10, 14, 27, 47].

In both the open-chest canine and intact neonatal porcine models, the tissue injury zone surrounding the perforations was small (around 3 mm), but the flanking injury zone was larger in the adult canine model, partly due to the higher motion of the atrial septum during the treatment in the adult dogs. In the canine model, the movement of atrial septum along the axial ultrasound beam was (8.1 ± 3.3) mm. There was also movement along the lateral beam estimated up to 4-5 mm. In comparison, the axial and lateral movement of human HLHS patients was measured to be approximately 1-2 mm. With less movement of the atrial septum in human neonates, it is expected that less flanking injury would be produced. At the same time, the critical structures in the neonatal heart surrounding the atrial septum are in closer proximity compared to the atrial septum in the adult canine heart or the ventricular septum in the neonatal porcine heart, and higher accuracy will be required. Similar to attempts using other forms of therapeutic ultrasound[48-50], innovative efforts to incorporate ultrafast motion-tracking[51] and to incorporate real-time color Doppler feedback are underway to further increase targeting and therapy accuracy and decrease flanking injury. These advances in combination with an optimized neonatal specific therapy transducer and other components leading to an integrated histotripsy cardiac system will allow approval from the Food and Drug Administrating enabling the initiation of an inaugural clinical trial of histotripsy for HLHS.

11.4.5 Conclusions

Under real-time ultrasound guidance, ASDs were created using extracardiac histotripsy in a live canine model, and VSDs were created using transcutaneous histotripsy in an intact neonatal pig model. Results demonstrated histotripsy as a safe and effective technique to create ASDs and VSDs, without acute and intermediate-term untoward systemic or clinical effects. These studies suggest that histotripsy has potential to become a novel, non-invasive clinical resource for not only ASD creation in treatment of HLHS and other conditions requiring ASD such as arterial-venous extracorporeal membrane oxygenation.

11.5 Thrombolysis Application

11.5.1 Introduction

Thrombosis is the medical term for the process of pathologic blood clot formation, the key mechanism behind many cardiovascular diseases. For example, deep vein thrombosis (DVT), generally described as clot formation in the deep veins in the legs, has an incidence rate of 1 in 1000 persons. DVT affects two million people[52, 53] and causes at least 100000 deaths annually in the United States alone[54].

Current treatment methods for thrombosis include thrombolytic drugs and catheter-based surgical procedures. Thrombolytic drugs (e.g, rt-PA) dissolve the blood clot by breaking down the cross-linked fibrin structures that solidify the clot[55]. As thrombolytic drugs are systemically administered and have a slow perfusion rate, this method is often ineffective, has a long treatment time, and can cause excessive bleeding, which may be fatal in a small number of cases[56]. Catheter-based procedures are more effective than thrombolytic drugs, but they are invasive and carry an increased risk of bleeding, vessel damage, and infection[57]. Ultrasound can be combined with thrombolytic drugs and/or contrast agents to accelerate thrombolysis by enhancing the delivery of drugs into the clot[58-63]. A multi-center retrospective evaluation of catheter-based ultrasound transducer (EKOS) showed that ultrasound combined with t-PA reduced infusion time and provided a greater incidence of complete clot lysis for DVT treatment[64, 65]. Despite these early signs of success, the thrombolysis rate is still quite slow, and this treatment carries the similar adverse effects of thrombolytic drugs. To further shorten the treatment time and eliminate the need for thrombolytic drugs, some researchers have started studying thrombolysis via inertial cavitation using ultrasound alone[66-69] or combined with contrast agents[70, 71].

Histotripsy can result in mechanical breakdown of the thrombus into acellular debris. Our study shows that clot lysis and blood flow restoration can be achieved using histotripsy at a speed an order of magnitude faster than any current methods, without vessel penetration or hazardous embolization[14, 72]. By eliminating thrombolytic drugs and catheters and achieving effective clot removal non-invasively with significantly reduced treatment time, histotripsy has the potential to substantially improve upon the standard of care for thrombosis treatment.

We have investigated the *in vitro* and *in vivo* feasibility of using histotripsy for non-invasive thrombolysis[13, 14, 72, 73]. First, the *in vitro* feasibility and efficacy of histotripsy thrombolysis was investigated, and cavitation was monitored and correlated to treatment results[13]. Second, the *in vivo* feasibility of histotripsy thrombolysis was studied in a porcine deep vein thrombosis model[14]. Third, we found that a cavitation cloud generated in a vessel can trap and simultaneously fractionate a clot particle (embolus) flowing near the cavitation zone[72, 73]. This trapping capability was used to develop a Non-invasive Embolus Trap (NET) to prevent embolization caused by escaping clot fragments.

11.5.2 Methods

First, to study the *in vitro* feasibility of histotripsy thrombolysis, *in-vitro* blood clots formed from fresh canine blood and $CaCl_2$ were treated by histotripsy[13]. The treatment was applied using a focused 1-MHz transducer, with 5-cycle pulses at a pulse repetition rate of 1 kHz. Acoustic pressures varying from 2-12 MPa peak negative pressure were tested. Acoustic backscatter from the cavitating bubble cloud was passively received using a 5-MHz focused single-element transducer confocally aligned with the therapy transducer. The initiation of cavitation cloud was detected as the initiation of the temporally changing scattered wave using the method detailed in our previous study[2]. A 5-MHz ultrasound imager (System Five, General Electric, USA) was used for targeting the clot and monitoring treatment progress, where the cavitation cloud was viewed as a hyperechoic zone on ultrasound imaging. The fractionated clot fragments from histotripys treatment were serially filtered through 1 mm, 100 μm, 20 μm, and 5 μm filters after treatment to measure the total weight of particles in each size category.

Second, to study the *in vivo* feasibility of histotripsy thrombolysis, acute thrombi were formed in the femoral vein of juvenile pigs weighing 30-40 kg by balloon occlusion with two catheters and thrombin infusion[14]. A 10-cm diameter 1-MHz focused transducer was used for therapy. An 8 MHz ultrasound imager was used to align the clot with the therapy focus. Therapy consisted of 5 cycle pulses delivered at a rate of 1 kHz and peak negative pressure between 14-19 MPa. The focus was scanned along the long axis of the vessel to treat the entire visible clot during ultrasound exposure. The targeted region identified by a hyperechoic cavitation bubble cloud was visualized via ultrasound during treatment. The treated vessel lumen was monitored using ultrasound and color Doppler to evaluate the clot removal and blood flow restoration. Histological samples where clots were formed but not treated showed that catheter insertion and balloon inflation caused damage to the vessel prior to treatment. To evaluate the vessel damage caused by histotripsy alone, vessels without clots were treated for 300 seconds using the same ultrasound parameters as those used to treat thrombi with peak negative pressure of 19 MPa, scanning the vessel twice at a rate of 0.1 mm/s.

Third, the acoustic trapping ability of the NET in a vessel-flow phantom was investigated[72, 73]. An optically transparent vessel phantom with a 6 mm inner diameter and 9 mm outer diameter, similar to that described by Ryan and Foster[74], was made to mimic a blood vessel and the surrounding soft tissue. An *in vitro* circulation system was constructed to produce steady, laminar flow of fluid to mimic blood flow in venous circulation, connecting a roller pump, a pulse dampener, and the vessel phantom. Clot particle phantoms were made by dissolving 4% w/v agarose powder into a 35 : 65 glycerol-water solution. A cavitation bubble cloud was generated inside the center of the vessel phantom using a focused PZT transducer with a center frequency of 1.063 MHz and an elliptical aperture (17.5 cm long axis and 15.5 cm short axis) with a 10 cm focal length. Pulse repetition frequencies (PRF) ranging from 30 Hz to 1000 Hz and pulse lengths from 5 to 50 cycles were tested. The trapping ability was evaluated by measuring the maximum mean fluid velocity of the circulation system for which a particle could remain trapped, termed here as the "maximum trapping velocity". Cavitation initiation was monitored by acoustic backscatter received by a passive 5 MHz single element transducer placed in the center hole of the therapy transducer.

11.5.3 Results

First, when treating the *in vitro* blood clots, histotripsy thrombolysis only occurred at peak negative pressure $\geqslant$6 MPa when initiation of a cavitating bubble cloud was detected using acoustic backscatter monitoring. Blood clots weighing 330 mg were completely broken down by histotripsy in 1.5-5 minutes. There was an increase in thrombolysis rate with peak negative pressure between 6-12 MPa (*t*-test, $P<0.05$). The mean rate was (0.21 ± 0.17) mg/sec at p of 6 MPa and (2.20 ± 0.85) mg/sec at p of 12 MPa. We measured the change in dry weight of the filters (5 μm-1 mm) to estimate the debris size distribution. All four filters' dry weights changed by $\leqslant$1 mg. No significant difference was found between control and any of the treated samples. These results suggest that at least 96% (96 mg of 100 mg) of the clot was broken down to particles smaller than 5 μm.

Second, in the vivo experiments, the bubble cloud generated at the focus of the therapy transducer was confined to the diameter vessel lumen as it appeared on a 2D image as a dynamic echogenic region on the surface of or within the thrombus[14]. As the focus was scanned over the clot following the predetermined route, the treated area underwent a reduction in echogenicity, indicating thrombolysis was occurring in that region. Thrombus breakdown was apparent as a decrease in echogenicity within the vessel in 10 of 12 cases, and in 7 cases, improved flow through the vein as measured by color Doppler (Figure 11.12, see the Color Inset, p.32). The mean ultrasound exposure time for a treatment was (10.5 ± 5.5) minutes (range from 2.6 min to 18.0 min) for clot length of (20.1 ± 6) mm and vein inner diameter of (5.5 ± 1.0) mm. Vessel histology showed denudation of vascular endothelium and small pockets of hemorrhage in the vessel adventitia and underlying muscle and fatty tissue, but perforation of the vessel wall was never observed.

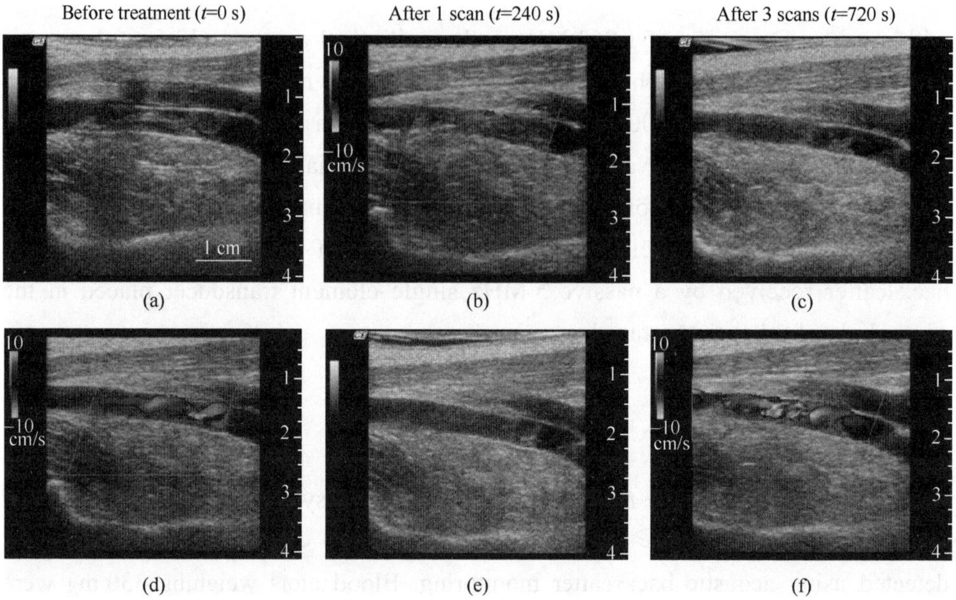

Figure 11.12 Ultrasound images of the femoral vein in the porcine DVT model captured by a linear array imaging probe between treatments of a thrombus. (a) and (b) show the original appearance of the vessel on 2D imaging and with color Doppler. (c) and (d) show the thrombus after 240 seconds of treatment (one scan). (e) and (f) show the final condition of the clot after 720 seconds (three scans). Note the decreased echogenicity in the lumen on (c) and (e) compared with (a). Also, a flow channel is clearly visible on (d) and (f), while none was present before treatment in (b). This figure is quoted from papers of Maxwell et al.[14]

Third, the acoustic trapping ability of the histotrips-induced cavitation cloud in the vessel-flow phantom was studied. A cavitation cloud generated in a vessel-flow phantom results in two vortices on either side of the cloud, and clot phantom particles were trapped in the vortices and simultaneously fractionated[72]. The maximum trapping velocity, defined by the maximum mean fluid velocity at which a 3-4 mm particle trapped in a 6 mm diameter vessel phantom, increased linearly with peak negative pressure (P–) and increased as the square root of pulse length and pulse repetition frequency (PRF)[73]. At 19.9 MPa P–, 1000 Hz PRF and 10 cycle pulse length, a 3 mm clot-mimicking particle could remain trapped under a background velocity of 9.7 cm/s. Clot fragments treated by NET resulted in debris particles $<$75 μm.

11.5.4 Discussion

The *in vitro* and *in vivo* feasibility of using histotripsy via cavitation alone for thormbolysis has been demonstrated. Histotripsy does not require drugs and is non-invasive, and thus has the potential to overcome the limitations of current thrombolysis methods requiring thrombolysis drugs and/or catheter-based procedures. In addition, our results show that histotripsy can dissolve clots at a speed significantly faster than those using drugs. Since histotripsy is non-invasive and does not involve a complex procedure to insert catheter into the treatment region, it would also require less time and lower cost than a surgical catheter. In the two instances of the *in vivo* experiments, where histotripsy could not successfully treat any part of the thrombus, a bubble cloud could not be generated within the vessel lumen. The likely cause of this phenomenon was air trapped either at the skin surface or within the vessel during catheterization to form the clots, which prevented transmission of ultrasound to the clot. These problems, however, are unrelated to the histotripsy treatment. In all cases where a bubble cloud was initiated in the vessel, at least partial breakdown of the thrombus was achieved. Flow restoration was not always achieved, in part because of limited resolution of the therapy guidance transducer, which was not optimized for the DVT study. A new transducer customized for human DVT patient anatomy has been designed and constructed. Further *in vivo* experiments will be needed to test the efficacy of the new apparatus.

As histotripsy fractionated the target tissue, the vessel wall may be damaged by histotripsy. While vessel perforation was not observed in any of the *in vivo* treatments, small areas of medial and adventitial damage were present in the part of the vessel. This damage was produced because the focal zone of the therapy transducer used was larger than the vessel lumen. This damage is expected to be avoided using the new transducer with a smaller focal zone.

There is a concern that clot fragments from histotripsy-induced fractionation may escape from the treatment zone and cause embolization downstream. The fractionated clot debris was measured *in vitro* to be no greater than 100 μm, a size unlikely to cause hazardous embolization. In addition, we found that the cavitating bubble cloud in the vessel can capture, trap, and simultaneously fractionate a clot fragment flowing through into or near the cavitation cloud. Two vortices were created on either side of the cavitation cloud in the vessel lumen, and the clot particles were trapped in the low-pressure regions of the vortices and further fractionated. This non-invasive embolus trapping (NET) ability may provide a novel tool to capture and eliminate any potential hazardous emboli by setting a secondary bubble cloud downstream of the histotripsy treatment region.

11.5.5 Conclusions

Both *in vitro* and *in vivo* experiments demonstrate that histotripsy alone is capable of non-invasively and rapidly fractionate blood clots without requiring thrombolytic drugs. The histotripsy thrombolysis treatment is guided by ultrasound imaging in real-time. The results indicate histotripsy has potential for development as a non-invasive treatment for thrombosis.

11.6 Liver Cancer Application

11.6.1 Introduction

Hepatocellular carcinoma (HCC) is a leading cause for cancer death, with 50% cases occurring in China. While liver transplantation may be curative, only a small patient population will receive this treatment[75-77]. Liver tumor surgical resection is a proven treatment option, but it is associated with high morbidity and death in 1%-5% of patients[78] and is not possible in patients with decompensated cirrhosis[78]. RFA has become a standard therapy for treatment of tumors less than 3 cm in diameter, with fewer than three tumor nodules[78]. Through a locally inserted electrode, RFA delivers heat to cause cell necrosis. Other thermal ablation techniques have also been developed, including microwave-[79], cryo-[80, 81], and laser-[82] therapy.

While these minimally invasive thermal-based therapies have shown some success, as blood flow through the highly vascular liver creates a natural source of heat dissipation or “heat sink”, RFA and other thermal-based minimally invasive ablation methods have the following inherent limitations: (1) often inconsistent ablation, leaving residual untreated tumor; (2) ineffective ablation for tumors near major vessels; (3) extensive treatment times for tumors larger than 3 cm in diameter. Further, these ablation modalities (4) do not have reliable imaging feedback during treatment and (5) are invasive, requiring device insertion. High intensity focused ultrasound (HIFU) thermal ablation has been investigated as a non-invasive ablation technique for liver cancer therapy[83, 84]. HIFU ablation does not require device insertion, is guided by real-time magnetic resonant imaging (MRI), and has potential to treat multiple and larger tumors. However, HIFU treatment still share the similar drawbacks as other thermal-based techniques due to the heat sink effect[44, 85]. A major challenge facing the non-invasive treatment of liver cancer using ultrasound is to overcome the rib obstruction. Skin burns and subcostal edema have been reported in clinical HIFU liver ablation cases[86-88]. Respiration-induced liver motion also poses significant technique challenges for HIFU liver ablation and reduces the treatment efficiency and safety.

As histotripsy non-invasively fractionates the target tissue through cavitation, it is not affected by the heat sink effect and has the potential to improve upon the limitations of thermal ablation. Histotripsy can produce consistent and fast fractionation of tissue with different heat dissipation patterns, even when the tissue is in proximity to major vessels. The fractionation is often self-limited at the boundaries of major vessels with surrounding tissue completely homogenized. As a non-invasive ablation method, the therapy focus can be scanned to treat a large tumor volume (＞3 cm) and multiple nodules. Further, the histotripsy cavitation bubble can be visualized with ultrasound imaging, allowing precise targeting. The change in tissue during treatment can be also be directly monitored using standard imaging modalities such as ultrasound and MRI, which allows histotripsy to be guided in real-time[89, 90]. For transthoracic liver ablation, ribs in the ultrasound pathway cause periodic blockage of ultrasound, resulting in a significantly decreased main lobe and increased grating lobes[91-93]. Histotripsy is more resistant to the grating lobes caused by rib aberration, as the cavitation cloud is only generated when the pressure exceeds a distinct threshold. By using an appropriate pressure where the main lobe is above the threshold while the grating lobes are not, a confined cloud within the main lobe and a precise lesion can be produced despite the intervening ribs[94, 95]. Thermal damage to the overlying and surrounding tissue is prevented by using prolonged cooling times between pulses.

We investigate the feasibility of developing histotripsy for non-invasive liver ablation in an *in vivo* porcine liver model with size and anatomy similar to human[31, 95]. First, histotripsy was used to generate consistent and complete fractionation of hepatic parenchyma through ribs and overlying tissue in various locations spanning all major regions of the liver. Second, the effectiveness of tissue-selective ablation using histotripsy to fractionate the liver surrounding major blood vessels and gallbladder while preserving these critical structures was tested. Third, the capability of ablating large regions in the liver was investigated. Finally, the thermal effect to the overlying ribs during histotripsy liver treatment without using any aberration correction was investigated.

11.6.2 Methods

A total of 16 healthy 60-90 pound mixed breed pigs were anesthetized. To ensure ultrasound propagation from the transducer to targeted tissue, a degassed water bolus was coupled to the skin with a thin plastic membrane and ultrasound coupling gel. The treatment targeting was guided by ultrasound imaging using an 8 MHz phased array ultrasonic imaging probe (Model S8, used with Sonos 7500 imaging system, Philips Electronics, Andover, MA) fixed coaxially with the histotripsy therapy transducer. The focal zone was viewed as a hyperechoic cavitaton bubble cloud on ultrasound imaging. The treated lesions were evaluated on ultrasound imaging and magnetic resonant imaging (MRI). Treated porcine liver samples were harvested after experiments for gross morphology and histological analysis using hematoxylin and eosin (H&E) staining.

Three sets of experiments were performed. First, to study the in vivo feasibility of using histotripsy to create precise lesions through ribs and overlying tissue in various locations spanning all major regions of the liver, twelve histotripsy lesions were created in the livers of six pigs through the intact chest. Histotripsy pulses of 10 cycles, 500 Hz pulse repetition frequency (PRF), and 14-17 MPa estimated *in situ* peak negative pressure were applied to the liver using a 1 MHz therapy transducer. Second, to study the tissue-selective ablation and large volume ablation, two larger volume lesions up to 60 cm^3 were generated in the livers of two pigs using the same histotripsy parameters above. The liver tissue, vessels, gallbladder, and other tissues within and surrounding the treatment region are evaluated histologically for any damage. Third, to study the effect of rib obstruction to histotripsy liver ablation without using aberration correction, non-invasive liver treatments were conducted in 8 pigs, with 4 lesions generated through transcostal windows with full ribcage obstruction and 4 lesions created through transabdominal windows without rib coverage. Treatments were performed by a 750 kHz focused transducer using 5 cycle pulses at 200 Hz PRF, with estimated *in situ* peak negative pressures of 13-17 MPa. Temperatures on overlying tissues including the ribs were measured with needle thermocouples inserted superficially beneath the skin. Treatments of approximately 40 minutes were applied, allowing overlying tissue temperatures to reach saturation.

11.6.3 Results

In the first experiment set, twelve ~1 cm^3 lesions were successfully created with locations spanning the entire liver, including the superior and inferior regions of the left, middle, and right lobes[31]. Treatments were performed in 16.7 minutes through

3-6.5 cm of overlying tissue with rib cage covering 30%-50% (including intercostal space) of the transducer aperture. Histological evaluation of histotripsy lesions showed complete fractionation of hepatic parenchyma inside the treated volume with sharp boundaries of partially ablated liver tissue <500 μm for all treatments.

In the second experiment set, two large liver volumes of 18 cm^3 and 60 cm^3 were fractionated *in vivo* within 60 minutes by mechanically scanning the focus to separate locations[31]. Histotripsy-induced liver fractionation was self-limited at the boundaries of critical structures including the major vessels and gallbladder, while surrounding liver tissue was completely fractionated. Liver tissue surrounding major vessels was completely fractionated, yet the vessels larger than 300 μm remained intact (Figure 11.13, see the Color Inset, p.32).

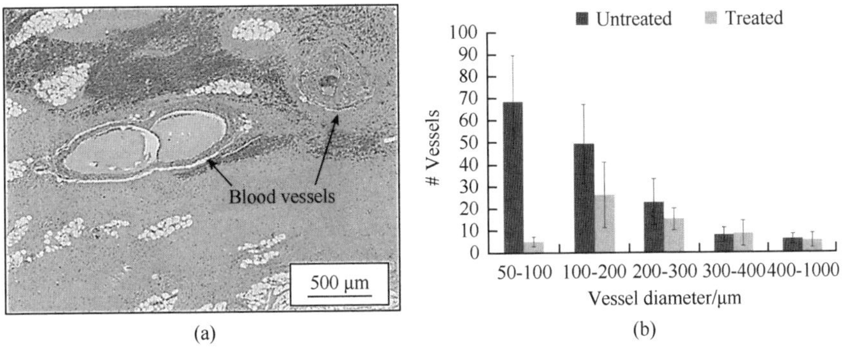

Figure 11.13 (a) H&E slide showing intact vessels (indicated by arrows) remained in the completely fractionated liver; (b) There is no statistical significance in the number of vessels above 300 μm diameter in the treated and control regions. This figure is quoted from papers by Vlaisavljevich et al.[31]

In the third experiment set to study the effect of rib obstruction on histotripsy liver treatment, lesions yielded statistically comparable ablation volumes of (3.6±1.7) cm^3 and (4.5±2.0) cm^3 in treatments with full ribcage coverage (transcostal) and no rib obstruction (transabdominal), respectively. The average temperature increase observed in transcostal treatments was (3.9±2.1)℃, while transabdominal treatments showed an increase of (1.7±1.3)℃. No damage was seen on the ribcage or other overlying tissues[95].

11.6.4 Discussion

In this study, we investigated the feasibility of using histotripsy for non-invasive liver fractionation in an *in vivo* porcine model. Results demonstrate that histotripsy created precise lesions at locations throughout the entire liver through ribs and overlying tissue without using aberration correction. Bubble clouds were successfully initiated and lesions were formed in all treatment attempts. The different heat dissipation patterns associated with regions containing different vasculatures did not affect the consistency of histotripsy liver ablation. As major vessels and gallbladder have higher mechanical strength and are more resistant to histotripsy[30], the liver surrounding these structures was completely fractionated while the major hepatic vessels and gallbladder remained intact.

This study also investigated the feasibility of using histotripsy therapy to generate lesions through rib obstacles without using aberration correction mechanisms. The results suggest that the therapeutic capabilities of histotripsy ablation are relatively unaffected by rib obstruction in transcostal treatment windows, as long as sufficient pressure is available at the focus of the treatment, with main lobe above the cavitation threshold and side lobes below the threshold. However, this threshold may vary between patients due to the different levels of rib obstruction and overlying tissue. Using ultrasound imaging to monitor the cavitation in real-time, the applied pressure can be gradually increased until a confined bubble cloud is seen. As such, an appropriate pressure can be tailored for individual treatment. Thermal damage to the overlying ribs and tissue can be prevented by using a prolonged cooling time between pulses[96].

11.6.5 Conclusions

This work demonstrates that histotripsy is capable of non-invasively and precisely fractionating liver tissue while preserving critical anatomical structures within the liver. Histotripsy can achieve effective treatment through the ribcage *in vivo* without requiring correction mechanisms, while inducing no substantial thermal effects or damage to overlying tissues. Results suggest histotripsy has potential for the non-invasive, non-thermal ablation of liver tumors.

11.7 Summary and Future Work

Histotripsy is a platform non-invasive, non-thermal ablation technique with many potential applications. The studies described here demonstrate the safety and feasibility of histotripsy for hypoplastic left heart syndrome (HLHS), deep vein thrombosis, and liver tumor. For HLHS application, clinical translation of histotripsy is well underway. An integrated, ultrasound image-guided, histotripsy pediatric cardiac system has been designed based on human HLHS patient anatomy and constructed following the regulation by Food Drug Administration (FDA). Validation of this clinical histotripsy pediatric cardiac system is being performed to achieve FDA approval to start a clinical trial in the near future. There are other common cardiac conditions that cab benefit from such a non-invasive tissue removal or remodeling approach, including septal ablation for hypertrophic cardiomyopathy, cardiac ablation for atrial fibrillation, and creation of an intracardiac pathway for pulmonary hypertension and d-transposition of the great arteries. For the thrombolysis study, blood flow was not restored in some cases and slight vessel damage was found because the prototype therapy transducer used in the study had a focal zone larger than the vessel lumen containing the clot. We have designed and constructed an integrated imaged guided histotripsy DVT system with a therapy transducer customized for human DVT patient anatomy and improved imaging guidance. This new and improved histotripsy DVT system will be tested *in vivo* to investigate the safety and efficacy. The same device can also be extended to treat peripheral arterial disease (PAD), which has even a higher prevalence than DVT. For liver cancer project, in addition to the studies included here, we have recently demonstrated that histotripsy can rapidly ablate large tumor volume using electric focal steering with phased array transducer. Future work includes design and construction of a phased array therapy transducer based on available acoustic window in liver cancer patients, parameter optimization for rapid and precise tumor ablation, and investigation of the histotripsy treatment response in rodent liver tumor models.

11.8 Acknowledgement

The authors would like to thank fellow co-inventors of histotripsy (Drs. Charles Cain, Tim Hall, Brian Fowlkes, and Will Roberts), clinical collaborators (Drs. Gabe Owens, Hitinder Gurm, and Theodore Welling, and other members in the histotripsy group at the University of Michigan for their contribution to the development of histotripsy. This work was supported by grants from National Institute of Health (R01 EB008998, R21 NS093121), National Science Foundation, The Hartwell Foundation, American Heart Association, The Focused Ultrasound Foundation, The Wallace Coulter Foundation, and a Research Scholar Grant from the American Cancer Society

(RSG-13-101-01-CCE).

References

[1] Parsons J E, Cain C A, Abrams G D, et al. Pulsed cavitational ultrasound therapy for controlled tissue homogenization. *Ultrasound in Medicine and Biology*, 2006, 32 (1): 115-129.

[2] Xu Z, Fowlkes J B, Rothman E D, et al. Controlled ultrasound tissue erosion: The role of dynamic interaction between insonation and microbubble activity. *The Journal of the Acoustical Socity of America*, 2005, 117 (1): 424-435.

[3] Roberts W W, Hall T J, Ives K, et al. Pulsed cavitational ultrasound: A non-invasive technology for controlled tissue ablation (histotripsy) in the rabbit kidney. *The Journal of Urology*, 2006, 175 (2): 734-738.

[4] Mrighifu H K. A tool for image-guided therapeutics. *Journal of Magnetic Resonance Imaging*, 2011, 34(3): 482-493.

[5] Kennedy J E, ter Haar G R, Cranston D. High intensity focused ultrasound: Surgery of the future? *The British Journal of Radiology*, 2003, 76 (909): 590-599.

[6] Wu F, Wang Z B, Zhu H, et al. Extracorporeal high intensity focused ultrasound treatment for patients with breast cancer. *Breast Cancer Research and Treatment*, 2005, 92 (1): 51-60.

[7] Xu Z, Raghavan M, Hall T L, et al. Evolution of bubble clouds induced in pulsed cavitational ultrasound therapy-histotripsy. *IEEE Transactions on Ultrasonics, Ferroelectrics and Frequency Control*, 2008, 55 (5): 1122-1132.

[8] Maxwell A D, Wang T Y, Cain C A, et al. Cavitation clouds created by shock scattering from bubbles during histotripsy. *The Journal of the Acoustical Society of America*, 2011, 130 (4): 1888-1898.

[9] Maxwell A D, Cain C A, Hall T L, et al. Probability of cavitation for single ultrasound pulses applied to tissues and tissue-mimicking materials. *Ultrasound in Medicine and Biology*, 2013, 39 (3): 449-465.

[10] Hall T L, Kieran K, Ives K, et al. Histotripsy of rabbit renal tissue *in vivo*: Temporal histologic trends. *Journal of Endourology*, 2007, 21 (10): 1159-1166.

[11] Xu Z, Fowlkes J B, Rothman E D, et al. Controlled ultrasound tissue erosion: The role of dynamic interaction between insonation and microbubble activity. *The Journal of the Acoustical Society of America*, 2005, 117 (1): 424-435.

[12] Hempel C R, Hall T L, Cain C A, et al. Histotripsy fractionation of prostate tissue: Local effects and systemic response in a canine model. *The Journal of Urology*, 2011, 185 (4): 1484-1489.

[13] Maxwell A D, Cain C A, Duryea A P, et al. Noninvasive thrombolysis using pulsed ultrasound cavitation therapy—histotripsy. *Ultrasound in Medicine and Biology*, 2009, 35 (12): 1982-1994.

[14] Maxwell A D, Owens G, Gurm H S, et al. Noninvasive treatment of deep venous thrombosis using pulsed ultrasound cavitation therapy (histotripsy) in a porcine model. *Journal of Vascular and Interventional Radiology*, 2011, 22(3): 369-377.

[15] Owens G E, Miller R M, Ensing G, et al. Therapeutic ultrasound to noninvasively create intracardiac communications in an intact animal model. *Catheterization and Cardiovascular Interventions*, 2011, 77: 580-588.

[16] Xu Z, Owens G, Gordon D, et al. Noninvasive creation of an atrial septal defect by histotripsy in a canine model. *Circulation*, 2010, 121 (6): 742-749.

[17] Styn N R, Wheat J C, Hall T L, et al. Histotripsy of vx-2 tumor implanted in a renal rabbit model. *Journal of Endourology*, 2010, 24 (7): 1145-1150.

[18] Duryea A P, Hall T L, Maxwell A D, et al. Histotripsy erosion of model urinary calculi. *Journal of Endourology*, 2011, 25(2): 341-344.

[19] Kim Y, Gelehrter S K, Fifer C G, et al. Non-invasive pulsed cavitational ultrasound for fetal tissue ablation: Feasibility study in a fetal sheep model. *Ultrasound in Obstetrics and Gynecology*, 2011, 37 (4): 450-457.

[20] Kim Y, Fifer C G, Gelehrter S K, et al. Developmental impact and lesion maturation of histotripsy-mediated non-invasive tissue ablation in a fetal sheep model. *Ultrasound in Medicine and Biology*, 2013, 39: 1047-1055.

[21] Xu Z, Ludomirsky A, Eun L Y, et al. Controlled ultrasound tissue erosion. *IEEE Transactions on Ultrasonics, Ferroelectrics and Frequency Control*, 2004, 51 (6): 726-736.

[22] Vlaisavljevich E, Maxwell A, Warnez M, et al. Histotripsy-induced cavitation cloud initiation thresholds in tissues of different mechanical properties. *IEEE Transactions on Ultrasonics, Ferroelectrics and Frequency Control*, 2014, 61: 341-352.

[23] Lin K W, Kim Y, Maxwell A D, et al. Histotripsy beyond the intrinsic cavitation threshold using very short ultrasound pulses: Microtripsy. *IEEE Transactions on Ultrasonics, Ferroelectrics and Frequency Control*, 2014, 61: 251-265.

[24] Xu Z, Raghavan M, Hall T L, et al. High speed imaging of bubble clouds generated in pulsed ultrasound cavitational therapy-histotripsy. *IEEE Transaction on Ultrasonics, Ferroelectrics and Frequency Control*, 2007, 54 (10): 2091-2101.

[25] Vlaisavljevich E, Maxwell A, Warnez M, et al. Histotripsy-induced cavitation cloud initiation thresholds in tissues of different mechanical properties. *IEEE Transactions on Ultrasonics, Ferroelectrics and Frequency Control*, 2014, 61: 341-352.

[26] Wang T Y, Xu Z, Winterroth F, et al. Ultrasound backscatter feedback for pulsed cavitational ultrasound therapy-histotripsy. *IEEE Transactions on Ultrason, Ferroelectrics and Frequency Control*, 2009, 56 (5): 995-1005.

[27] Hall T L, Hempel C R, Wojno K, et al. Histotripsy of the prostate: Dose effects in a chronic canine model. *Urology*, 2009, 74 (4): 932-937.

[28] Winterroth F, Xu Z, Wang T Y, et al. Examining and analyzing subcellular morphology of renal tissue treated by histotripsy. *Ultrasound in Medicine and Biology*, 2011, 37 (1): 78-86.

[29] Xu Z, Fan Z Z, Hall T L, et al. Size measurement of tissue debris particles generated from pulsed ultrasound cavitational therapy-histotripsy. *Ultrasound in Medicine & Biology*, 2009, 35 (2): 245-255.

[30] Vlaisavljevich E, Kim Y, Owens G, et al. Effects of tissue mechanical properties on susceptibility to histotripsy-induced tissue damage. *Physics in Medicine and Biology*, 2014, 59: 253-270

[31] Vlaisavljevich E, Kim Y, Allen S, et al. Image-guided non-invasive ultrasound liver ablation using histotripsy: Feasibility study in an *in vivo* porcine model. *Ultrasound in Medicine and Biology*, 2013, 39 (8): 1398-1409.

[32] Owens G E, Miller R M, Ensing G, et al. Therapeutic ultrasound to non invasively create intracardiac

communications in an intact animal model. *Catheterization and Cardiovascular Interventions*, 2011, 77: 580-588.

[33] Hall T L, Fowlkes J B, Cain C A. A real-time measure of cavitation unduced tissue disruption by ultrasound imaging backscatter reduction. *IEEE Transactions on Ultrasonics, Ferroelectrics, and Frequency Control*, 2007, 54: 569-575.

[34] Wang T Y, Hall T L, Xu Z. Imaging feedback of histotripsy treatments using ultrasound shear wave elastography. *IEEE Transactions on Ultrasonics, Ferroelectrics*, and *Frequency Control*, 2012, 59 (6): 1167-1181.

[35] Wang T Y, Hall T L, Xu Z, et al. Imaging feedback for histotripsy by characterizing dynamics of acoustic radiation force impulse (arfi)-induced shear waves excited in a treated volume. *IEEE Transactions on Ultrasonics Ferroelectrics and Frequecy Control*, 2014, 61: 1137-1151.

[36] Zhang X, Miller R M, Lin K W, et al. Real-time feedback of histotripsy thrombolysis using bubble-induced color doppler. *Ultrasound in Medicine and Biology*, 2015, 41 (5): 1386-1401.

[37] Allen S P, Hall T L, Cain C A, et al. Controlling cavitation-based image contrast in focused ultrasound histotripsy surgery. *Magnetic Resonance in Medicine*, 2015, 73 (1): 204-213.

[38] Abu-Harb M, Hey E, Wren C. Death in infancy from unrecognised congenital heart disease. *Archives of Disease in Childhood*, 1994, 71 (1): 3-7.

[39] Fyler D C. Report of the new england regional infant cardiac program. *Pediatrics*, 1980, 65 (suppl): 375-461.

[40] Fry W J, Fry F J. Fundamental neurological research and human neurosurgery using intense ultrasound. *IRE Transactions on Medical Electronics*, 1960, 7: 166-181.

[41] Strickberger S A, Tokano T, Kluiwstra J U, et al. Extracardiac ablation of the canine atrioventricular junction using high intensity focused ultrasound. *Circulation*, 1999, 100 (2): 203-208.

[42] Miller R M, Owens G, Ensing G, et al. Histotripsy for pediatric cardiac applications: *In vivo* neonatal pig model. *9th International Symposium on Therapeutic Ultrasound: ISTU*, 2010, 1215 (1): 15-18.

[43] Ninet J, Roques X, Seitelberger R, et al. Surgical ablation of atrial fibrillation with off-pump, epicardial, high-intensity focused ultrasound: Results of a multicenter trial. *Journal of Thoracic and Cardiovascular Surgery*, 2005, 130 (3): 803-809.

[44] Leslie T A, Kennedy J E, Illing R O, et al. High-intensity focused ultrasound ablation of liver tumours: Can radiological assessment predict the histological outcome? *British Journal of Radiology*, 2008, 81 (967): 564-571.

[45] Owens G E, Miller R M, Owens S T, et al. Intermediate-term effects of intracardiac communications created noninvasively by therapeutic ultrasound (histotripsy) in a porcine model. *Pediatric Cardiology*, 2012, 33: 83-89.

[46] Xu Z, Fan Z, Hall T L, et al. Size measurement of tissue debris particles generated from mechanical tissue fractionation by pulsed cavitational ultrasound therapy-histotripsy. *Ultrasound in Medicine and Biology*, 2009, 35 (2): 245-255.

[47] Duryea A P, Maxwell A D, Roberts W W, et al. Non-invasive comminution of renal calculi using pulsed cavitational ultrasound therpy-histotripsy//*Ultrasonics Symposium* (IUS), 2009 *IEEE International*. IEEE, 2009: 73-76.

[48] Curiel L, Chopra R, Hynynen K. *In vivo* monitoring of focused ultrasound surgery using local harmonic motion. *Ultrasound in Medicine and Biology*, 2008, 35 (1): 65-78.

[49] Tanter M, Pernot M, Aubry J F, et al. Compensating for bone interfaces and respiratory motion in high-intensity

focused ultrasound. *International Journal of Hyperthermia*, 2007, 23 (2): 141-151.

[50] Pernot M, Tanter M, Fink M. 3-d real-time motion correction in high-intensity focused ultrasound therapy. *Ultrasound in Medicine and Biology*, 2004, 30 (9): 1239-1249.

[51] Miller R M, Kim Y, Lin K W, et al. Histotripsy cardiac therapy system integrated with real-time motion correction. *Ultrasound Medicine and Biology*, 2013, 39: 2362-2373.

[52] Nordström M, Lindblad B, Bergqvist D, et al. A prospective study of the incidence of deep-vein thrombosis within a defined urban population. *Journal of Internal Medicine*, 1992, 232: 155-160.

[53] Bulger C M, Jacobs C, Patel N H. Epidemiology of acute deep vein thrombosis. *Techniques in Vascular and Interventional radiology*, 2004, 7: 50-54.

[54] Goldhaber S Z. Pulmonary embolism. *New Engl and Journal of Medicine*, 1998, 339: 93-104.

[55] Collen D, Stump D C, Gold H K. Thrombolytic therapy. *Annual Review Medicine*, 1988, 39: 405-423.

[56] Rogers L Q, Lutcher C L. Streptokinase therapy for deep vein thrombosis: A comprehensive review of the English literature. *American Journal of Medicine*, 1990, 88: 389-395.

[57] Sharafuddin M J, Sun S, Hoballah J J, et al. Endovascular management of venous thrombotic and occlusive diseases of the lower extremities. *Journal of Vascular and Interventional Radiology*, 2003, 14: 405-423.

[58] Harpaz D, Chen X, Francis C W, et al. Ultrasound enhancement of thrombolysis and reperfusion *in vitro*. *Journal of the American College of Cardiology*, 1993, 21: 1507-1511.

[59] Luo H, Birnbaum Y, Fishbein M, et al. Enhancement of thrombolysis *in vivo* without skin and soft tissue damage by transcutaneous ultrasound. *Thrombosis Research*, 1998, 89: 171-177.

[60] Siegel R, Atar S, Fishbein M, et al. Noninvasive, transthoracic, low-frequency ultrasound augments thrombolysis in a canine model of acute myocardial infarction. *Circulation*, 2000, 101: 2026-2029.

[61] Francis C W, Suchkova V N. Ultrasound and thrombolysis. *Vascular Medicine*, 2001, 6: 181-187.

[62] Frenkel V, Oberoi J, Stone M, et al. Pulsed high-intensity focused ultrasound enhances thrombolysis in an *in vitro* model. *Radiology*, 2006, 239 (1): 86-93.

[63] Meunier J, Holland C, Lindsell C, et al. Duty cycle dependence of ultrasound enhanced thrombolysis in a human clot model. *Ultrasound in Medicine and Biology*, 2007, 33 (4): 576-583.

[64] Parikh S, Motarjeme A, McNamara T, et al. Ultrasound-accelerated thrombolysis for the treatment of deep vein thrombosis: Initial clinical experience. *Journal of Vascular and Interventional Radiology*, 2008, 19 (4): 521-528.

[65] Dumantepe M, Tarhan I A, Ozler A. Treatment of chronic deep vein thrombosis using ultrasound accelerated catheter-directed thrombolysis. *European Journal of Vascular and Endovascular Surgery*, 2013, 46(3): 366-371.

[66] Rosenschein U, Roth A, Rassin T, et al. Analysis of coronary ultrasound thrombolysis endpoints in acute myocardial. *Circulation*, 1997, 95 (6): 1411-1416.

[67] Rosenschein U, Furman V, Kerner E, et al. Ultrasound imaging-guided noninvasive ultrasound thrombolysis: Preclinical results. *Circulation*, 2000, 102 (2): 238-245.

[68] Schafer S, Kliner S, Klinghammer L, et al. Influence of ultrasound operating parameters on ultrasound-induced thrombolysis *in vitro*. *Ultrasound in Medicine and Biology*, 2005, 31 (6): 841-847.

[69] Wright C, Hynynen K, Goertz D. *In vitro* and *in vivo* high-intensity focused ultrasound thrombolysis. *Investigative Radiology*, 2012, 47 (4): 217-225.

[70] Deng C X, Xu Q, Apfel R E, et al. *In vitro* measurements of inertial cavitation thresholds in human blood. *Ultrasound in Medicine and Biology*, 1996, 22 (7): 939-948.

[71] Daniels S, Kodama T, Price D J. Damage to red blood cells induced by acoustic cavitation. *Ultrasound in Medicine and Biology*, 1995, 21 (1): 113-119.

[72] Maxwell A D, Park S, Vaughan B L, et al. Trapping of embolic particles in a vessel phantom by cavitation-enhanced acoustic streaming. *Physics in Medicine and Biology*, 2014, 59: 4927-4943.

[73] Park S, Maxwell A D, Owens G E, et al. Non-invasive embolus trap using histotripsy-an acoustic parameter study. *Ultrasound in Medicine and Biology*, 2013, 39: 611-619.

[74] Ryan L K, Foster F S. Tissue equivalent vessel phantom for intravascular ultrasound. *Ultrasound in Medicine and Biology*, 1997, 23 (2): 261-273.

[75] Pelletier S J, Fu S, Thyagarajan V, et al. An intention-to-treat analysis of liver transplantation for hepatocellular carcinoma using organ procurement transplant network data. *Liver Transplantation*, 2009, 15 (8): 859-868.

[76] Bosch F X, Ribes J, Diaz M, et al. Primary liver cancer: Worldwide incidence and trends. *Gastroenterology*, 2004, 127 (5): S5-S16.

[77] El-Serag H B, Mason A C. Rising incidence of hepatocellular carcinoma in the United States. *The New England Journal of Medicine*, 1999, 340 (10): 745-750.

[78] Livraghi T, Makisalo H, Line P D. Treatment options in hepatocellular carcinoma today. *Scandinavian Journal of Surgery*, 2011, 100 (1): 22-29.

[79] Boutros C, Somasundar P, Garrean S, et al. Microwave coagulation therapy for hepatic tumors: Review of the literature and critical analysis. *Surgical Oncology*, 2010, 19 (1): e22-e32.

[80] Gage A A. Cryosurgery in the treatment of cancer. *Surgery, Gynecology and Obstetrics*, 1992, 174 (1): 73-92.

[81] Gage A A, Baust J G. Cryosurgery for tumors. *Journal of the American College of Surgeons*, 2007, 205 (2): 342-356.

[82] Dick E A, Taylor-Robinson S D, Thomas H C, et al. Ablative therapy for liver tumours. *Gut*, 2002, 50 (5): 733-739.

[83] Leslie T A, Kennedy J E. High-intensity focused ultrasound principles, current uses, and potential for the future. *Ultrasound Quarterly*, 2006, 22 (4): 263-272.

[84] ter Haar G. High intensity ultrasound. *Seminars in Laparoscopic Surgery*, 2001, 8 (1): 77-89.

[85] Okada A, Murakami T, Mikami K, et al. A case of hepatocellular carcinoma treated by mr-guided focused ultrasound ablation with respiratory gating. *Magnetic Resonance in Medical Sciences*, 2006, 5 (3): 167-171.

[86] Illing R O, Kennedy J E, Wu F, et al. The safety and feasibility of extracorporeal high-intensity focused ultrasound (hifu) for the treatment of liver and kidney tumours in a western population. *The British Journal of Cancer*, 2005, 93: 890-895.

[87] Jung S E, Cho S H, Jang J H, et al. High-intensity focused ultrasound ablation in hepatic and pancreatic cancer: Complications. *Abdominal Imaging*, 2010, 36 (2): 185-195.

[88] Wu F, Wang Z B, Chen W Z, et al. Extracorporeal high intensity focused ultrasound ablation in the treatment of 1038 patients with solid carcinomas in china: An overview. *Ultrasonics-Sonochemistry*, 2004, 11 (3): 149-154.

[89] Hall T L, Lee G R, Hernandez L, et al. Relaxation properties of cavitation induced tissue lesions. Joint Annual Meeting ISMRM (International Society for Magnetic Resonance in Medicine)-ESMRMB (European Society for

Magnetic Resonance in Medicine and Biology) 2007b: Poster. 2007, 1118.

[90] Wang T Y, Xu Z, Winterroth F, et al. Quantitative ultrasound backscatter for pulsed cavitational ultrasound therapy-histotripsy. *IEEE Transactions on Ultrasonics, Ferroelectrics and Frequency Control*, 2009, 56 (5): 995-1005.

[91] Bobkova S, Gavrilov L, Khokhlova V, et al. Focusing of high-intensity ultrasound through the rib cage using a therapeutic random phased array. *Ultrasound in Medicine and Biology*, 2010, 36 (6): 888-906.

[92] Khokhlova V A, Bobkova S M, Gavrilov L R. Focus splitting associated with propagation of focused ultrasound through the rib cage. *Acoustical Physics*, 2010, 56 (5): 665-674.

[93] Wang H, Ebbini E S, O'Donnell M, et al. Phase abberation correction and motion compensation for ultrasonic hyperthermia phased arrays: Experimental results. *Ultrasonics Ferroelectrics & Frequency Control IEEE Transactions on*, 1994, 41: 34-43.

[94] Kim Y, Wang T Y, Xu Z, et al. Lesion generation through ribs using histotripsy therapy without aberration correction. *IEEE Transactions on Ultrasonics, Ferroelectrics and Frequency Control*, 2011, 58 (11): 2334-2343.

[95] Kim Y, Vlaisavljevich E, Owens G E, et al. *In vivo* transcostal histotripsy therapy without aberration correction. *Physics in Medicine and Biology*, 2014, 59 (1): 2553-2568.

[96] Kieran K, Hall T L, Parsons J E, et al. Refining histotripsy: Defining the parameter space for the creation of non-thermal lesions with high intensity, pulsed focused ultrasound of the *in vitro* kidney. *The Journal of Urology*, 2007, 178 (2): 672-676.

Color Inset

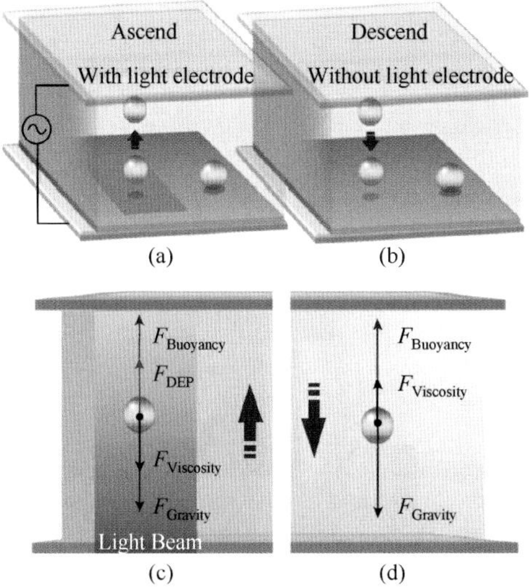

Figure 1.2

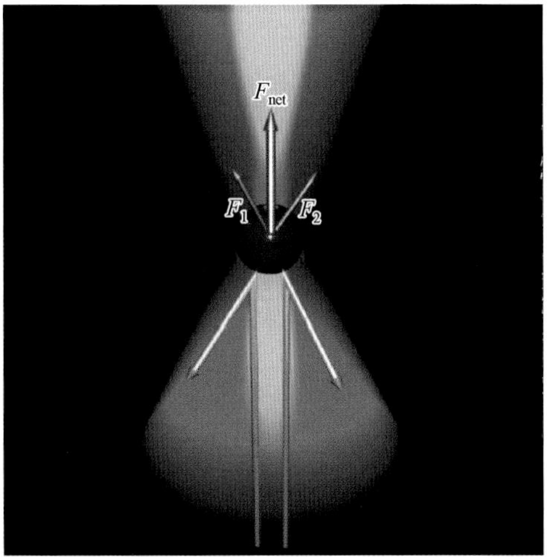

Figure 1.3

Figure 1.4

1

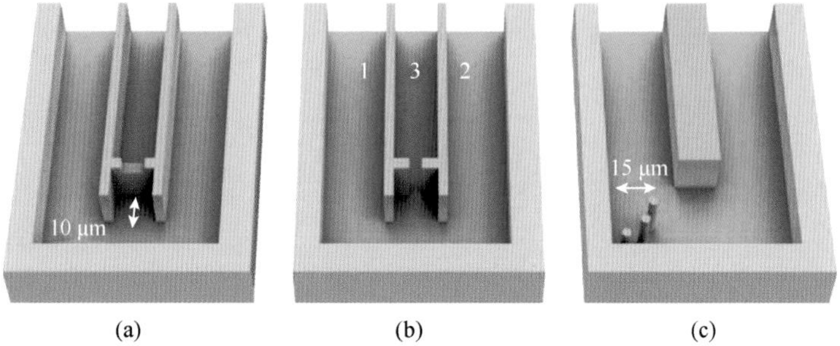

Figure 1.7

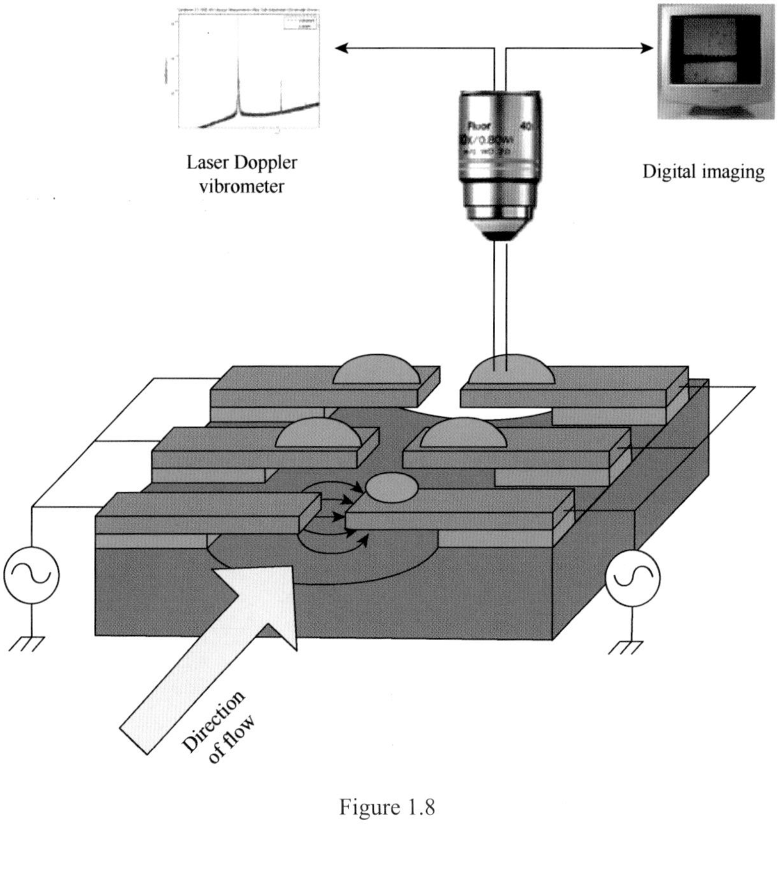

Figure 1.8

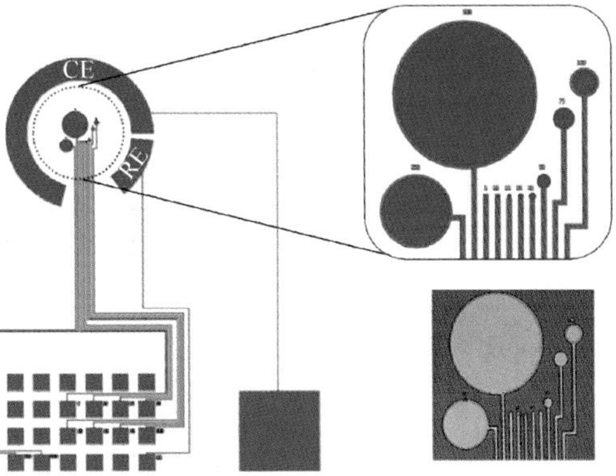

Figure 1.10

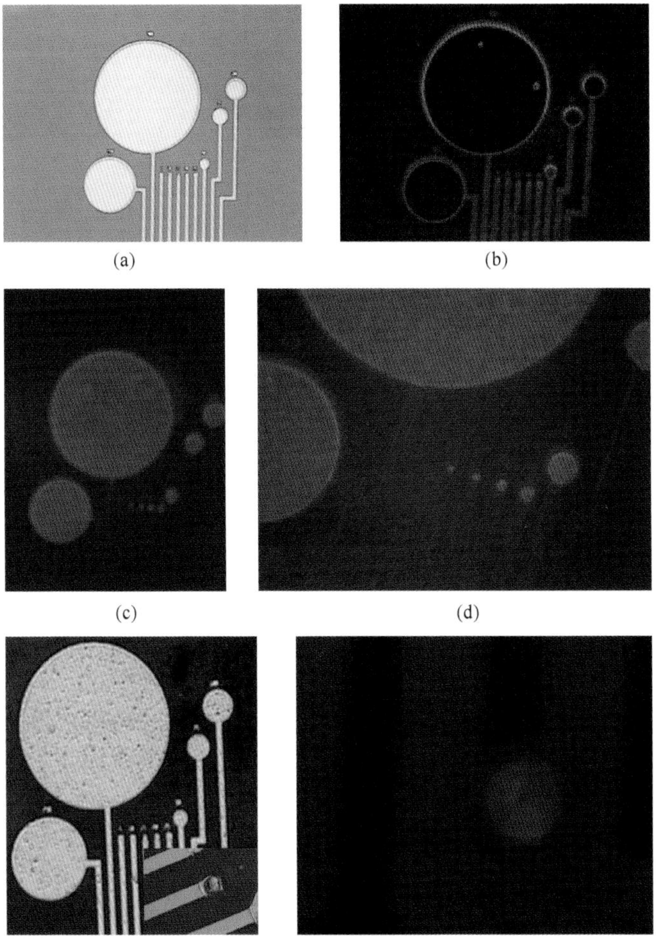

(a) (b)

(c) (d)

(e) (f)

Figure 1.11

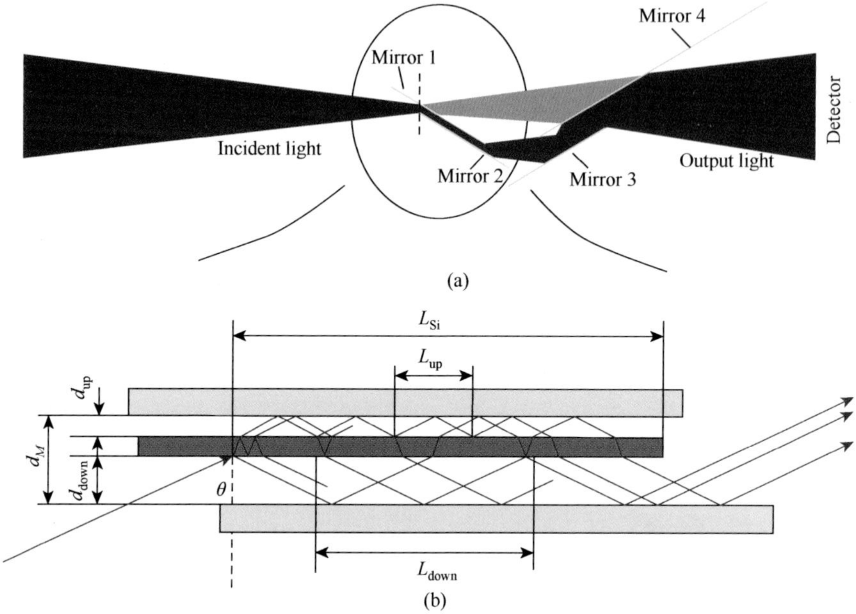

Figure 2.13

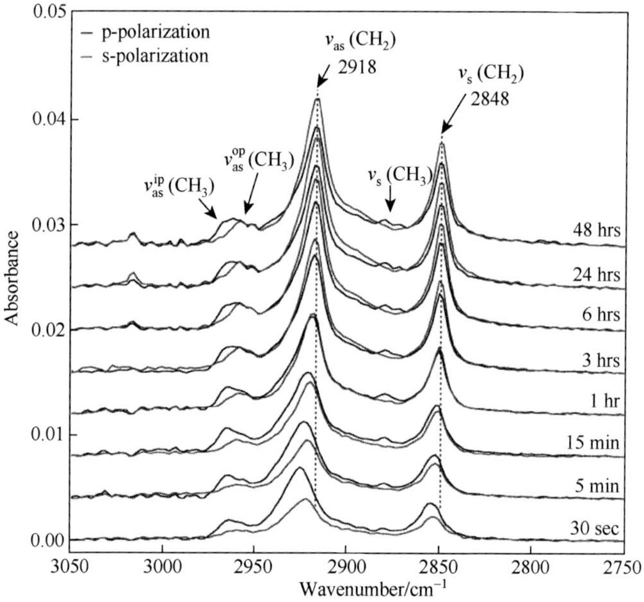

Figure 2.26

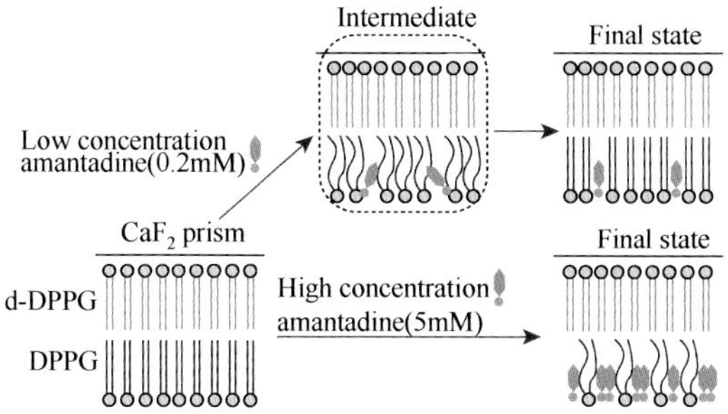

Figure 3.5

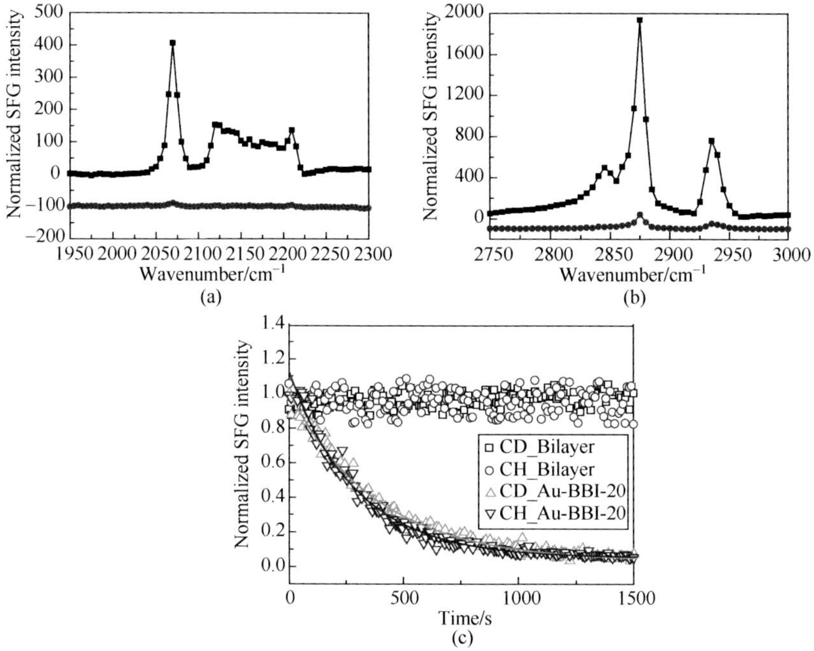

Figure 3.6

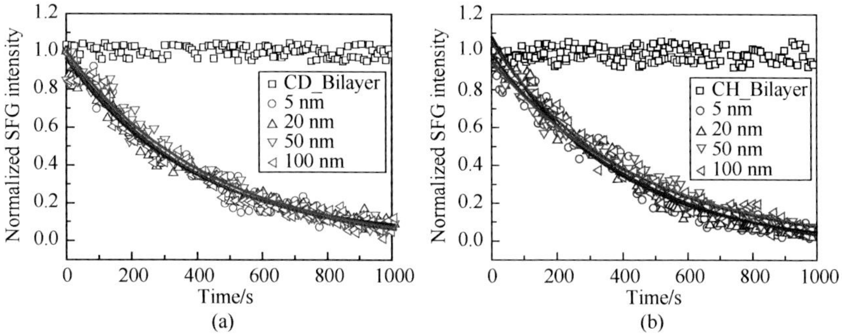

Figure 3.7

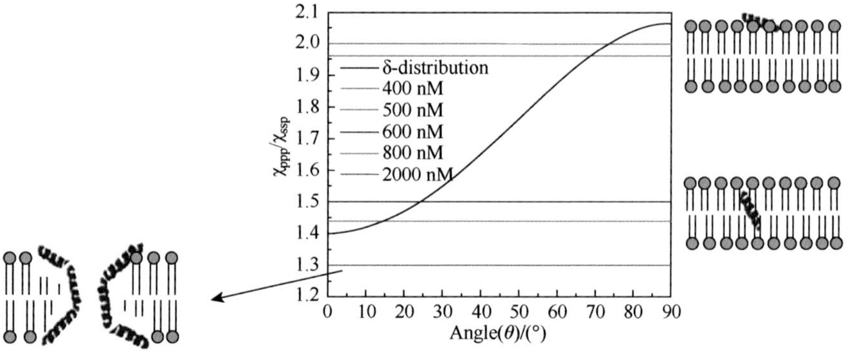

Figure 3.11

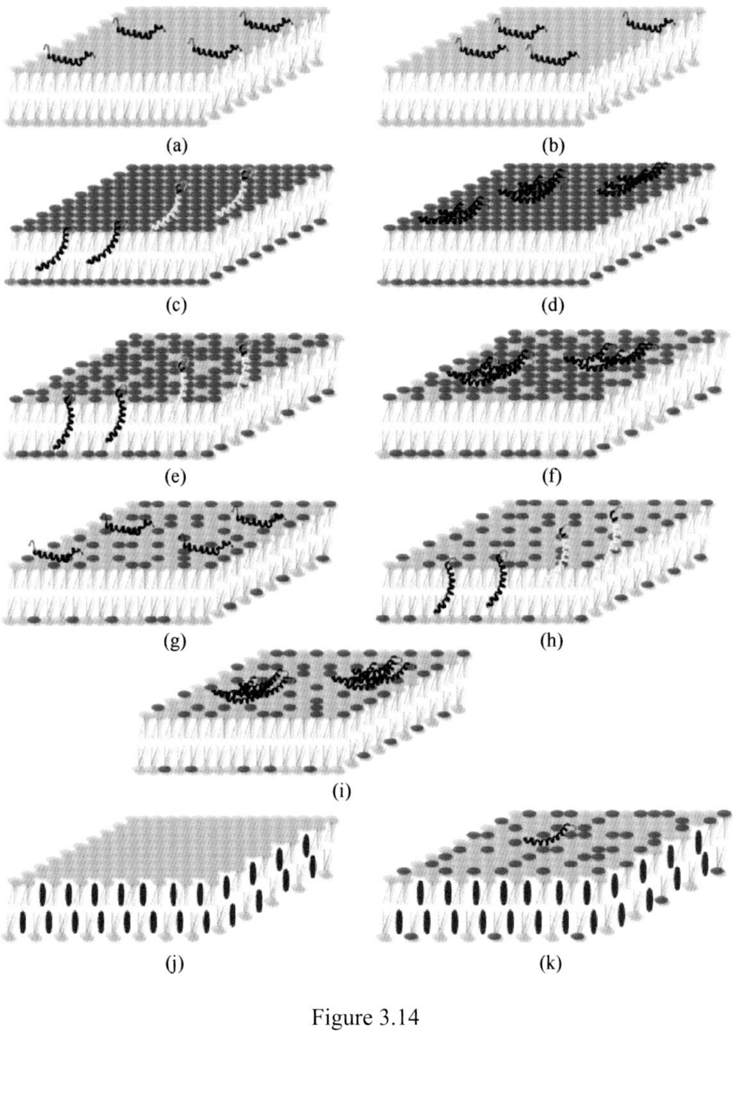

Figure 3.14

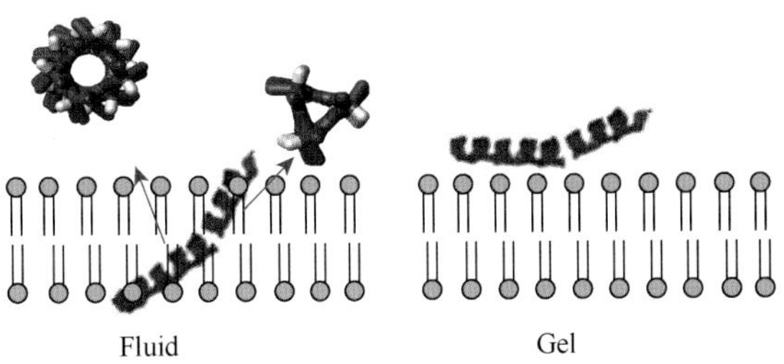

Figure 3.15

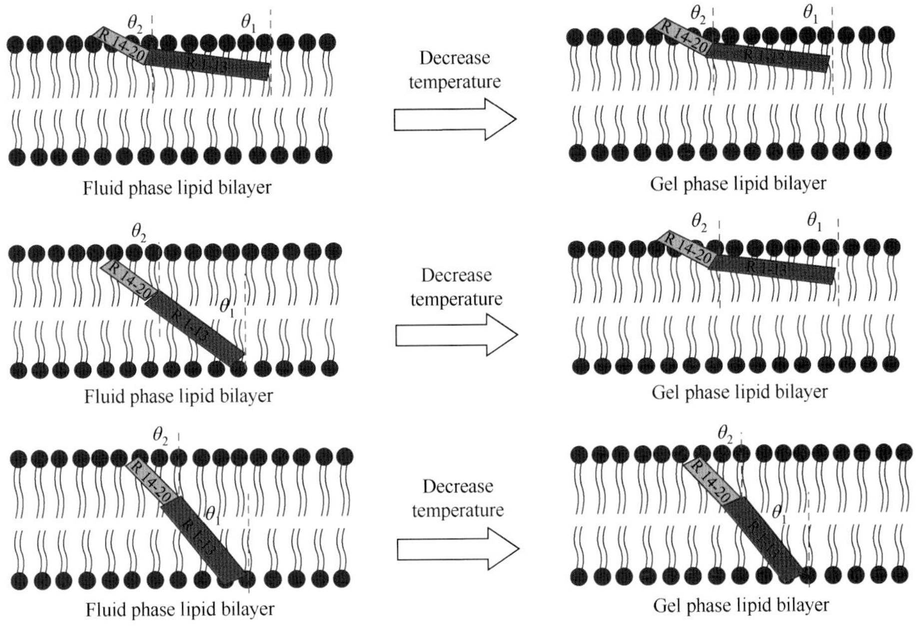

Figure 3.16

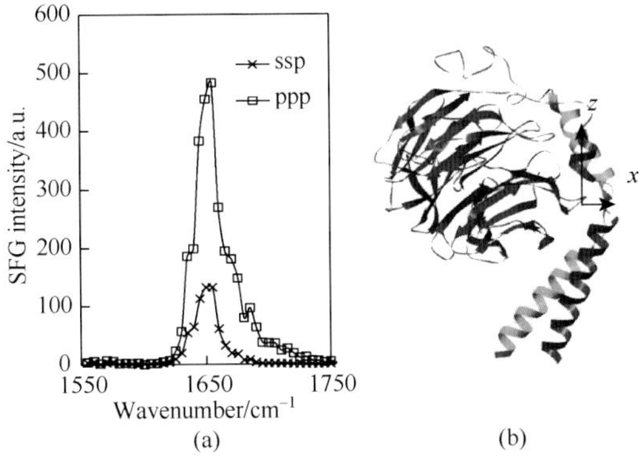

Figure 3.17

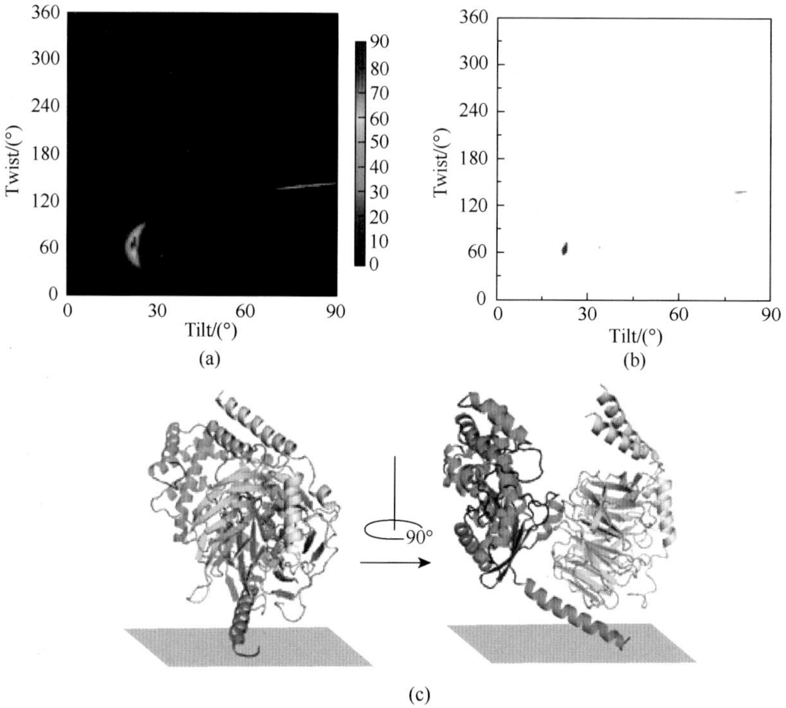

Figure 3.18

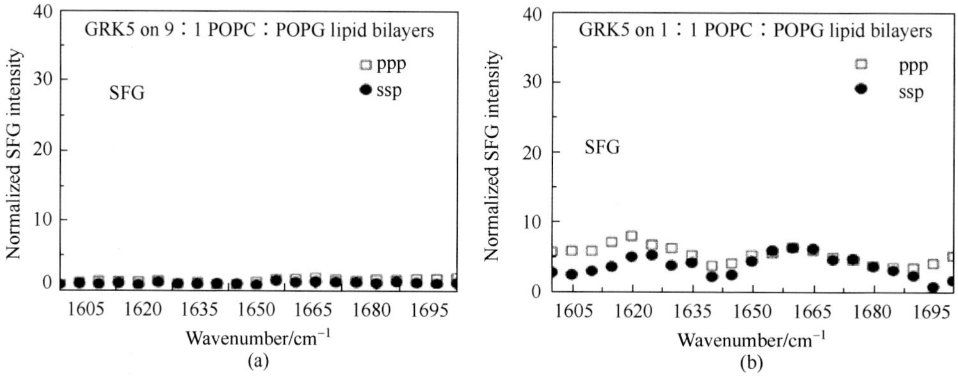

Figure 3.19

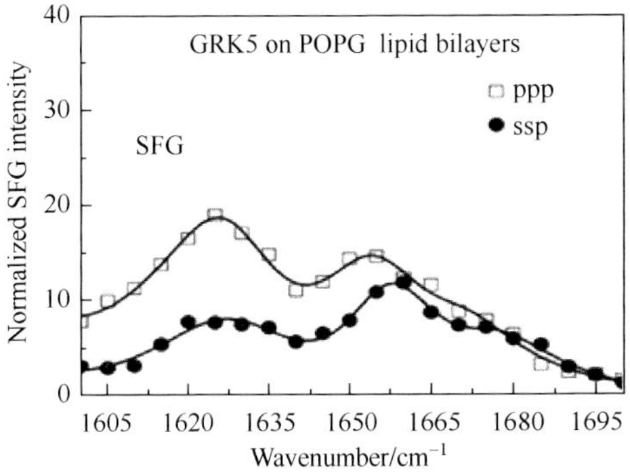

(c)

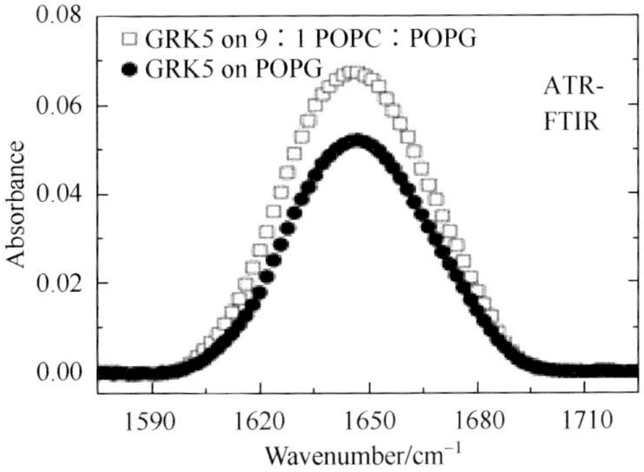

(d)

Figure 3.19 (Continued)

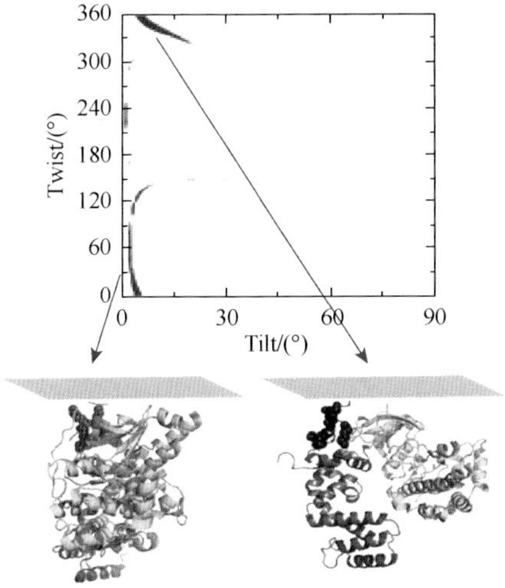

(a)

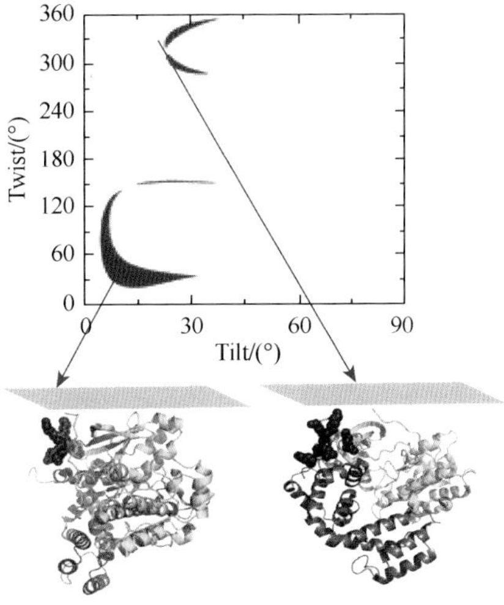

(b)

Figure 3.20

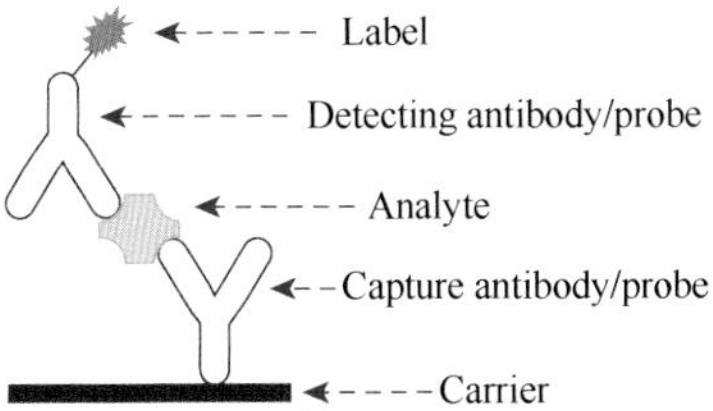

(a)

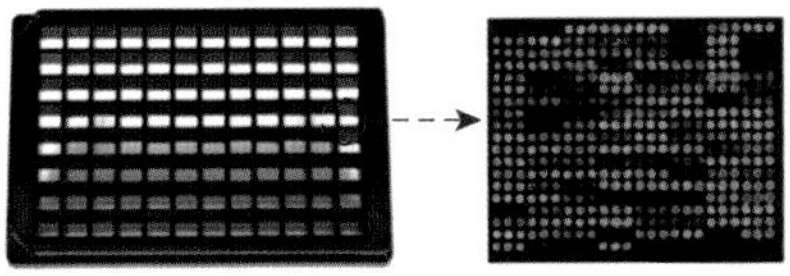

(b)

Figure 4.1

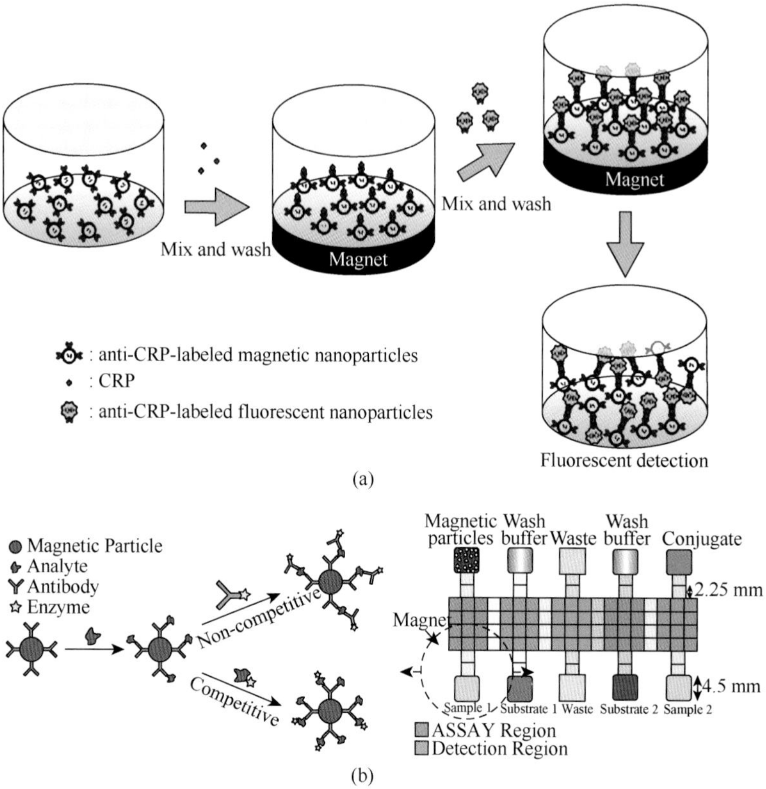

Figure 4.2

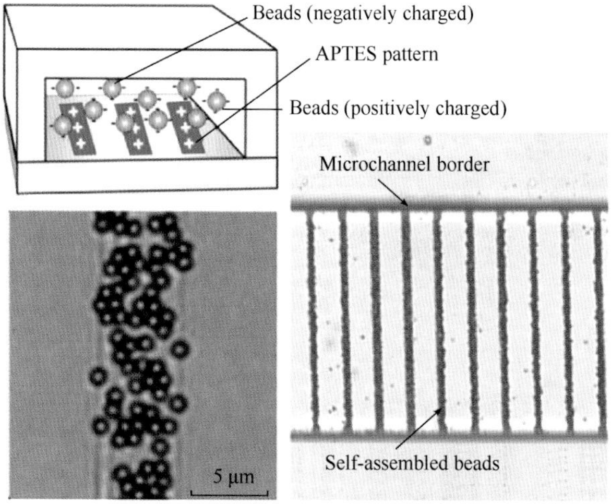

(a)

Figure 4.3

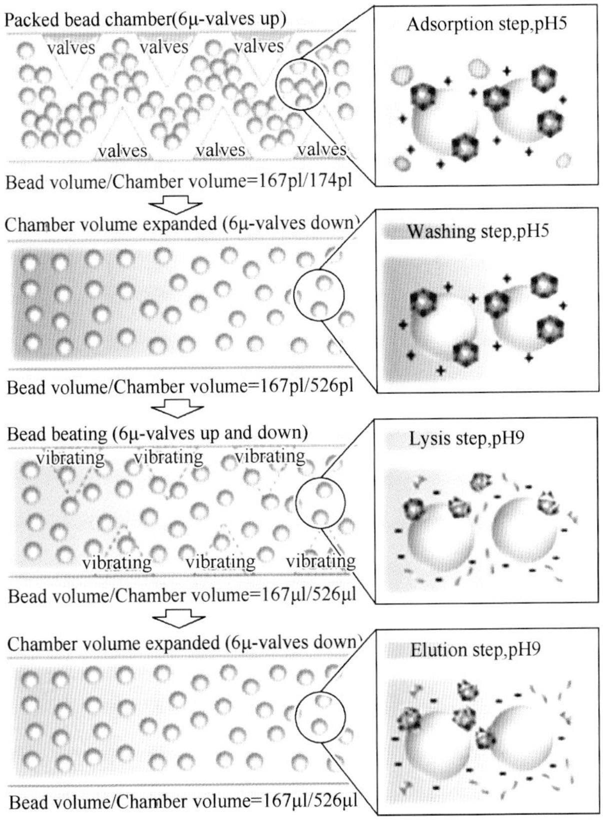

(b)

Figure 4.3 (Continued)

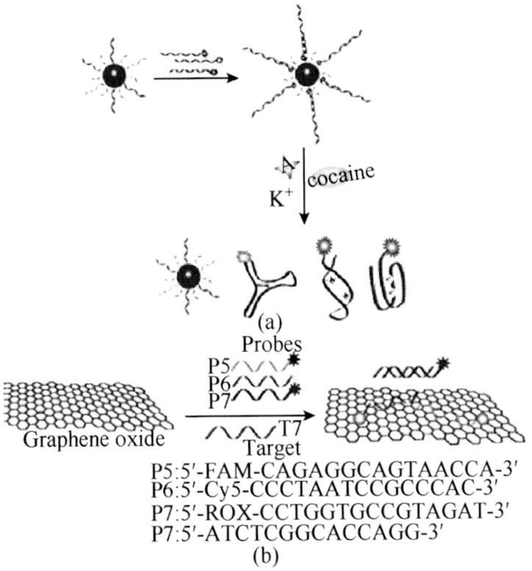

Figure 4.4

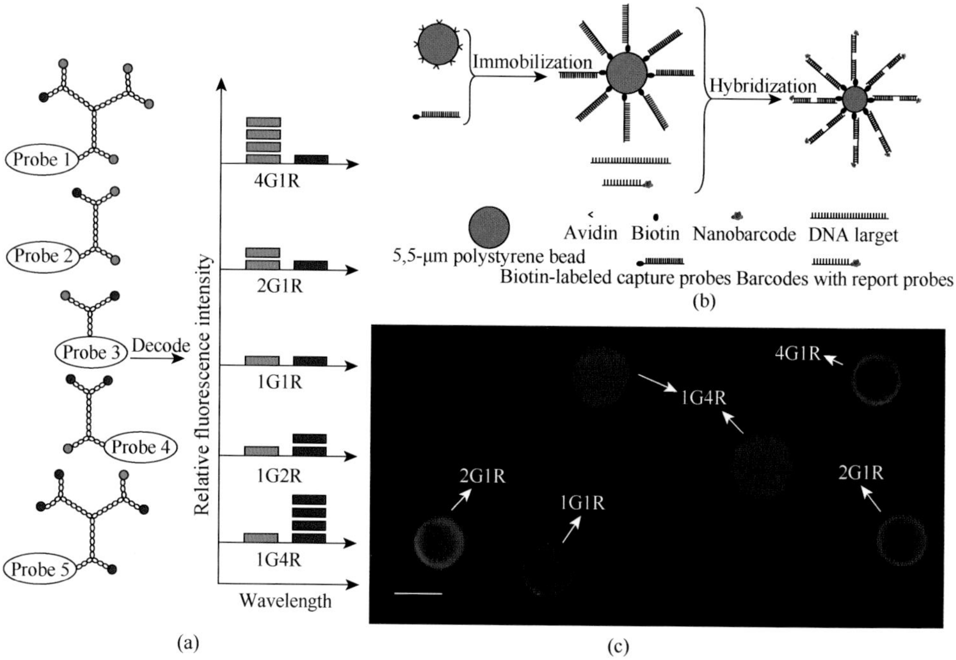

Figure 4.5

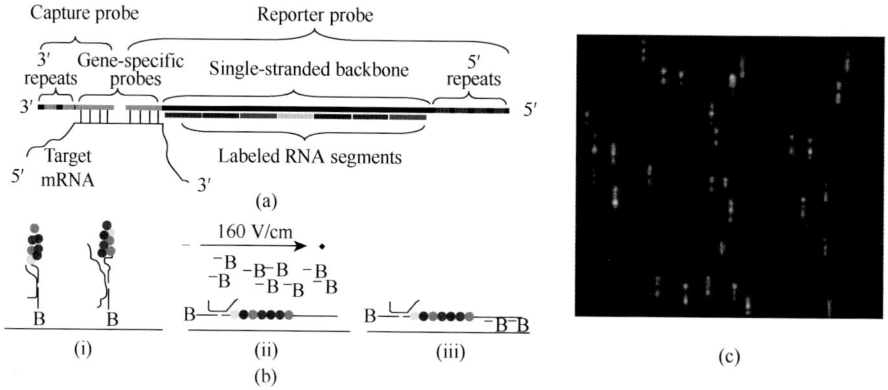

Figure 4.6

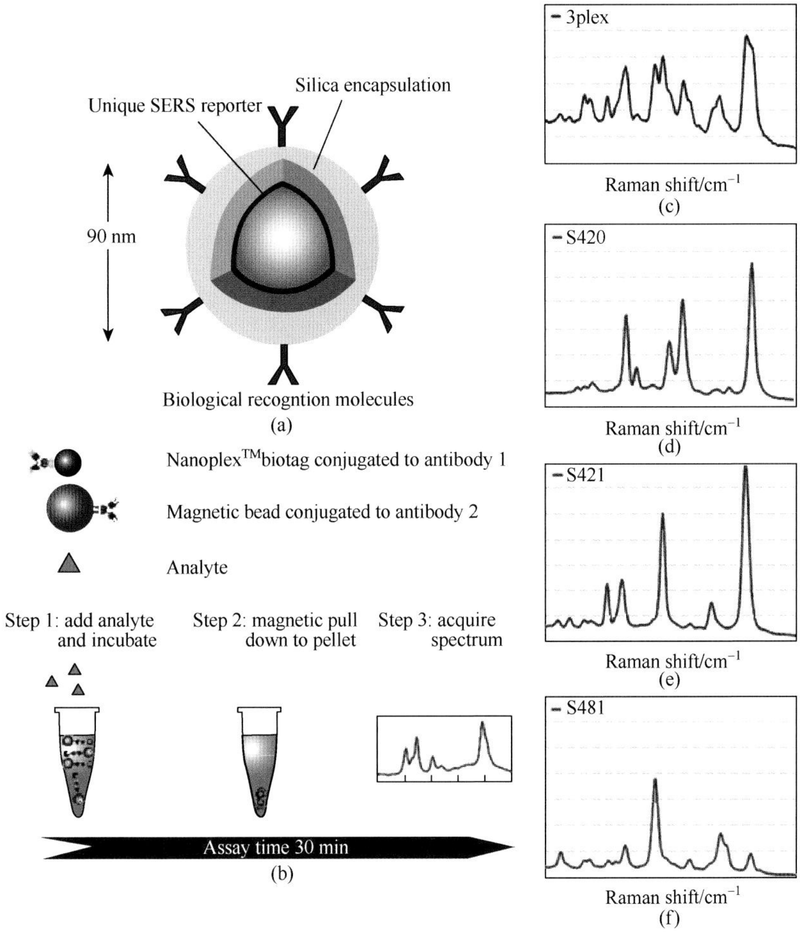

Figure 4.7

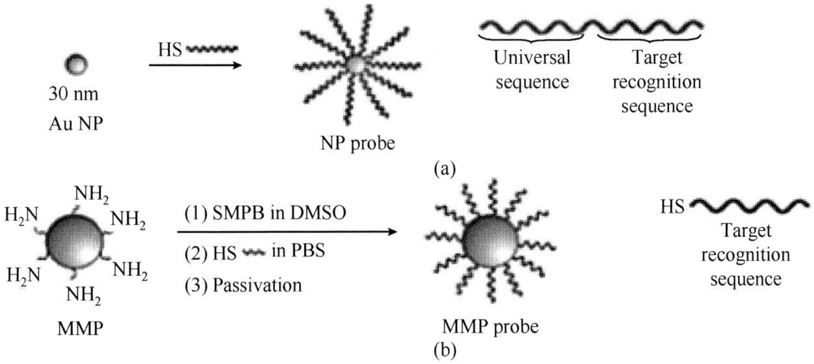

Figure 4.8

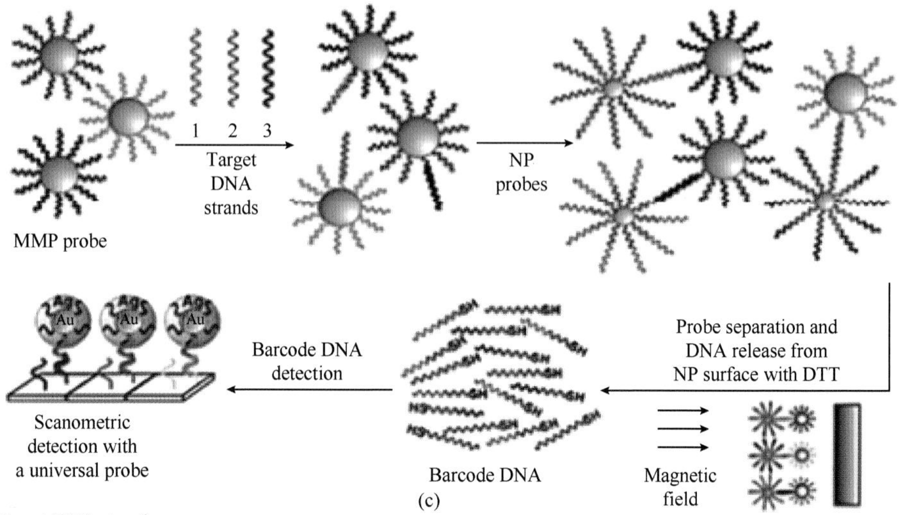

(c)

Target DNA strands:

HBV: 5′TTGGCTTTCAGTTAT-ATGGATGTGGTA-3′

VV: 5′AGTTGTAACGGAAGA-TGCAATAGTAATCAG-3′

EV: 5′-GGAGTAAATGTTGGA-GAACAGTATAACAA-3′

HIV: 5′-AGAAGATATTTGGAATAA-GATGACCTGGATGCA-3′

un=TACGAGTTGAGAATC

PEG_{15}=$(\cdot OCH_2CH_2\cdot)_6$

MMP DNA strands:

MMP-HBA: 5′ATAACTGAAAGCCAA-A_{10}-SH-3′

MMP-W: 5′-TCTTCCGTTACAACT-A_{10}-SH-3′

MMP-EV: 5′-TCCAACATTTACTCC-A_{10}-SH-3′

MMP-HIV: 5′-TTATTCCAAATATCTTCT-A_{10}-SH-3′

Barcode DNA strands:

B-HBV: 5′-HS-un-PEG_{10}-TACCACATCATCCAT-3′

B-VV: 5′-HS-un-PEG_{10}-CTGATTACTATTGCA-3′

B-EV: 5′-HS-un-PEG_{10}-TTGTTGATACTGTTC-3′

B-HIV: 5′-HS-un-PEG_{10}-TGACTCCAGGTCATG-3′

Figure 4.8 (Continued)

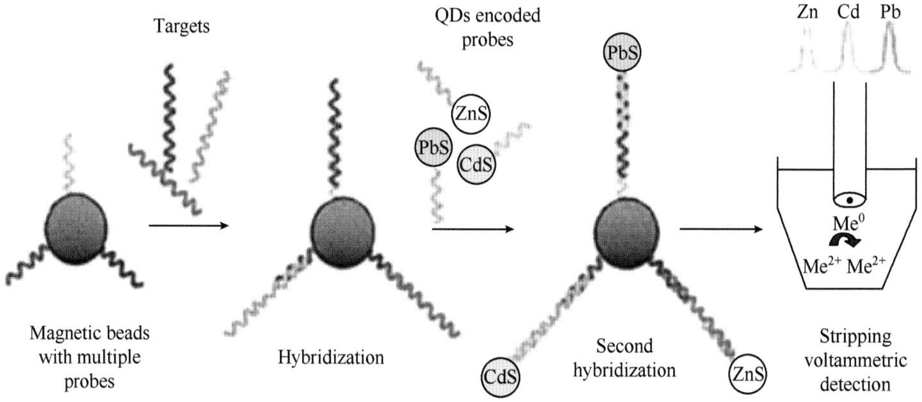

Figure 4.9

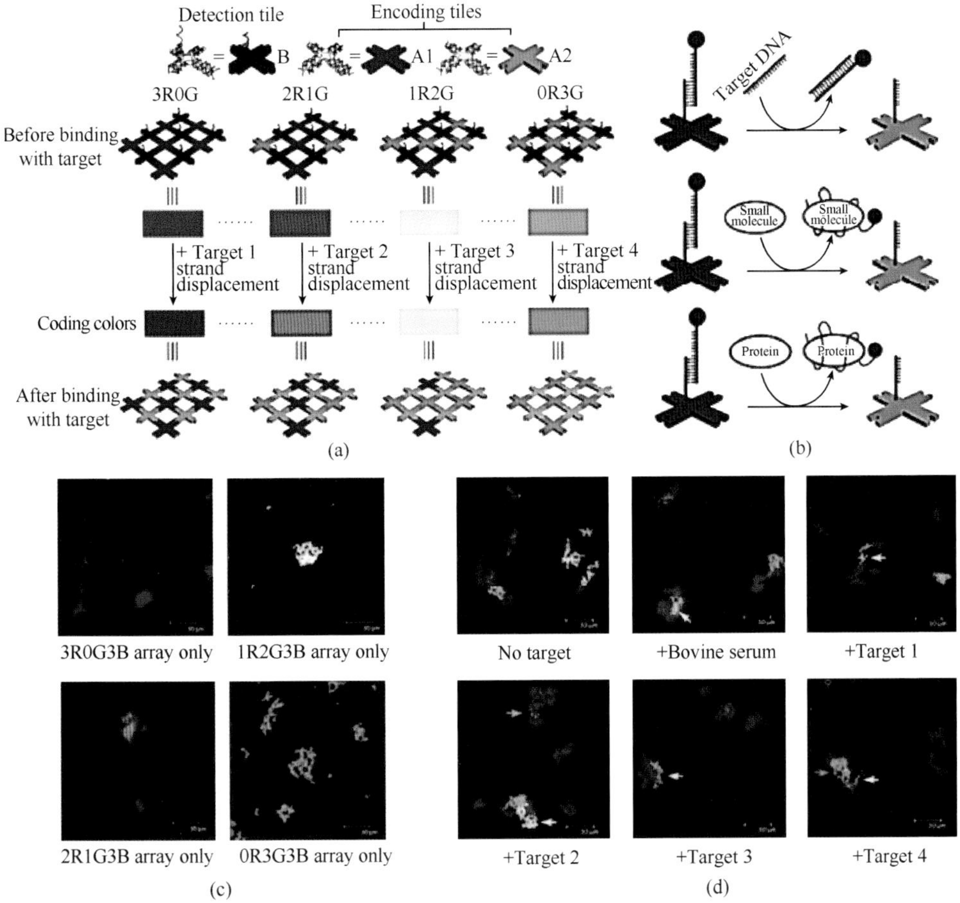

Figure 4.10

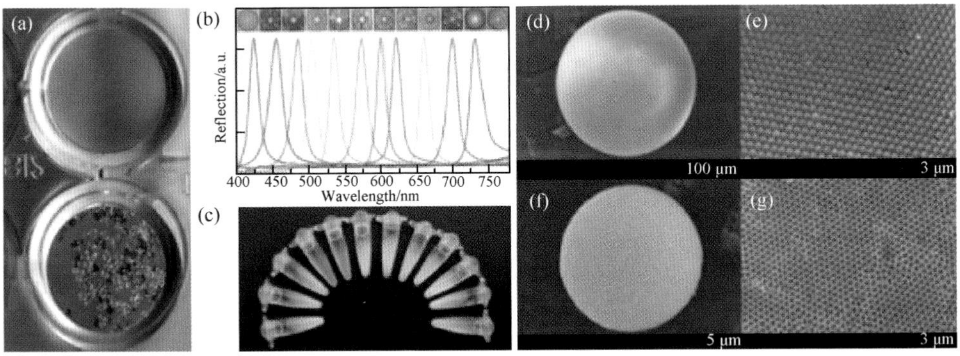

Figure 4.11

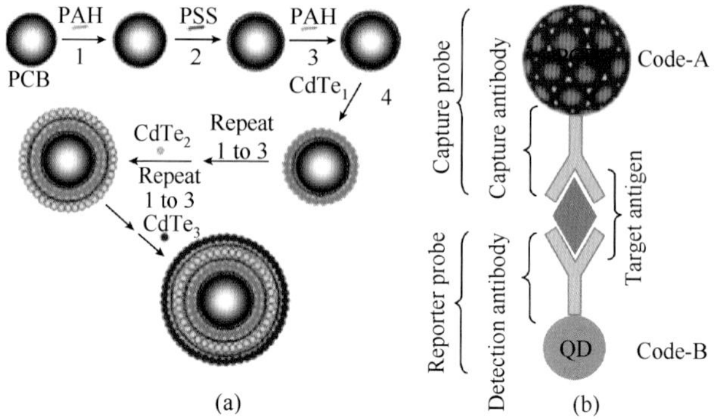

Figure 4.12

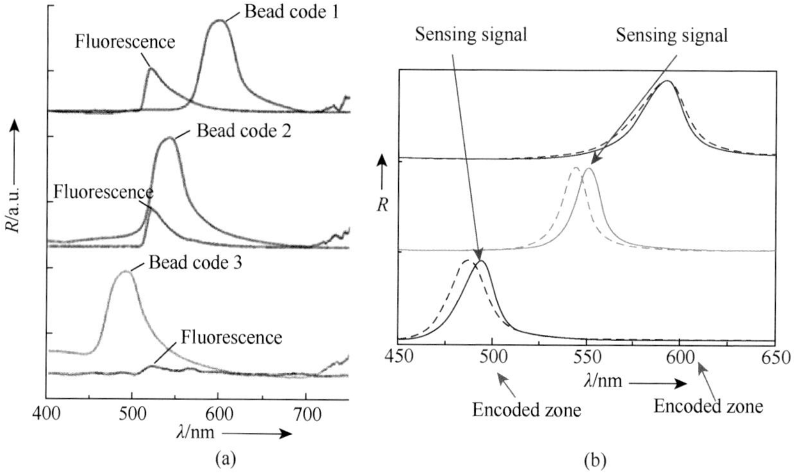

Figure 4.13

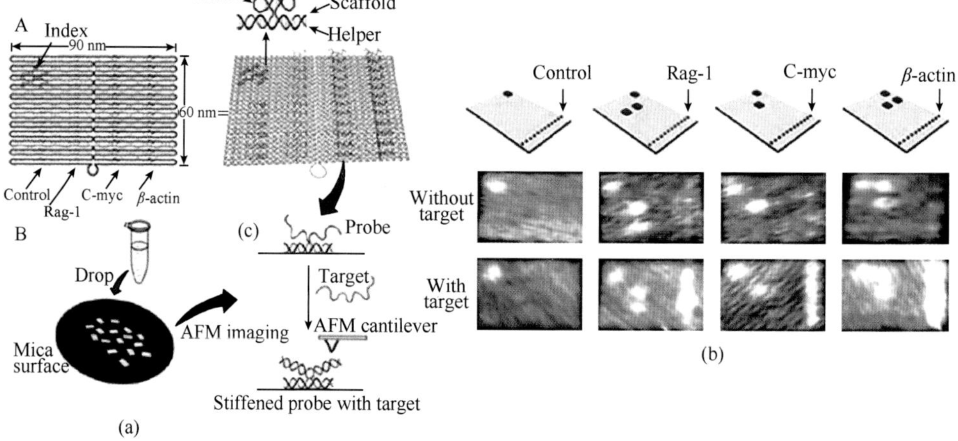

Figure 4.14

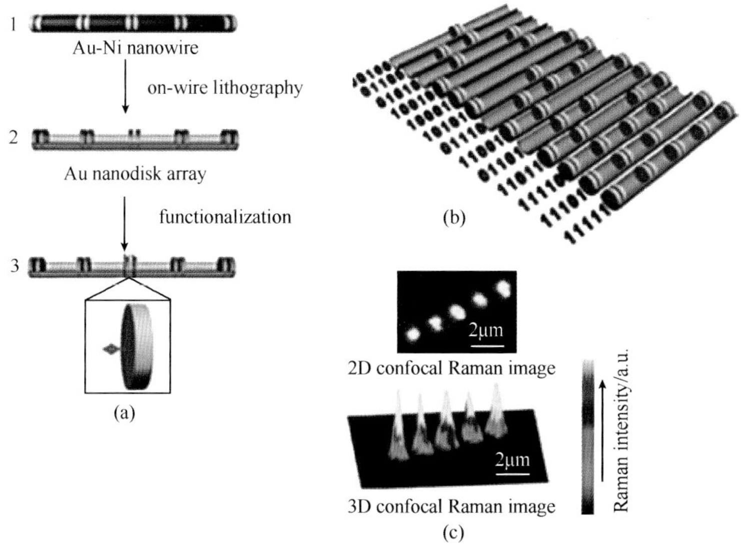

Figure 4.15

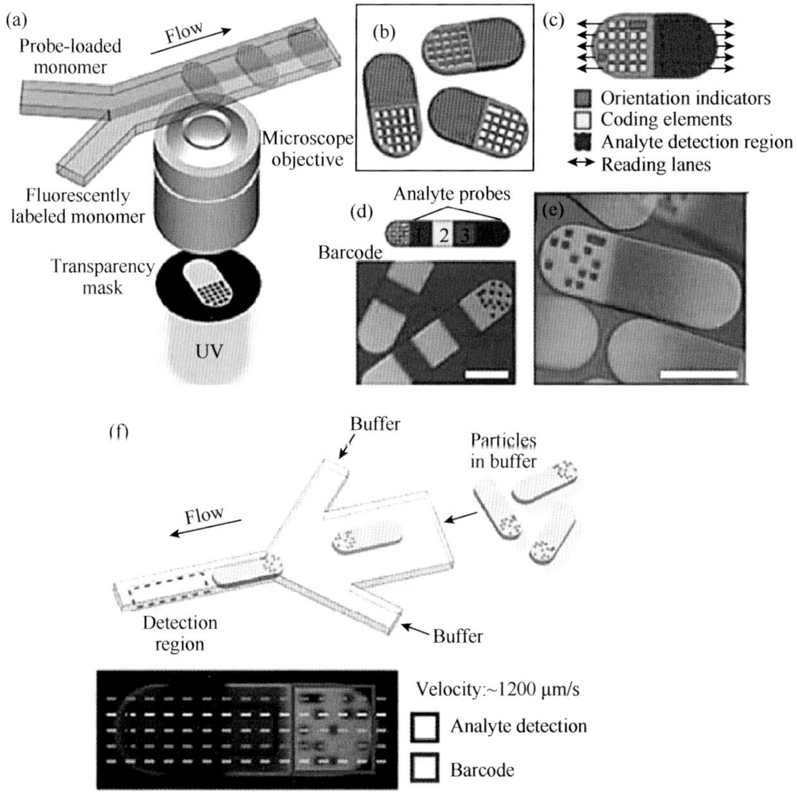

Figure 4.16

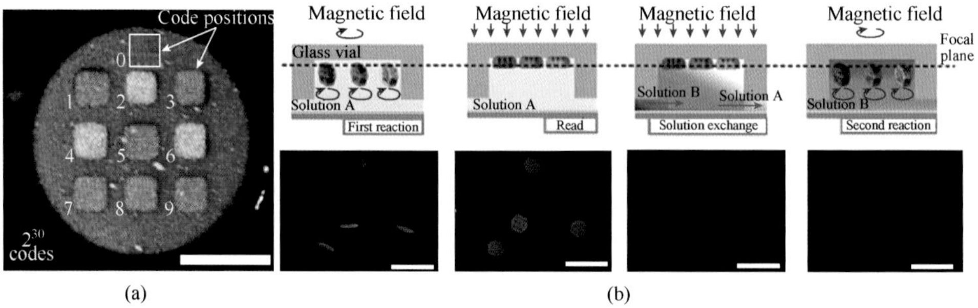

Figure 4.17

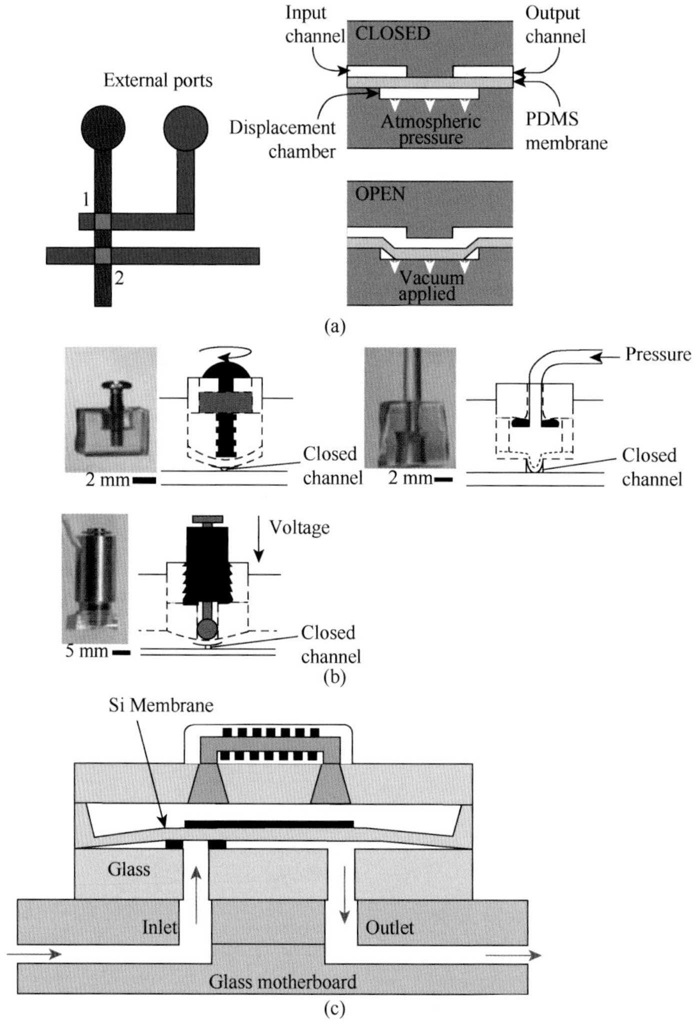

Figure 4.18

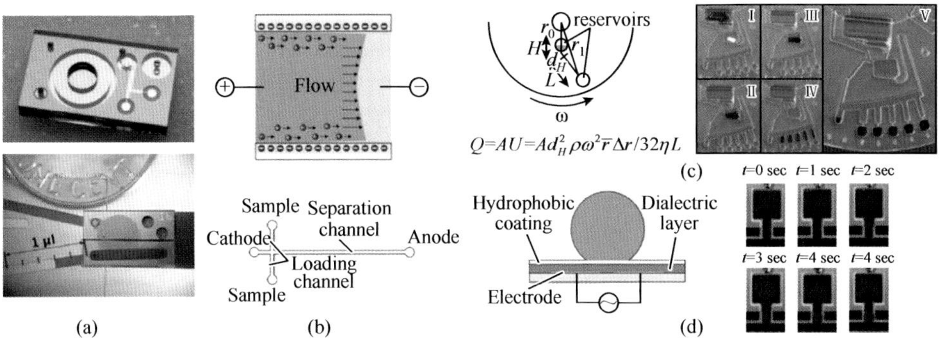

Figure 4.20

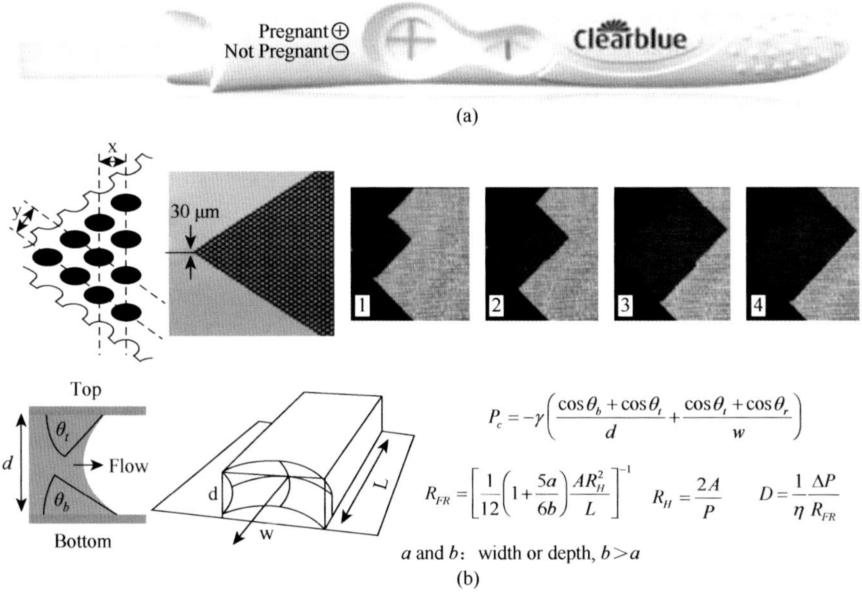

Figure 4.21

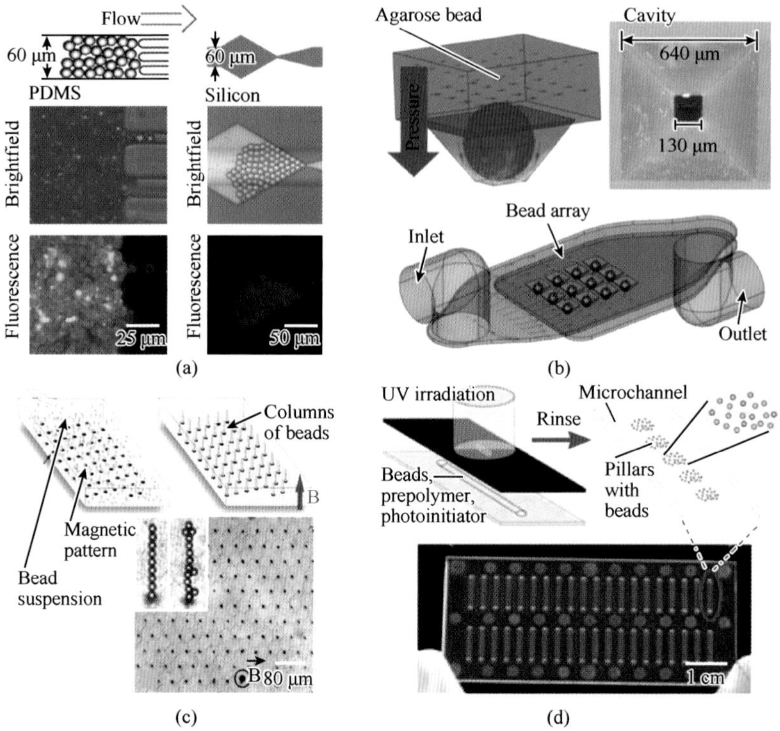

Figure 4.22

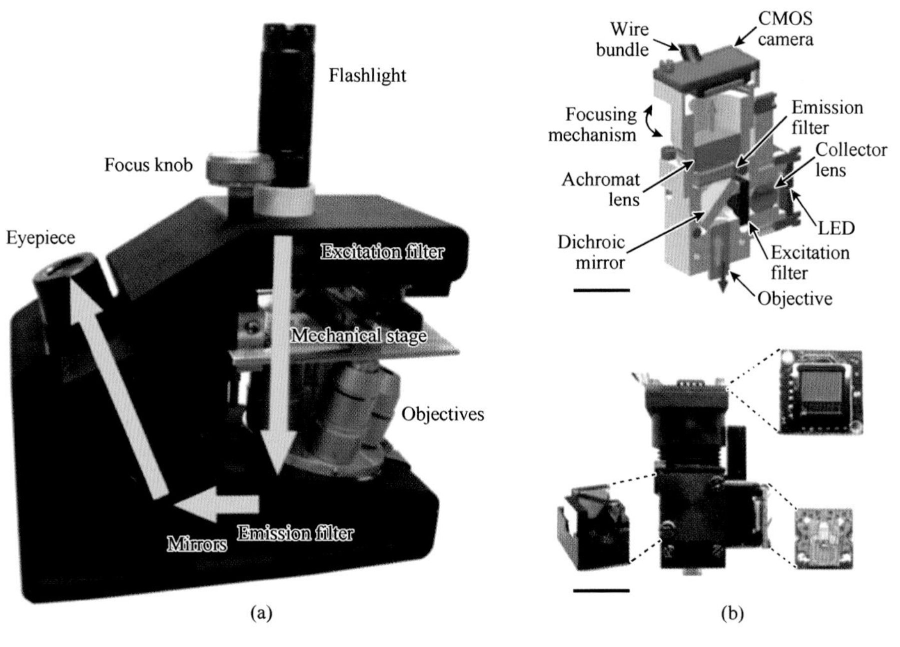

Figure 4.23

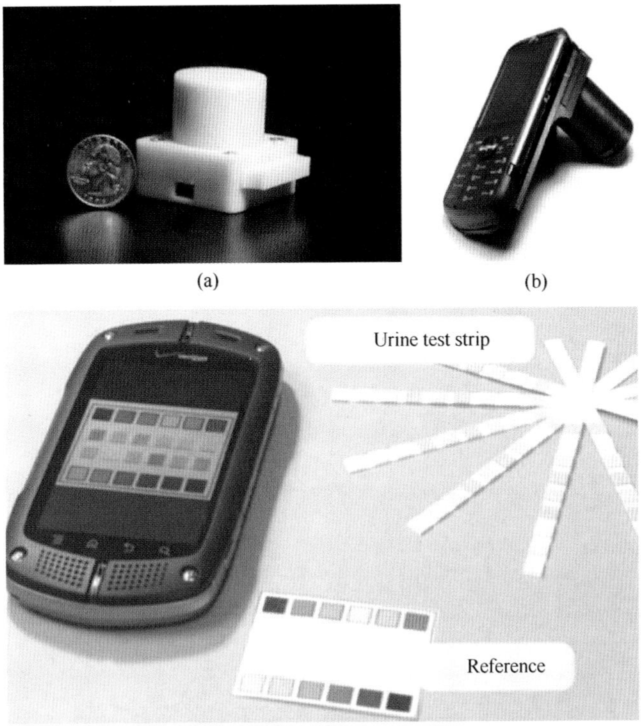

Figure 4.24

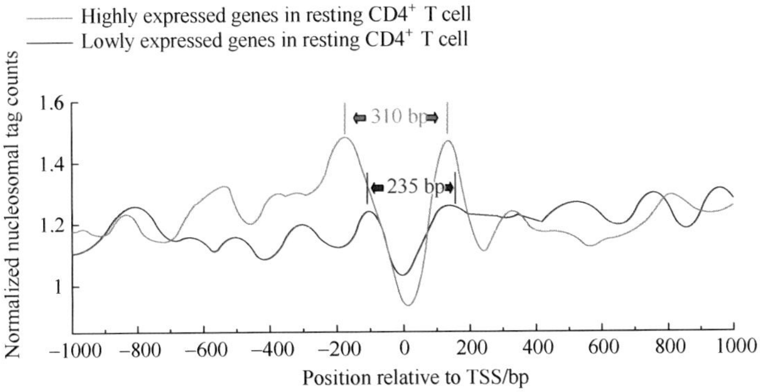

Figure 6.1

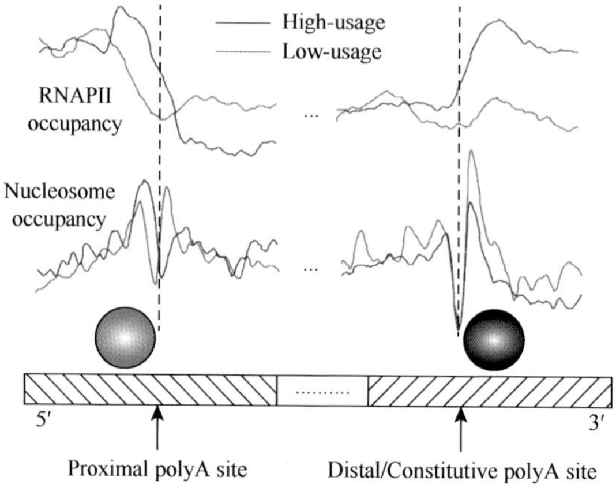

Figure 6.3

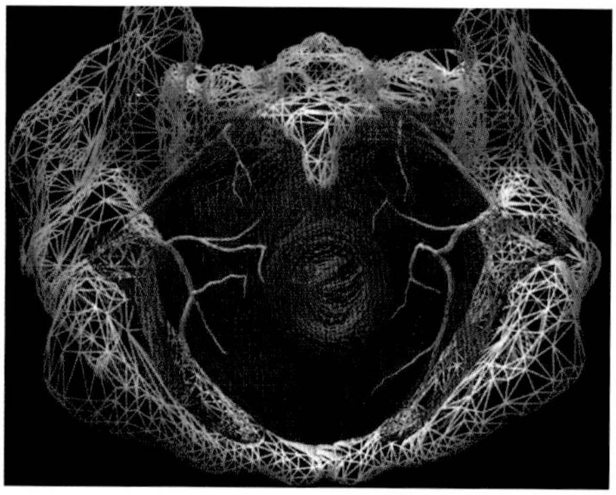

Figure 7.1

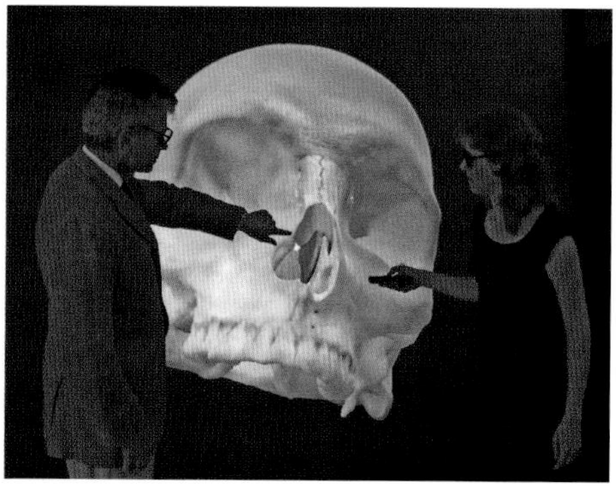

Figure 7.2

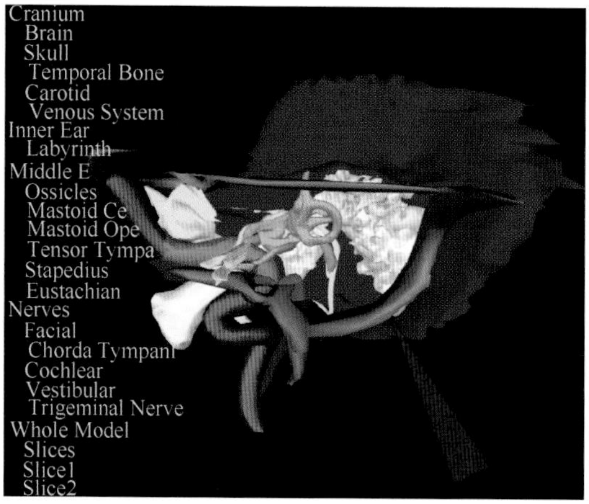

Figure 7.3

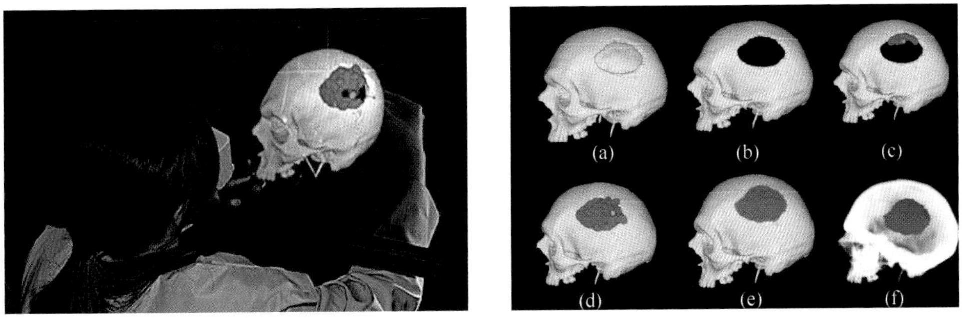

Figure 7.4

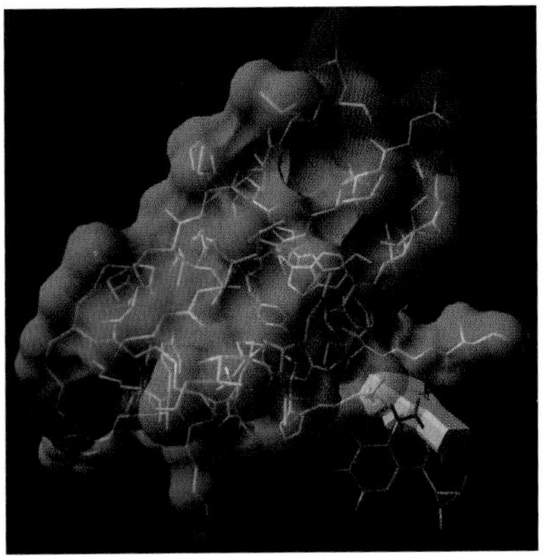

Figure 7.5

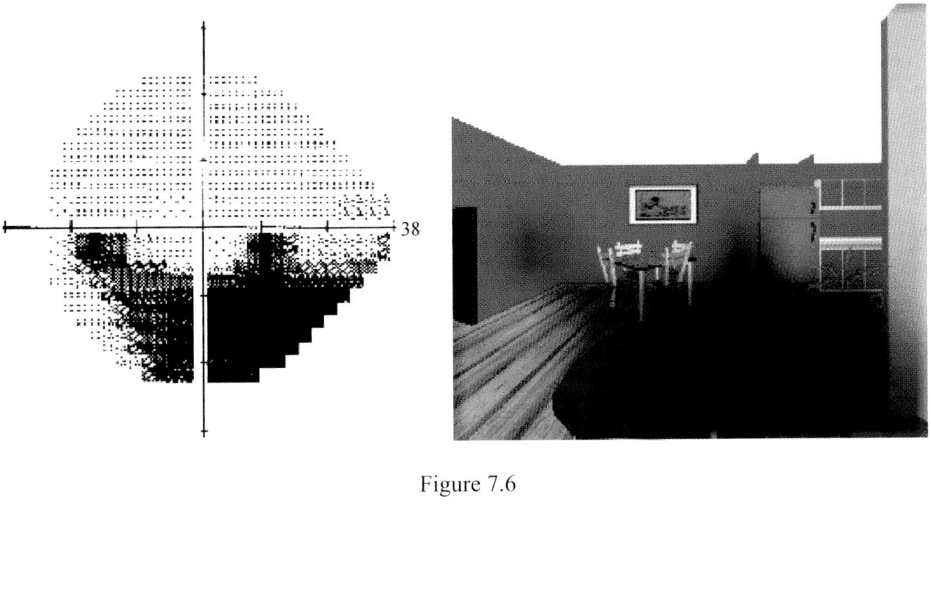

Figure 7.6

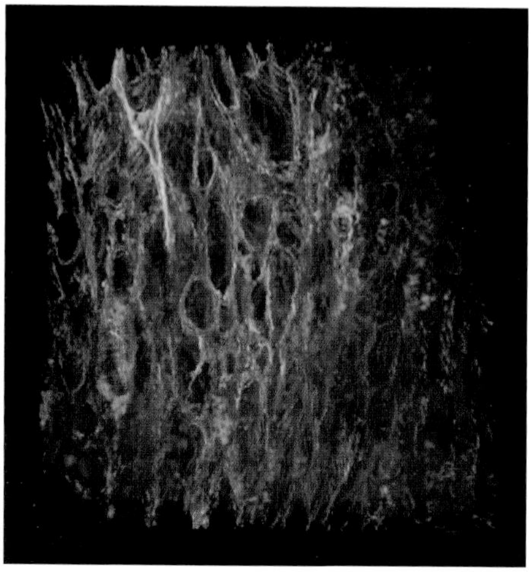

Figure 7.7

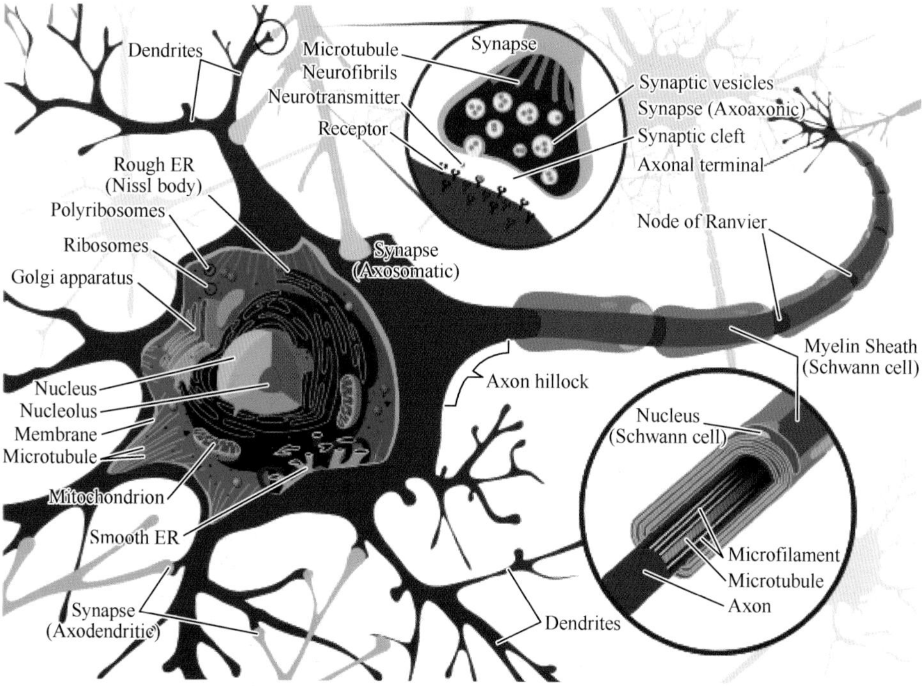

Figure 9.3

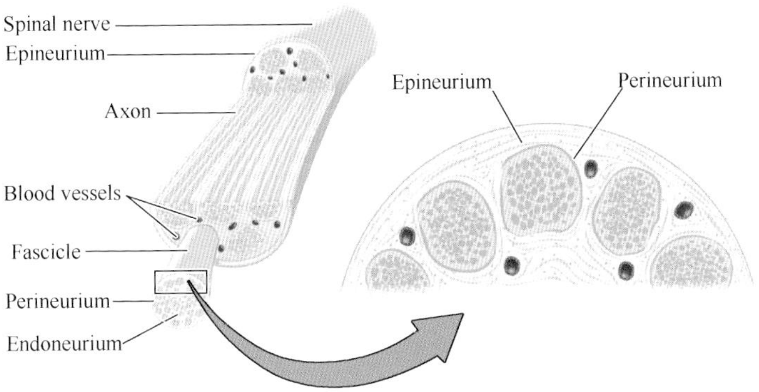

Figure 9.12

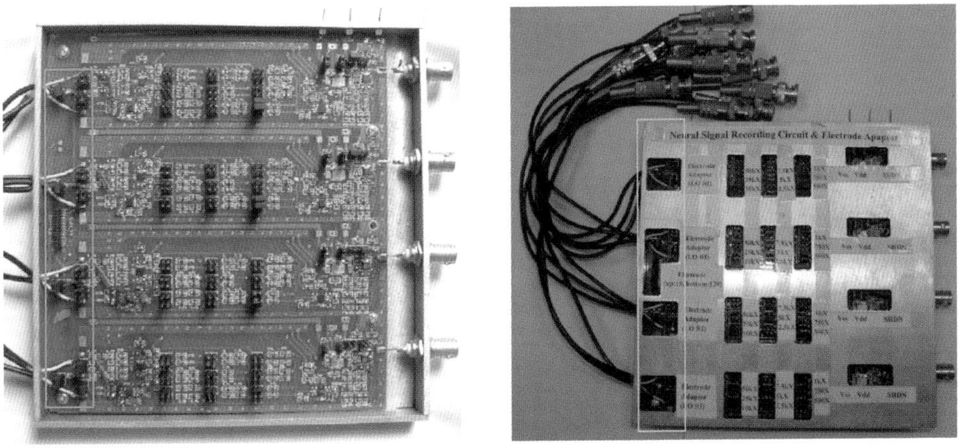

(a) (b)

Figure 9.20

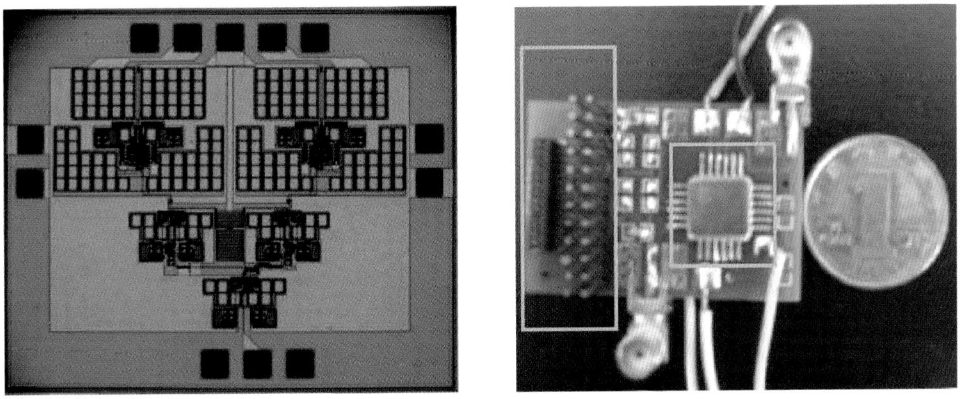

(a) (b)

Figure 9.25

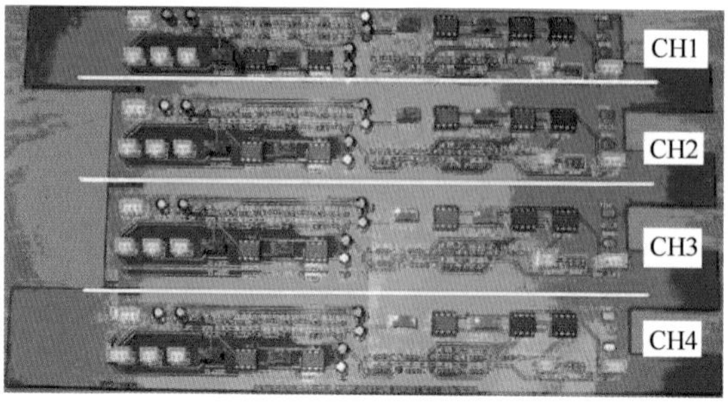

Figure 9.35

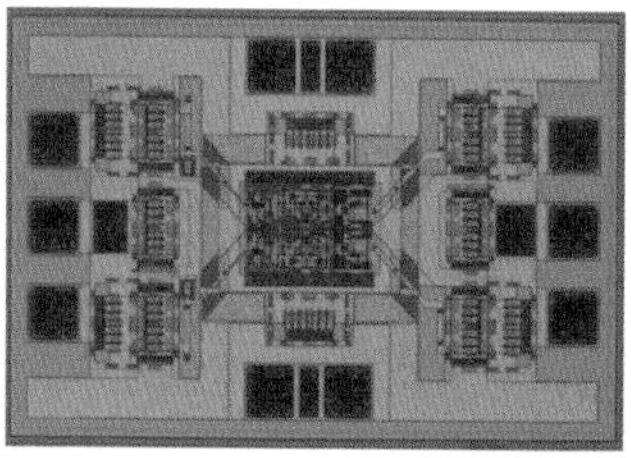

(a)

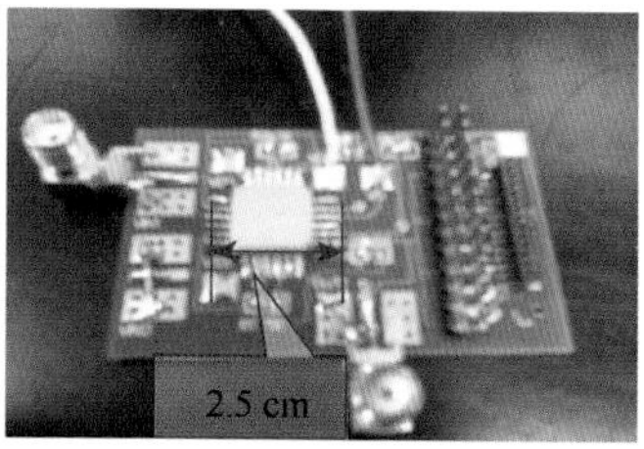

(b)

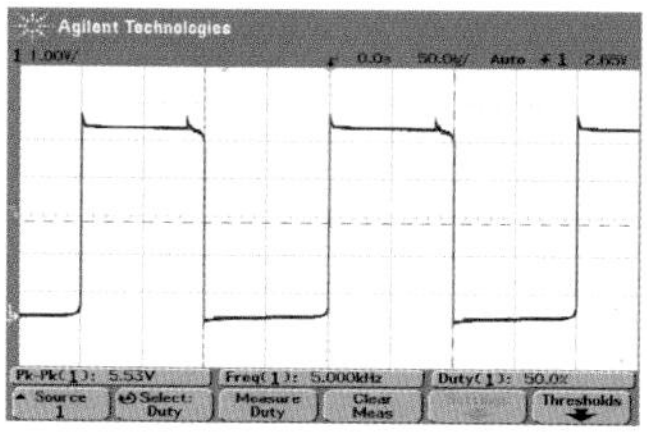

(c)

Figure 9.38

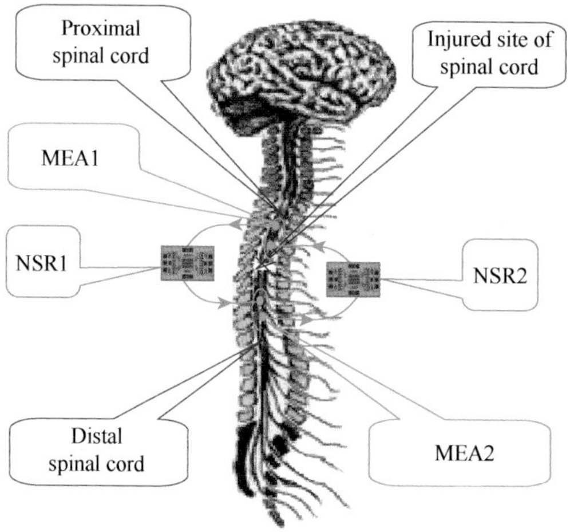

Figure 9.39

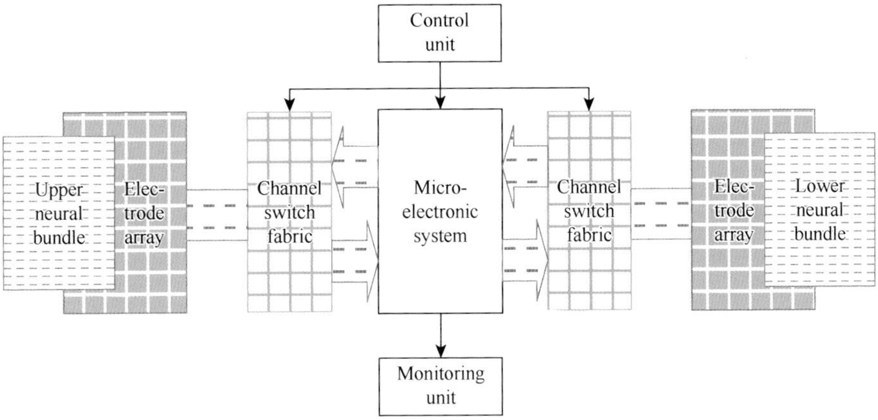

Figure 9.40

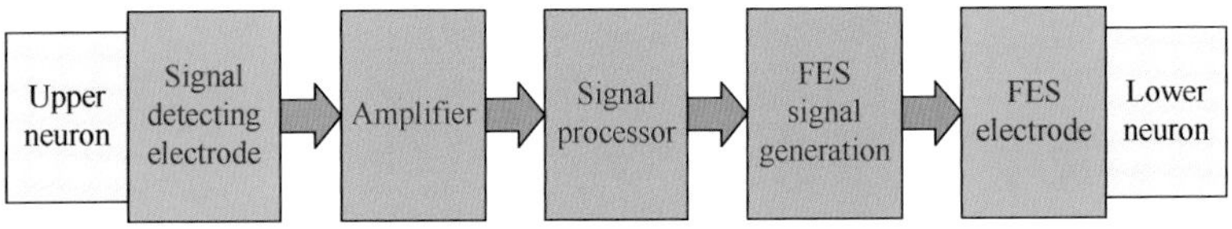

Figure 9.41

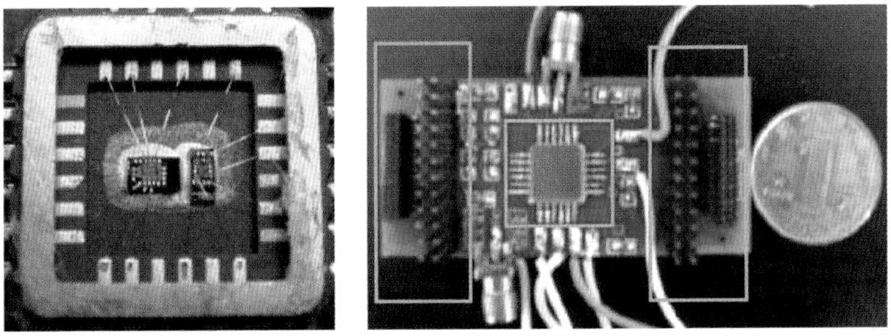

(a) (b)

Figure 9.42

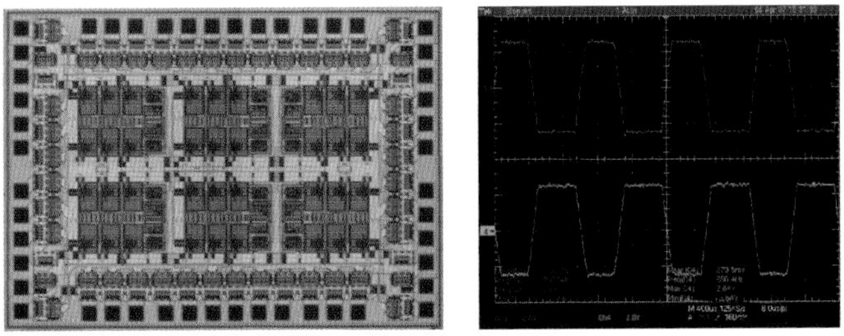

(a) (b)

Figure 9.44

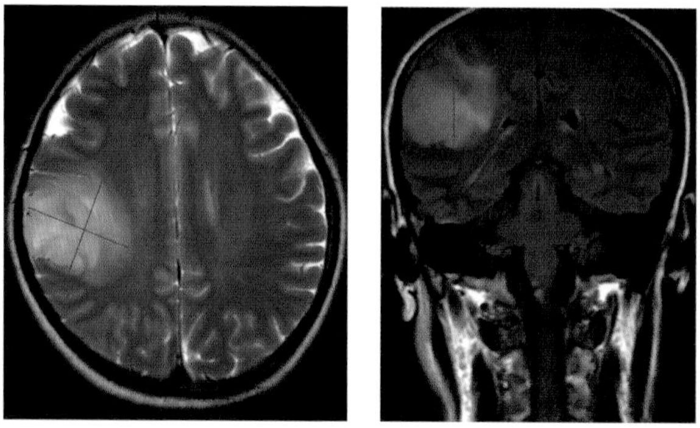

Figure 10.1

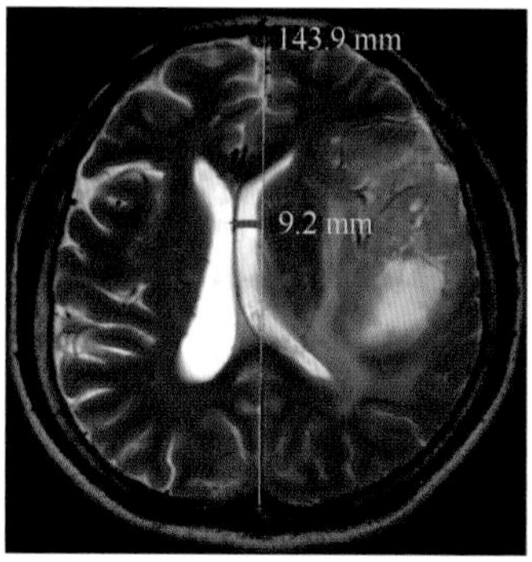

Figure 10.2

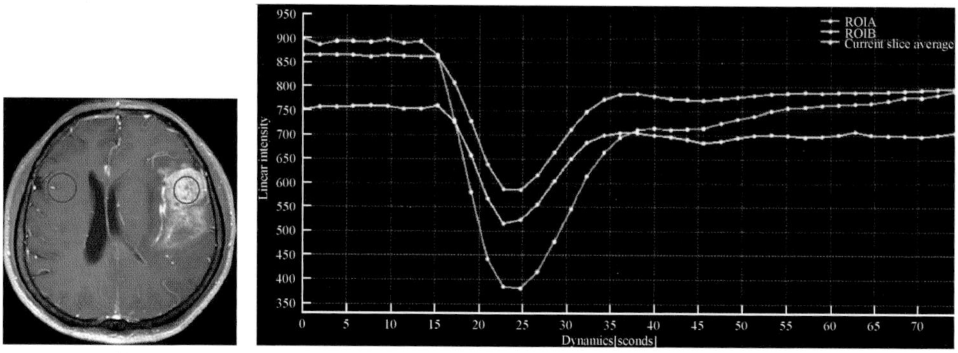

(a) (b)

Figure 10.4

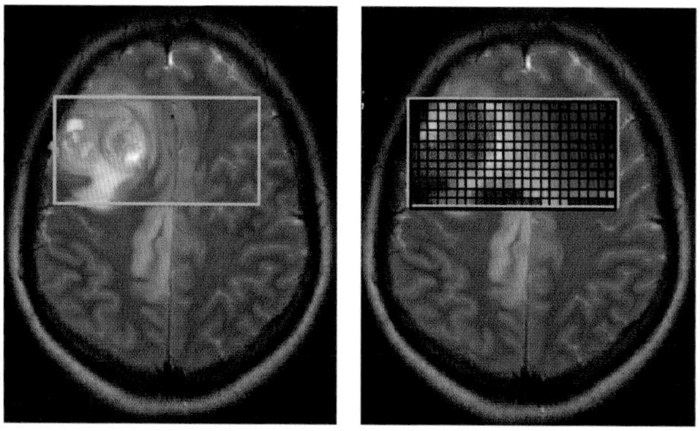

(a) (b)

Figure 10.5

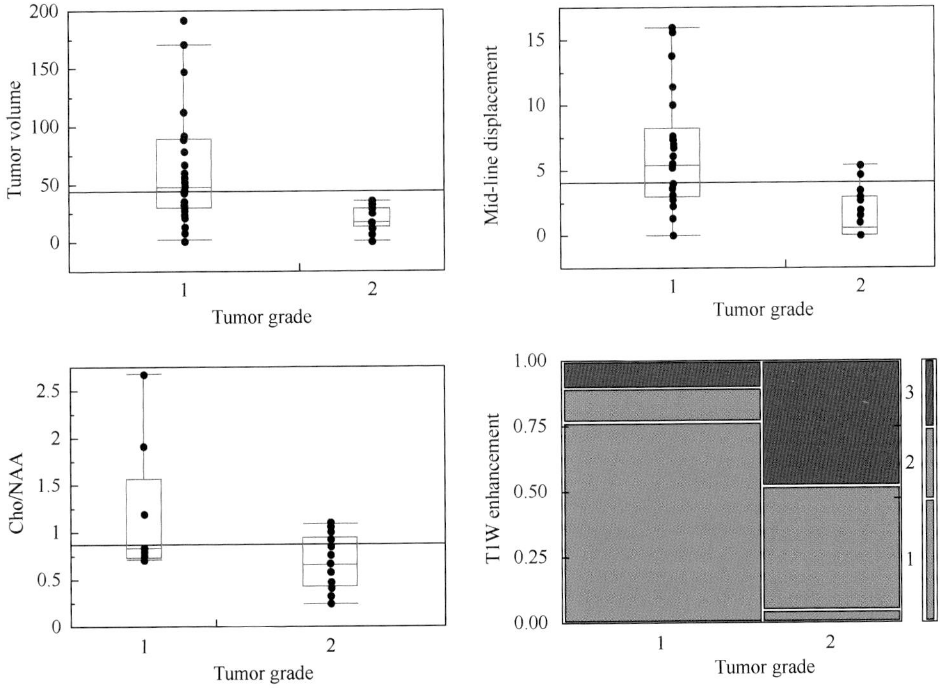

Figure 10.7

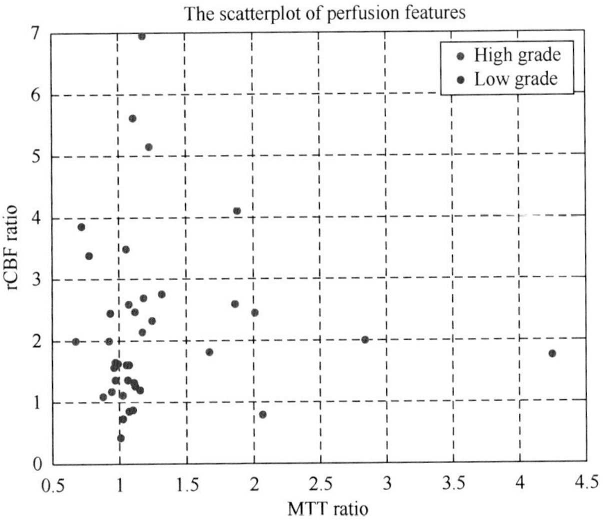

Figure 10.8

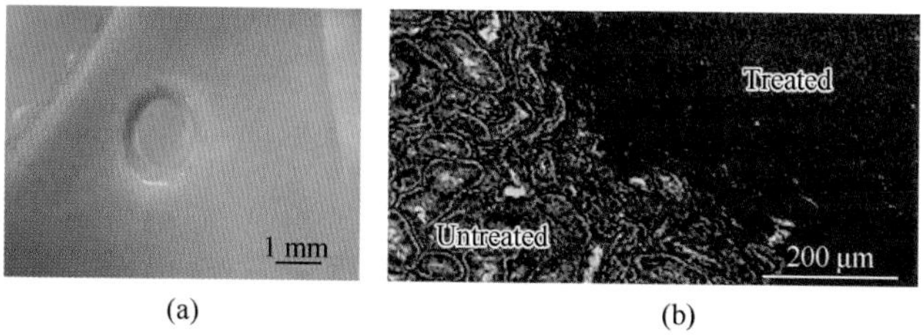

Figure 11.1

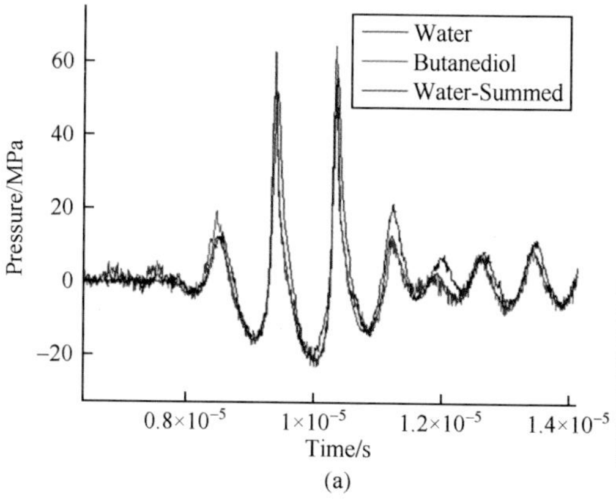

Figure 11.3

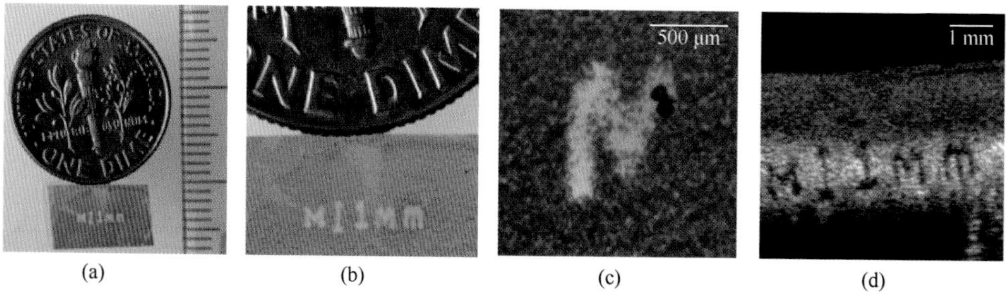

Figure 11.4

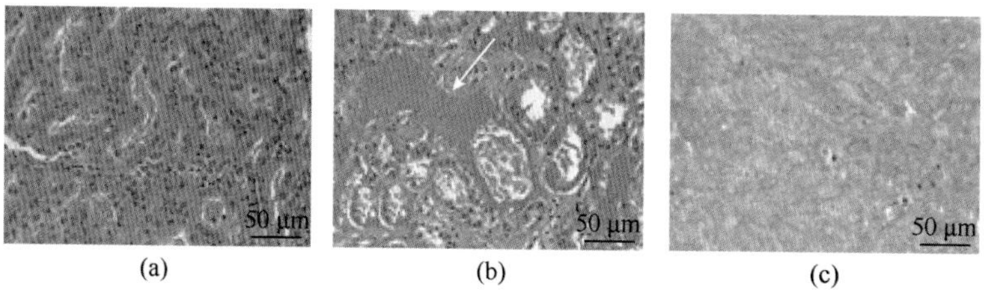

Figure 11.6

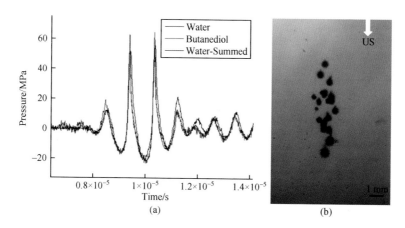

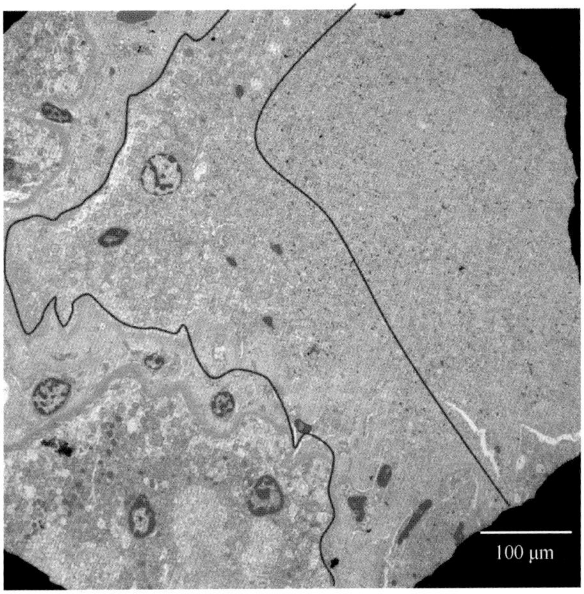

Figure 11.7

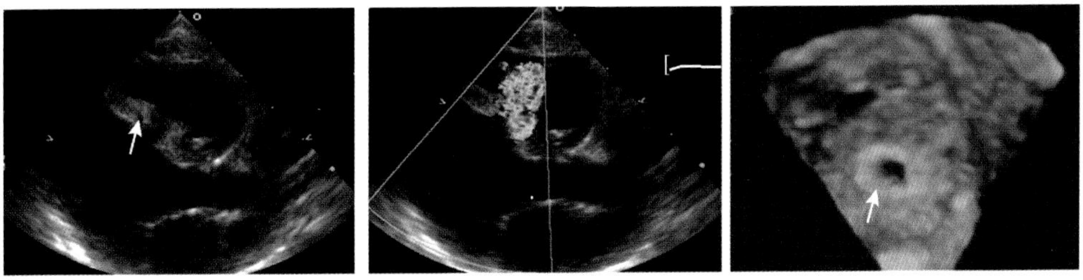

(a) (b) (c)

Figure 11.10

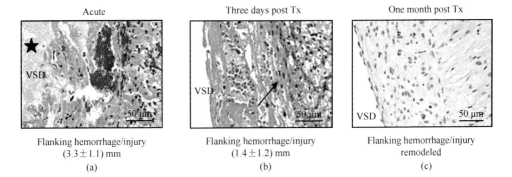

Acute | Three days post Tx | One month post Tx

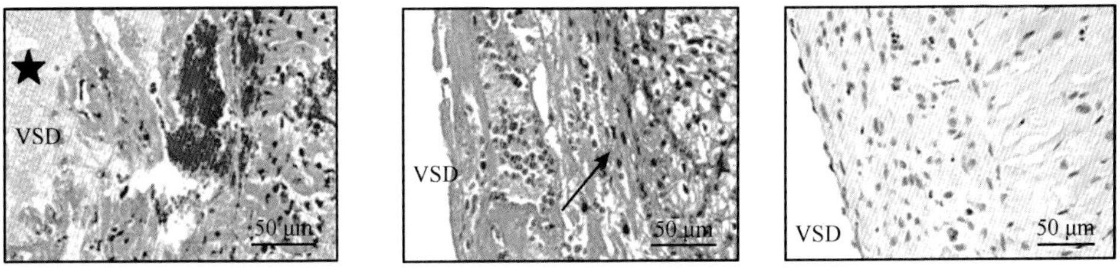

Flanking hemorrhage/injury (3.3±1.1) mm (a)

Flanking hemorrhage/injury (1.4±1.2) mm (b)

Flanking hemorrhage/injury remodeled (c)

Figure 11.11

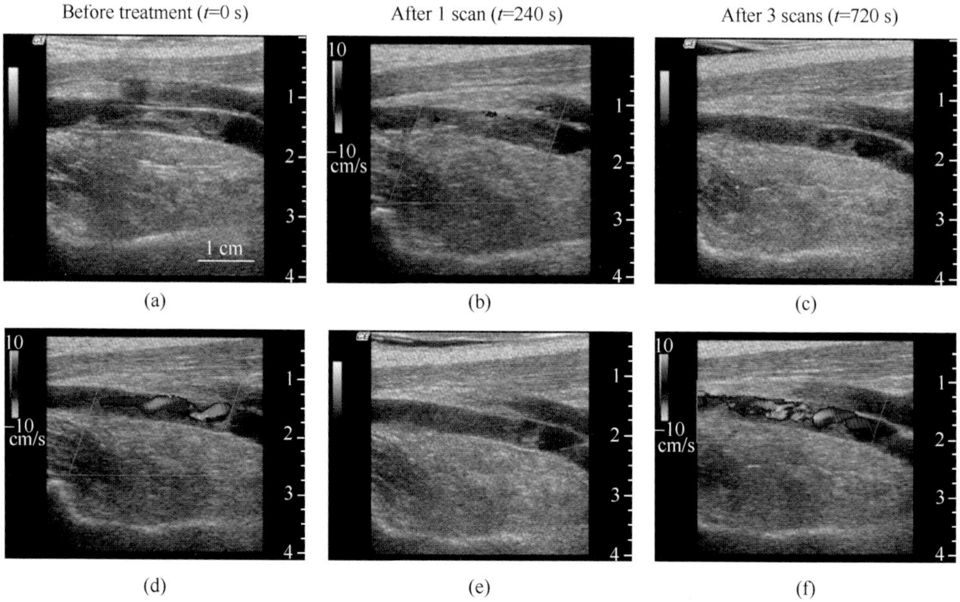

Figure 11.12

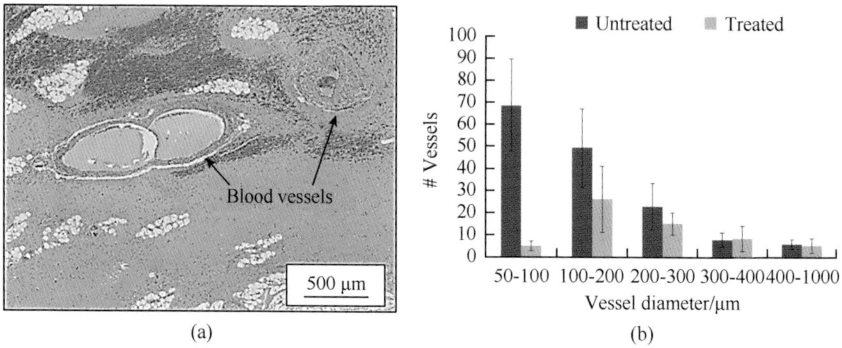

Figure 11.13